Charles Lyell

The geological evidences of the antiquity of man with an outline of glacial and post-tertiary geology; and remarks on the origin of species

with special reference to mans first appearance on the earth

Charles Lyell

The geological evidences of the antiquity of man with an outline of glacial and post-tertiary geology; and remarks on the origin of species
with special reference to mans first appearance on the earth

ISBN/EAN: 9783742823137

Manufactured in Europe, USA, Canada, Australia, Japa

Cover: Foto ©Klaus-Uwe Gerhardt /pixelio.de

Manufactured and distributed by brebook publishing software
(www.brebook.com)

Charles Lyell

The geological evidences of the antiquity of man with an outline of glacial and post-tertiary geology; and remarks on the origin of species

THE

ANTIQUITY OF MAN

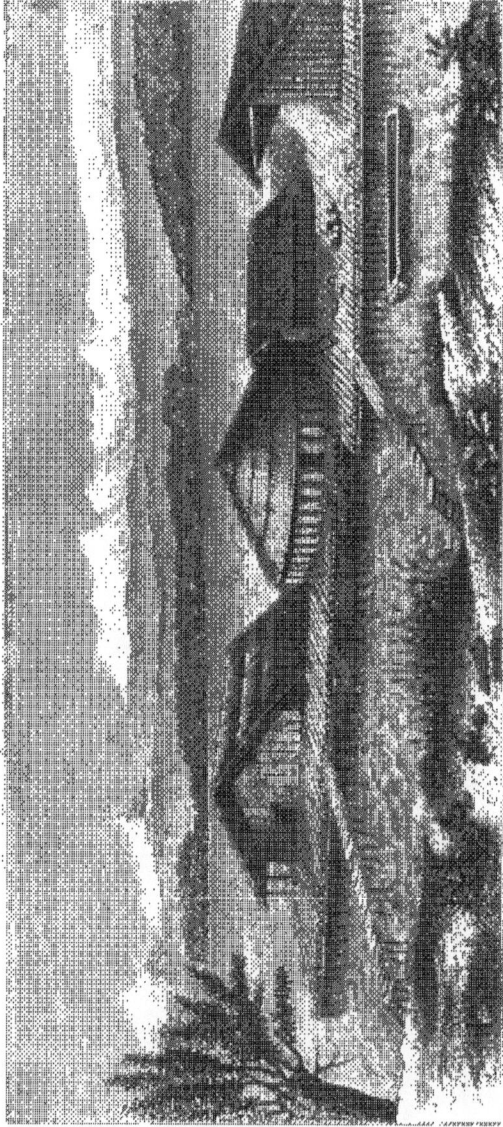

(Frontispiece.)

A VILLAGE BUILT ON PILES IN A SNUG LAKE.

(Restored by Dr. F. Keller, partly from Dumont D'Urville's Sketch of similar habitations in New Guinea.)

THE GEOLOGICAL EVIDENCES

OF

THE ANTIQUITY OF MAN

WITH AN OUTLINE OF

GLACIAL AND POST-TERTIARY GEOLOGY

AND REMARKS ON

THE ORIGIN OF SPECIES

WITH SPECIAL REFERENCE TO MAN'S FIRST APPEARANCE ON THE EARTH

By Sir CHARLES LYELL, Bart. M.A. F.R.S.

AUTHOR OF 'PRINCIPLES OF GEOLOGY' 'ELEMENTS OF GEOLOGY' ETC.

FOURTH EDITION, REVISED

ILLUSTRATED BY WOODCUTS

LONDON

JOHN MURRAY, ALBEMARLE STREET

1873

PREFACE.

THE First Edition of the 'Antiquity of Man' was published in 1863, and was the first work in which I expressed my opinion of the prehistoric age of man and also my belief in Mr. Darwin's theory of the 'Origin of Species' as the best explanation yet offered of the connection between man and those animals which have flourished successively on the earth.

The Second Edition appeared two months later with a few alterations and corrections, and a short Appendix, and the Third Edition with a long Appendix followed before the end of the year.

A space of ten years has now elapsed since the publication of the Third Edition, during which time I have been occupied with two editions of the 'Principles of Geology,' and two editions of the 'Elements of Geology,' in which I have embodied incidentally many of the daily accumulating proofs of man's antiquity. A fourth edition of this work has now long been called for, and I have therefore carefully revised it and introduced such new matter as seemed to me necessary to bring it up to the present state of our knowledge.

Finding that many persons have failed to recognise the natural connection of the three separate parts of this work, I have now placed them under three distinct titles, the first of which, called the 'Antiquity of Man,' might with more propriety have been distinguished as the 'Geological Memorials of Man.'

The second part on the Glacial Period, which was formerly the subject of much criticism, will no longer be regarded as irrelevant to the main subject of man's antiquity now that so much discussion is going on whether man is pre-glacial or post-glacial.

The third part, treating of the Origin of Species with reference to Man's place in Nature, has too evident a bearing on the preceding parts to need any special comment.

Among other corrections and improvements made in the present edition, I may mention—

1st. That in the earlier chapters I have profited in many places by the works of Sir John Lubbock on 'Pre-historic Man,' and that of Mr. John Evans on 'Ancient Stone Implements.'

2ndly. I have reconsidered the question of the rise of land in Scotland, and have brought forward the many facts which lead me to conclude that the opinion I formerly adopted, on the evidence of Mr. Geikie, that there had been a rise of twenty-five feet in Post-Roman times, is no longer tenable (p. 50, et seq.)

3rdly. I have added an account of the exploration of the caverns of the Lesse in Belgium by M. Dupont, and some interesting conclusions which he has drawn from them (p. 77).

4thly. I have completely recast the chapter relating to Brixham Cavern and Kent's Hole, and have added some new information respecting the age of the deposits in the latter cavern (p. 99, et seq.)

5thly. In the spring of 1872 I visited the cave of Aurignac with my friend Mr. T. McK. Hughes, now Woodwardian Professor at Cambridge, and we convinced ourselves that the geological proofs of the remoteness of the era to which the existence of sepulchral rites may be carried back are more doubtful than M. Lartet and I had formerly supposed (p. 131, *et seq.*).

6thly. At the same time I visited with Mr. Hughes the caverns of the Dordogne, of which, together with the carvings contained in them, I have given some account (p. 139).

7thly. I have also given a description of a skeleton found by M. Rivière in a cave at Mentone, which, from the unpolished implements and extinct animals associated with it, I am inclined to consider as of Paleolithic age (p. 143). Since the sheets were printed, a second skeleton has been brought to light by M. Rivière in a neighbouring cavern under similar conditions. He informs me in a letter (April 17, 1873) that he found with this second human fossil a flint lance and flint hatchet, both unpolished. Around the arms, wrists, and knees were bracelets of Mediterranean shells, Nassa, Cypraea, and Buccinum; and the skeleton and surrounding earth were stained red by oxide of iron, as was the case with the skeleton discovered in 1872. Extinct animals were found also at a higher level than this second skeleton, but I infer from letters received from Mr. Charles Moore, now at Mentone, that the time of inhumation of these remains of elephant, rhinoceros, and cave-bear in subaerial breccias at different altitudes in the cliffs will have to be critically ascertained before their geological bearing on the age of the human skeletons can be finally settled.

8thly. In Part II., in dealing with the period immediately preceding that in which we have positive traces of
man, I have found it necessary entirely to recast the twelfth
chapter relating to the crags of Norfolk and Suffolk, and
have taken the opportunity of adding new information
from the recent memoirs of Mr. Prestwich, aided by Mr.
Gwyn Jeffreys, and those of MM. Searles Wood, sen., S.
Wood, jun., and Harmer (Chapter XII.)

9thly. In regard to the question of ice-action in the
Glacial Period, both in the British Islands and in the Alps
and elsewhere, I have discussed the merits of the rival
theories proposed to account for the transportation of erratic
blocks and the erosion of lake-basins, and have added some
new facts respecting the continental ice of Greenland derived
from the observations of MM. Nordenskiöld and Richard
Brown (Chapters XIII. to XV.)

10thly. In the recapitulation of the proofs of man's
antiquity, I have dwelt more fully than in former editions
on the remote dates of civilisation afforded us by the
monuments and traditions of Egypt and other Oriental
countries (p. 425).

11thly. In Part III. I have been enabled to make one
very important addition. At the time when the third edition
of this work was published, the absence of intermediate links
was one of the greatest difficulties experienced by the advocates of transmutation. Since then, three intermediate fossil
forms have been discovered, linking together the two classes,
Aves and Reptilia (p. 483); and among the Mammalia, two
extremely ancient and less specialised forms of the horse
have been found in the Upper and Lower Miocene for-

mations, affording evidence of the gradual modification of this animal from a different ancestral type (p. 491).

12thly. In the concluding chapter on the theory of transmutation as applied to man I have considered the bearing of some of the new facts and conclusions in Mr. Darwin's 'Descent of Man,' and in reference to the origin of races, that of the Greeks in particular, I have considered Mr. Francis Galton's speculations on the causes of their preeminence as treated of in his book on 'Hereditary Genius' (p. 545).

List of the Dates of Publication of successive Editions of the 'Antiquity of Man,' and other Geological Works of Intermediate Date.

Antiquity of Man, 1st edition	Feb. 1863
Antiquity of Man, 2nd edition	April 1863
Antiquity of Man, 3rd edition	Nov. 1863
Elements of Geology, 6th edition . . .	1865
Principles of Geology, 10th edition . . .	1866-68
Student's Elements of Geology . . .	1871
Principles of Geology, 11th edition . . .	1872
Antiquity of Man, 4th edition . . .	May 1873

CHARLES LYELL.

73 Harley Street:
April 21, 1873.

CONTENTS.

PART I.

GEOLOGICAL MEMORIALS OF MAN.

CHAPTER I.

INTRODUCTORY.

CHAPTER II.

RECENT PERIOD—DANISH PEAT AND SHELL-MOUNDS— SWISS LAKE-DWELLINGS.

CHAPTER III.

FOSSIL HUMAN REMAINS AND WORKS OF ART OF THE RECENT PERIOD—*continued.*

CHAPTER IV.

PLEISTOCENE PERIOD.

BONES OF MAN AND EXTINCT MAMMALIA IN BELGIAN CAVERNS.

CHAPTER V.

PLEISTOCENE PERIOD.

FOSSIL HUMAN SKULLS OF THE NEANDERTHAL AND ENGIS CAVES.

CHAPTER VI.

FLINT IMPLEMENTS IN ENGLISH CAVE-DEPOSITS.

CHAPTER XIII.

CHRONOLOGICAL RELATIONS OF THE GLACIAL PERIOD AND THE EARLIEST SIGNS OF MAN'S APPEARANCE IN EUROPE.

CHAPTER XIV.

CHRONOLOGICAL RELATIONS OF THE GLACIAL PERIOD AND THE EARLIEST SIGNS OF MAN'S APPEARANCE IN EUROPE—continued.

CHAPTER XV.

EXTINCT GLACIERS OF THE ALPS AND THEIR CHRONOLOGICAL RELATION TO THE HUMAN PERIOD.

PART III.

THE ORIGIN OF SPECIES AS BEARING ON MAN'S PLACE IN NATURE

CHAPTER XX.

THEORIES OF PROGRESSION AND TRANSMUTATION.

CHAPTER XXI.

ON THE ORIGIN OF SPECIES BY VARIATION AND NATURAL SELECTION.

CHAPTER XXII.

OBJECTIONS TO THE HYPOTHESIS OF TRANSMUTATION CONSIDERED.

CHAPTER XXIII.

ORIGIN AND DEVELOPMENT OF LANGUAGES AND SPECIES COMPARED.

CHAPTER XXIV.

BEARING OF THE DOCTRINE OF TRANSMUTATION ON THE ORIGIN OF MAN, AND HIS PLACE IN THE CREATION.

GEOLOGICAL EVIDENCE

OF

THE ANTIQUITY OF MAN.

PART I.

THE ANTIQUITY OF MAN.

—◦—

CHAPTER I.

INTRODUCTORY.

PRELIMINARY REMARKS ON THE SUBJECTS TREATED OF IN THIS WORK—
DEFINITION OF THE TERMS RECENT, PLEISTOCENE, AND POST-TER-
TIARY—TABULAR VIEW OF THE ENTIRE SERIES OF FOSSILIFEROUS
STRATA.

NO subject has during the last ten years excited more
curiosity and general interest among geologists and the
public than the question of the Antiquity of the Human Race,
—or, in other words, whether or no we have sufficient evidence
in caves, or in the superficial deposits commonly called drift
or 'diluvium,' to prove the former co-existence of man with
certain extinct mammalia. For the last half-century, the
occasional occurrence, in various parts of Europe, of the
bones of Man or the works of his hands, in cave-breccias and
stalagmites, associated with the remains of the extinct hyæna,
bear, elephant, or rhinoceros, has given rise to a suspicion

B

that the date of Man must be carried further back than we had heretofore imagined. On the other hand, extreme reluctance was naturally felt, on the part of scientific reasoners, to admit the validity of such evidence, seeing that so many caves have been inhabited by a succession of tenants, and have been selected by Man, as a place not only of domicile, but of sepulture, while some caves have also served as the channels through which the waters of occasional land-floods or engulfed rivers have flowed, so that the remains of living beings which have peopled the district at more than one era have subsequently been mingled in such caverns and confounded together in one and the same deposit. But the facts brought to light in 1858, during the systematic investigation of the Brixham cave, near Torquay in Devonshire, which will be described in the sequel, excited anew the curiosity of the British public, and prepared the way for a general admission that scepticism in regard to the bearing of cave evidence in favour of the antiquity of Man had previously been pushed to an extreme.

Since that period, many of the facts formerly adduced in favour of the co-existence in ancient times of Man with certain species of mammalia long since extinct have been re-examined in England and on the Continent, and new cases bearing on the same question, whether relating to caves or to alluvial strata in valleys, have been brought to light. To qualify myself for the appreciation and discussion of these cases, I visited, between the years 1860 and 1873, many parts of England, France, and Belgium, and have communicated personally or by letter with not a few of the geologists, English and foreign, who have taken part in these researches. Besides explaining in the present volume the results of this enquiry, I shall give a description of the glacial formations of Europe and North America, that I may allude to the theories entertained respecting their origin, and consider their probable relations in a chronological point of view to the human epoch, and why throughout a great part of the northern hemisphere they so often interpose an abrupt barrier to all attempts to trace farther back into the past the signs of the existence of Man upon the earth.

In the concluding chapters I shall offer a few remarks on the recent modifications of the Lamarckian theory of progressive development and transmutation, which are suggested by Mr. Darwin's work on the 'Origin of Species, by Variation and Natural Selection,' and the bearing of this hypothesis on the different races of mankind and their connection with other parts of the animal kingdom.

Nomenclature.—Some preliminary explanation of the nomenclature adopted in the following pages will be indispensable, that the meaning attached to the terms Recent, Pleistocene or Post-Pliocene, and Post-Tertiary may be correctly understood. In the first edition of my 'Principles of Geology,' in 1830, I divided the whole of the Tertiary formations into three groups; the Eocene, Miocene, and Pliocene, characterised by the percentage of extinct shells, or shells then unknown as living, which they contained. I then again subdivided the Pliocene into Older and Newer Pliocene. All strata of later age than these, or such as contained none but recent shells, I termed 'Recent,' and these strata were subsequently subdivided into Recent and Post-Pliocene, united under the general term Post-Tertiary.

In 1839 I proposed the term Pleistocene as an abbreviation for Newer Pliocene, and it soon became popular, because adopted by the late Edward Forbes in his admirable essay 'On the Geological Relations of the existing Fauna and Flora of the British Isles;' but he applied the term almost precisely in the sense in which I have hitherto used Post-Pliocene, and not as short for Newer Pliocene. In order, therefore, to prevent confusion, I thought it best entirely to abstain from the use of Pleistocene in future; but in a note to my 'Elements of Geology,' I advised such geologists as wished to retain Pleistocene to use it as strictly synonymous with Post-Pliocene. This was done by many, and has been found so convenient that it has now been very generally adopted; therefore, as the term Post-Pliocene has many inconveniences, especially that of being often confounded with Post-Tertiary, I propose, in this volume, to adopt the term Pleistocene for the lower subdivision of the Post-Tertiary, retaining only sometimes in brackets the word Post-Pliocene, to remind the reader who is accustomed to that term that Pleistocene is used as its synonym.

The annexed tabular view of the whole series of fossiliferous strata will enable the reader to see at a glance the chronological relation of the Recent and Pleistocene to the antecedent periods.

ABRIDGED GENERAL TABLE OF FOSSILIFEROUS STRATA.

1. RECENT.	POST-TERTIARY.		
2. PLEISTOCENE (POST-PLIOCENE).			
3. NEWER PLIOCENE.	PLIOCENE.		
4. OLDER PLIOCENE.			
5. UPPER MIOCENE.	MIOCENE.	TERTIARY or CAINOZOIC.	
6. LOWER MIOCENE.			
7. UPPER EOCENE.			
8. MIDDLE EOCENE.	EOCENE.		
9. LOWER EOCENE.			
10. MAESTRICHT BEDS.			NEOZOIC.
11. WHITE CHALK.			
12. CHLORITIC SERIES.	CRETACEOUS.		
13. GAULT.			
14. NEOCOMIAN.			
15. WEALDEN.			
16. PURBECK BEDS.			
17. PORTLAND STONE.			
18. KIMMERIDGE CLAY.			
19. CORAL RAG.	JURASSIC.	SECONDARY or MESOZOIC.	
20. OXFORD CLAY.			
21. GREAT or BATH OOLITE.			
22. INFERIOR OOLITE.			
23. LIAS.			
24. UPPER TRIAS.			
25. MIDDLE TRIAS.	TRIASSIC.		
26. LOWER TRIAS.			
27. PERMIAN.	PERMIAN.		
28. COAL-MEASURES.	CARBONIFEROUS.		
29. CARBONIFEROUS LIMESTONE.			
30. UPPER			
31. MIDDLE DEVONIAN.	DEVONIAN.	PRIMARY or PALÆOZOIC.	PALÆOZOIC.
32. LOWER			
33. UPPER SILURIAN.	SILURIAN.		
34. LOWER			
35. UPPER CAMBRIAN.	CAMBRIAN.		
36. LOWER			
37. UPPER LAURENTIAN.	LAURENTIAN.		
38. LOWER			

CHAPTER II.

RECENT PERIOD—DANISH PEAT AND SHELL MOUNDS—SWISS LAKE DWELLINGS.

WORKS OF ART IN DANISH PEAT-MOSSES—REMAINS OF THREE PERIODS OF VEGETATION IN THE PEAT—AGES OF STONE, BRONZE, AND IRON—COMPARATIVE ANTIQUITY OF THESE AGES—SHELL-MOUNDS OR ANCIENT REFUSE-HEAPS OF THE DANISH ISLANDS—CHANGE IN GEOGRAPHICAL DISTRIBUTION OF MARINE MOLLUSCA SINCE THEIR ORIGIN—EMBEDDED REMAINS OF MAMMALIA OF RECENT SPECIES—HUMAN SKULLS OF THE SAME PERIOD—PALÆOLITHIC AND NEOLITHIC DIVISIONS OF THE STONE AGE—SWISS LAKE-DWELLINGS BUILT ON PILES—STONE AND BRONZE IMPLEMENTS FOUND IN THEM—FOSSIL CEREALS AND OTHER PLANTS—REMAINS OF MAMMALIA, WILD AND DOMESTICATED—NO EXTINCT SPECIES—CHRONOLOGICAL COMPUTATIONS OF THE DATE OF THE BRONZE AND STONE PERIODS IN SWITZERLAND—LAKE-DWELLINGS, OR ARTIFICIAL ISLANDS CALLED 'CRANNOGES,' IN IRELAND.

Works of Art in Danish Peat.

WHEN treating in the 'Principles of Geology' of the changes of the earth which have taken place in comparatively modern times, I have spoken (chap. xliv.) of the embedding of organic bodies and human remains in peat, and explained under what conditions the growth of that vegetable substance is going on in northern and humid climates. Of late years, since I first alluded to the subject, more extensive investigations have been made into the history of the Danish peat-mosses. Of the results of these enquiries I shall give a brief abstract in the present chapter, that we may afterwards compare them with deposits of older date, in which the bones of extinct animals are not unfrequent, and which throw light on the antiquity of the human race.

The deposits of peat in Denmark,[*] varying in depth from ten to thirty feet, have been formed in hollows or depressions in the northern drift or boulder formation hereafter to be described. The lowest stratum, two to three feet thick, consists of swamp-peat composed chiefly of a kind of moss called sphagnum, above which lies another growth of peat, not made up exclusively of aquatic or swamp plants. Around the borders of the bogs, and at various depths in them, lie trunks of trees, especially of the Scotch fir (*Pinus sylvestris*), often three feet in diameter, which must have grown on the margin of the peat-mosses, and have frequently fallen into them. This tree is not now, nor has ever been in historical times, a native of the Danish Islands, and when introduced there has not thriven: yet it was evidently indigenous in the human period, for Steenstrup has taken out with his own hands a flint instrument from below a buried trunk of one of these pines. It appears clear that the same Scotch fir was afterwards supplanted by the sessile variety of the common oak, of which many prostrate trunks occur in the peat at higher levels than the pines: and still higher the pedunculated or stalked variety of the same oak (*Quercus Robur* L.) occurs with the alder (*Alnus glutinosa* L.), birch (*Betula verrucosa* Ehrh.), and hazel (*Corylus Avellana* L.). The oak has now in its turn been almost superseded in Denmark by the common beech (*Fagus sylvatica* L.). Other trees, such as the white birch (*Betula alba*), characterise the lower part of the bogs, and disappear from the higher; while others again, like the aspen (*Populus tremula*), occur at all levels, and still flourish in Denmark. All the land and fresh-water shells, and all the mammalia as well as the plants, whose

* An excellent account of these researches of Danish naturalists and
antiquaries has been drawn up by an
able Swiss geologist, the late M. A.

Morlot, and will be found in the Bulletin de la Société Vaudoise des Sci.
Nat., t. vi. Lausanne, 1860.

remains occur buried in the Danish peat, are of recent species.

It has been stated above, that a stone implement was found under a buried Scotch fir at a great depth in the peat. By collecting and studying a vast variety of such implements, and other articles of human workmanship preserved in peat and in sand-dunes on the coast, as also in certain shell-mounds of the aborigines presently to be described, the Danish and Swedish antiquaries and naturalists, MM. Nilsson, Steenstrup, Forchhammer, Thomsen, Worsaae and others, have succeeded in establishing a chronological succession of periods, which they have called the ages of stone, of bronze, and of iron, named from the materials which have each in their turn served for the fabrication of implements.

The age of stone in Denmark coincided with the period of the first vegetation, or that of the Scotch fir, and in part at least with the second vegetation, or that of the oak. But a considerable portion of the oak epoch coincided with ' the age of bronze,' for swords and shields of that metal, now in the museum of Copenhagen, have been taken out of peat in which oaks abound. The age of iron corresponds more nearly with that of the beech tree.[*] Although it will probably never be possible to fix exact dates for these three ages, as in fact they must differ in different parts of the world, yet in any one country, as in Denmark for example, we may perhaps hope to arrive approximately at some idea of the periods they represent. Some writers have imagined that because the lines of division are very ill-defined, and all the three ages may even be said to exist in the present day,—the Fuegians being now, as Sir J. Lubbock[†] points out, in the age of stone,—therefore any attempt at classification is useless. But Mr. Evans has well said

* Morlot, Bulletin de la Société Vaudoise des Sci. Nat., t. vi. p. 292.
† Prehistoric Times, 2nd ed. 1869, p. 4.

that these periods in no way imply an exact chronology, but only a succession of different stages of civilisation, and that though, 'like the three principal colours of the rainbow, these three stages of civilisation overlap, intermingle, and shade off the one into the other, yet their succession, so far as Western Europe is concerned, appears to be equally well defined with that of the prismatic colours.'[*]

The late M. Morlot, to whom we are indebted for a masterly sketch of the early progress of this new line of research, followed up with so much success in Scandinavia and Switzerland, observed that the introduction of the first tools made of bronze among a people previously ignorant of the use of metals, implies a great advance in the arts, for bronze is an alloy of about nine parts of copper and one of tin ; and although the former metal, copper, is by no means rare, and is occasionally found pure or in a native state, tin is not only scarce but never occurs native. To detect the existence of this metal in its ore, then to disengage it from the matrix, and finally, after blending it in due proportion with copper, to cast the fused mixture in a mould, allowing time for it to acquire hardness by slow cooling, all this bespeaks no small sagacity and skilful manipulation. Accordingly, the pottery found associated with weapons of bronze in Switzerland is of a more ornamental and tasteful style than any which belongs to the age of stone; but this is not universally the case, some of the pottery of the bronze age being very little advanced beyond that found with stone implements. Some of the moulds in which the bronze instruments were cast, and 'tags,' as they are called, of bronze, which are formed in the hole through which the fused metal was poured, have been found. The number and variety of objects belonging to the age of bronze indicate its long duration, as does the progress in the arts implied by the rudeness of the

* Evans, Ancient Stone Implements of Great Britain, 1872, p. 2.

earlier tools, often mere repetitions of those of the stone age, as contrasted with the more skilfully worked weapons of a later stage of the same period. (See fig. 1.)

Hatchets of copper have been found in the Danish peat, and it has been suggested that an age of copper must always have intervened between that of stone and bronze; but if so, the interval seems to have been short in Europe, owing apparently to the territory occupied by the aboriginal inhabitants having been invaded and conquered by a people coming from the East, to whom the use of swords, spears, and other weapons of bronze was familiar, and whose presence in Scandinavia M. Nilsson considers proved by the small size of the sword-handles, which would not fit the hand of the present European race. The abundance of bronze in the East is proved both by many passages in Homer, and also by the frequent mention of the use of brass in the early books of the Bible. For it is admitted that the word translated 'brass' in our version should be 'bronze,' the present mixture of copper and zinc, which we call brass, being then unknown.

Fig. 1.

Highly ornamented bronze dagger from the Lake of Bienne. In the collection of Herr Schwab, of Biel.[*]

The next stage of improvement, or that manifested by the substitution of iron for bronze, indicates another stride in the progress of the arts. Iron never presents itself, except in meteorites, in a native state; so that to recognise

* Keller, Pfahlbauten, Zweiter Bericht. Plate I. fig. 60.

its ores, and then to separate the metal from its matrix, demands no inconsiderable exercise of the powers of observation and invention. To fuse the ore requires an intense heat, not to be obtained without artificial appliances, such as pipes inflated by the human breath, or bellows, or some other suitable machinery.

It is very difficult to obtain any positive dates, even for the more recent transition between the iron and bronze ages, still less for the period when in any particular country bronze was superseding stone implements. We find, however, that Hesiod, writing about B.C. 850, speaks of a time when bronze had not been superseded by iron,* and Homer mentions iron but rarely, while he makes frequent reference to weapons and implements of bronze. Mr. Evans also points out that the continuance of the use of bronze cutting implements in certain religious rites, as when the Tuscans, on founding a city, ploughed the pomœrium with a bronze ploughshare, and the priests of the Sabines cut their hair with bronze knives, affords evidence of comparative antiquity; in the same way as we find the use of stone implements surviving in the early Egyptian ceremonies, and in the rite of circumcision among the Jews. We have also some indication of the great antiquity of the use of stone in the fact, that stone implements were already regarded, in very distant ages, with superstitious awe, showing that the knowledge of their construction and use must long have died out. Solacus, says Mr. Evans, who must have written at least two thousand years before our time, ' speaks, even at that remote period, of the use of stone halberds and axe-heads having so long ceased in Greece, that when found they were regarded as of superhuman origin, and invested with magical virtues.'†

* Op. et Di. I. 150.
† Evans, Stone Implements of Great Britain, pp. 4, 58.

Danish Shell-mounds, or Kjökkenmödding.

In addition to the peat-mosses, another class of memorials found in Denmark has thrown light on the pre-historic age. At certain points along the shores of nearly all the Danish islands, mounds may be seen, consisting chiefly of thousands of cast-away shells of the oyster, cockle, and other mollusks of the same species as those which are now eaten by Man. These shells are plentifully mixed up with the bones of various quadrupeds, birds and fish, which served as the food of the rude hunters and fishers by whom the mounds were accumulated. I have seen similar large heaps of oysters, and other marine shells, with interspersed stone implements, near the sea-shore, both in Massachusetts and in Georgia, U.S., covering in the latter state ten acres of ground to an average height of five feet. These mounds were left by the native North American Indians, at points near to which they were in the habit of pitching their wigwams, for centuries before the white man arrived.*

Such accumulations are called by the Danes, Kjökken-mödding, or 'kitchen-middens,' midden being a name still used in the North of England for refuse-heaps. Scattered all through them are flint knives, hatchets, and other instruments of stone, horn, wood, and bone, with fragments of coarse pottery, mixed with charcoal and cinders, but never any implements of bronze, nor of iron. The stone hatchets and knives had been polished and sharpened by rubbing, and in this respect are less rude than those of an older date, associated in France with the bones of extinct mammalia, of which more in the sequel. The mounds vary in height from 3 to 10 feet, and in area are some of them 1,000 feet long, and from 150 to 200 wide. They are rarely placed more than 10 feet above the level of the sea, and are confined to its

* Second Visit to the United States, p. 338, vol. I. 1845.

immediate neighbourhood, or if not (and there are cases
where they are several miles from the shore), the distance is
ascribable to the entrance of a small stream, which has
deposited sediment, or to the growth of a peaty swamp, by
which the land has been made to advance on the Baltic, as
it is still doing in many places, aided, according to M.
Puggaard, by a very slow upheaval of the whole country at
the rate of two or three inches in a century.

There is also another geographical fact equally in favour
of the antiquity of the mounds, viz., that they are wanting
on those parts of the coast which border the Western Ocean,
or exactly where the waves are now slowly eating away the
land. There is every reason to presume that originally there
were stations along the coast of the German Ocean as well
as that of the Baltic, but by the gradual undermining of
the cliffs they have all been swept away.

Another striking proof, perhaps the most conclusive of
all, that the 'kitchen-middens' are very old, is derived
from the character of their embedded shells. These consist entirely of living species; but, in the first place, the
common eatable oyster is among them, attaining its full
size, whereas the same *Ostrea edulis* cannot live at present
in the brackish waters of the Baltic, except near its entrance,
where, whenever a north-westerly gale prevails, a current
setting in from the ocean pours in a great body of salt
water. Yet it seems that during the whole time of the
accumulation of the 'shell-mounds' the oyster flourished in
places from which it is now excluded. In like manner,
the cockle, mussel, and periwinkle (*Cardium edule, Mytilus edulis*, and *Littorina littorea*), which are met with
in great numbers in the 'refuse-heaps,' are of the ordinary
dimensions which they acquire in the ocean, whereas the
same species now living in the adjoining parts of the Baltic
only attain a third of their natural size, being stunted and

dwarfed in their growth by the quantity of fresh water
poured by rivers into that inland sea.* Hence we may
confidently infer that in the days of the aboriginal hunters
and fishers, the ocean had freer access than now to the
Baltic, communicating probably through the peninsula of
Jutland, Jutland having been at no remote period an
archipelago. Even in the course of the present century,
the salt waters have made one irruption into the Baltic by
the Lymfiord, although they have been now again excluded.
It is also affirmed that other channels were open in historical times which are now silted up.†

If we next turn to the remains of vertebrata preserved
in the mounds, we find that here also, as in the Danish
peat-mosses, all the quadrupeds belong to species known to
have inhabited Europe within the memory of Man. No
remains of the mammoth, or rhinoceros, or of any extinct
species appear, except those of the wild bull (*Bos Urus*
Linn., or *Bos primigenius* Bojanus), which are in such
numbers as to prove that the species was a favourite food of
the ancient people. But as this animal was seen by Julius
Cæsar, and survived long after his time, its presence alone
would not go far to prove the mounds to be of high antiquity. The Lithuanian aurochs or bison (*Bos Bison* L.,
Bos priscus Boj.), which has escaped extirpation only because
protected by the Russian Czars, surviving in one forest in
Lithuania, has not yet been met with, but will no doubt be
detected hereafter, as it has been already found in the
Danish peat. The beaver, long since destroyed in Denmark,
occurs frequently, as does the seal (*Phoca Gryppus* Fab.),
now very rare on the Danish coast. With these are mingled
bones of the red deer and roe, but the reindeer has not yet
been found. There are also the bones of many carnivora,

* See Principles of Geology, 11th   † See Morlot, Bulletin de la Société
edition, vol. ii. p. 193.   Vaudoise des Sci. Nat., t. vi.

such as the lynx, fox, and wolf, but no signs of any domesticated animals except the dog. The long bones of the
larger mammalia have been all broken as if by some
instrument, in such a manner as to allow of the extraction
of the marrow, and the gristly parts have been gnawed off,
as if by dogs, to whose agency is also attributed the almost
entire absence of the bones of young birds and of the
smaller bones and softer parts of the skeletons of birds
in general, even of those of large size. In reference to
the latter, it has been proved experimentally by Professor
Steenstrup, that if the same species of birds are now given
to dogs, they will devour those parts of the skeleton which
are missing, and leave just those which are preserved in the
old ' kitchen-middens.'

The dogs of the mounds, the only domesticated animals,
are of a smaller race than those of the bronze period, as
shown by the peat-mosses, and the dogs of the bronze age
are inferior in size and strength to those of the iron age.
The domestic ox, horse, and sheep, which are wanting in the
mounds, are confined to that part of the Danish peat which
was formed in the ages of bronze and iron.

Among the bones of birds, scarcely any are more frequent
in the mounds than those of the auk (*Alca impennis*),
now extinct in Europe, having but lately died out in Iceland, but said still to survive in Greenland, where, however,
its numbers are fast diminishing. The Capercailzie (*Tetrao
Urogallus*) is also met with, and may, it is suggested,
have fed on the buds of the Scotch fir in times when that
tree flourished around the peat-bogs. The different stages of
growth of the roe-deer's horns, and the presence of the wild
swan, now only a winter visitor, have been appealed to as
proving that the aborigines resided in the same settlements
all the year round. That they also ventured out to sea in
canoes such as are now found in the peat-mosses, hollowed

out of the trunk of a single tree, to catch fish far from land,
is testified by the bony relics of several deep-sea species, such
as the herring, cod, and flounder. The ancient people were
not cannibals, for no human bones are mingled with the spoils
of the chase. Skulls, however, have been obtained not only
from peat, but from tumuli of the stone period believed to be
contemporaneous with the mounds. These skulls are small
and round, and have a prominent ridge over the orbits of
the eyes, showing that the ancient race was of small stature,
with round heads and overhanging eyebrows,—in short, they
bore a considerable resemblance to the modern Laplanders.
The human skulls of the bronze age found in the Danish peat,
and those of the iron period, are of an elongated form and
larger size. There appear to be very few well-authenticated examples of crania referable to the bronze period,—a
circumstance no doubt attributable to the custom prevalent
among the people of that era of burning their dead and
collecting their bones in funeral urns. Sir John Lubbock
has collected together the evidence as to the different modes
of burial prevailing in the three ages of stone, bronze, and
iron, and he arrives at the conclusion that interments where
the corpse is in a sitting or contracted posture belong to the
stone age, those in which the body has been burnt, to the age
of bronze, while cases in which the skeleton is extended may
be referred, with little hesitation, to the age of iron.[*]

No traces of grain of any sort have hitherto been discovered, nor any other indication that the men of the
Kjökken-möddings had any knowledge of agriculture. The
only vegetable remains in the mounds are burnt pieces of
wood and some charred substance referred by Dr. Forchhammer to the *Zostera marina*, a sea plant which was perhaps
used in the production of salt.[†]

---

[*] Prehistoric Times. 1865, p. 148.
[†] Lubbock, on the Kjökken-möddings, Nat. Hist. Rev. 1861, p. 490.

What may be the antiquity of the earliest human remains preserved in the Danish peat cannot be estimated in centuries with any approach to accuracy. In the first place, in going back to the bronze age, we already find ourselves beyond the reach of history or even of tradition. In the time of the Romans the Danish Isles were covered, as now, with magnificent beech forests. Nowhere in the world does this tree flourish more luxuriantly than in Denmark, and eighteen centuries seem to have done little or nothing towards modifying the character of the forest vegetation. Yet in the antecedent bronze period there were no beech trees, or at most but a few stragglers, the country being then covered with oak. In the age of stone again, the Scotch fir prevailed (see p. 9), and already there were human inhabitants in those old pine forests. How many generations of each species of tree flourished in succession before the pine was supplanted by the oak, and the oak by the beech, can be but vaguely conjectured, but the minimum of time required for the formation of so much peat must, according to the estimate of Steenstrup and other good authorities, have amounted to at least 4,000 years; and there is nothing in the observed rate of the growth of peat opposed to the conclusion that the number of centuries may have been four times as great.

As to the 'shell-mounds,' they correspond in date to the older portion of the peaty record, or to the earliest part of the age of stone as known in Denmark. Sir John Lubbock believes them to be referable to the beginning of that division of the stone age, called by him 'Neolithic,' when the art of polishing flint implements was known but not far advanced, and in which we find no trace of metal except gold. The earlier or 'Palæolithic' division of the stone age was that of the 'Drift' of Amiens and Abbeville, when man was contemporary in Europe with the mammoth, rhinoceros, and other extinct animals, and when

c

the weapons were of a ruder and unpolished type; whether
some of the less perfect weapons of the shell-mounds may
belong to this earlier period, is a point not yet clearly ascertained.

Ancient Swiss Lake-dwellings, built on Piles.

In the shallow parts of many Swiss lakes, where there is
a depth of no more than from five to fifteen feet of water,
ancient wooden piles are observed at the bottom sometimes
worn down to the surface of the mud, sometimes projecting
slightly above it.   These have evidently once supported
villages, nearly all of them of unknown date, but the most
ancient of which certainly belonged to the age of stone, for
hundreds of implements resembling those of the Danish
shell-mounds and peat-mosses have been dredged up from
the mud into which the piles were driven.

The earliest historical account of such habitations is that
given by Herodotus of a Thracian tribe, who dwelt, in the
year 520 B.C., in Prasias, a small mountain-lake of Pæonia,
now part of Modern Roumelia.*

Their habitations were constructed on platforms raised
above the lake, and resting on piles.   They were connected
with the shore by a narrow causeway of similar formation.
Such platforms must have been of considerable extent, for
the Pæonians lived there with their families and horses.
Their food consisted largely of the fish which the lake
produced in abundance.

In rude and unsettled times, such insular sites afforded
safe retreats, all communication with the main land being
cut off, except by boats, or by such wooden bridges as could
be easily removed.

The Swiss lake-dwellings seem first to have attracted

* Herodotus, lib. v. cap. 16.—Rediscovered by M. Deville, Nat. Hist. Rev
Oct. 1862, vol. ii. p. 486.

attention during the dry winter of 1853–4, when the lakes
and rivers sank lower than had ever been previously known,
and when the inhabitants of Meilen, on the lake of Zurich,
resolved to raise the level of some ground and turn it into
land, by throwing mud upon it obtained by dredging in the
adjoining shallow water. During these dredging operations
they discovered a number of wooden piles deeply driven into
the bed of the lake, and among them a great many hammers,
axes, celts, and other instruments. All these belonged to the
stone period with two exceptions, namely, an armlet of thin
brass wire, and a small bronze hatchet.

Fragments of rude pottery fashioned by the hand were
abundant, also masses of charred wood, supposed to have
formed parts of the platform on which the wooden cabins
were built. Of this burnt timber, on this and other sites
subsequently explored, there was such an abundance as to
lead to the conclusion that many of the settlements must
have perished by fire. Herodotus has recorded that the
Pæonians, above alluded to, preserved their independence
during the Persian invasion, and defied the attacks of Darius
by aid of the peculiar position of their dwellings. 'But
their safety,' observes Mr. Wylie,* 'was probably owing to
their living in the middle of the lake, ἐν μέσῃ τῇ λίμνῃ, whereas
the ancient Swiss settlers were compelled by the rapidly
increasing depth of the water near the margins of their lakes
to construct their habitations at a short distance from the
shore, within easy bowshot of the land, and therefore not out
of reach of fiery projectiles, against which thatched roofs and
wooden walls could present but a poor defence.' To these
circumstances and to accidental fires we are probably indebted for the frequent preservation, in the mud around the
site of the old settlements, of the most precious tools and

* W. M. Wylie, M.A., Archæologia, vol. xxxviii., 1859, a valuable paper on
the Swiss and Irish lake habitations.

works of art, such as would never have been thrown into the
Danish 'shell-mounds,' which have been aptly compared to a
modern dusthole.

Dr. Ferdinand Keller of Zurich has drawn up a series
of most instructive memoirs, illustrated with well-executed
plates, of the treasures in stone, bronze, and bone brought to
light in these subaqueous repositories, and has given an ideal
restoration of part of one of the old villages (see plate 1,
Frontispiece)[*], such as he conceives may have existed on the
Lakes of Zurich and Bienne. In this view, however, he has not
simply trusted to his imagination, but has availed himself of
a sketch published by M. Dumont d'Urville, of similar habitations of the Papoos in New Guinea in the Bay of Dorei.
It is also stated by Dr. Keller, that on the river Limmat,
near Zurich, so late as the last century, there were several
fishing-huts constructed on the same plan.[†] It will be
remarked, that one of the cabins is represented as circular.
That such was the form of many in Switzerland is inferred
from the shape of pieces of clay which lined the interior, and
which owe their preservation apparently to their having been
hardened by fire when the village was burnt. In the sketch
some fishing-nets are seen spread out to dry on the wooden
platform. The Swiss archæologist has found abundant
evidence of fishing gear, consisting of pieces of cord, hooks,
and stones used as weights. A canoe also is introduced,
such as are occasionally met with. One of these, made of
the trunk of a single tree, fifty feet long, and three and a
half feet wide, was found capsized at the bottom of the Lake of
Bienne. It appears to have been laden with stones, such as
were used to raise the foundation of some of the artificial
islands.

It is believed that as many as 300 wooden huts were

* Keller, Pfahlbauten, Antiqua-   xiii. 1858-61.
rische Gesellschaft in Zürich, Bd. xii.   † Keller, ibid. Bd. ix. p. 81, note.

sometimes comprised in one settlement, and that they may
have contained about 1,000 inhabitants. At Wangen, M.
Lohle has calculated that 40,000 piles were used, probably
not all planted at one time nor by one generation. Among
the works of great merit devoted specially to a description of
the Swiss lake-habitations is that of M. Troyon, published in
1860.* The number of sites which he and other authors
have already enumerated in Switzerland is truly wonderful.
They occur on the large lakes of Constance, Zurich, Geneva,
and Neufchatel, and on most of the smaller ones. Some are
exclusively of the stone age, others of the bronze period. Of
these last more than twenty are spoken of on the Lake of
Geneva alone, more than forty on that of Neufchatel, and
twenty on the small Lake of Bienne.

One of the sites first studied by the Swiss antiquaries was
the small lake of Moosseedorf, near Berne, where implements
of stone, horn, and bone, but none of metal, were obtained.
Although the flint here employed must have come from a
distance (probably from the South of France), the chippings
of the material are in such profusion as to imply that there
was a manufactory of implements on the spot. Here also,
as in several other settlements, hatchets and wedges of jade
have been observed of a kind said not to occur in Switzerland
or the adjoining parts of Europe, and which some mineralo-
gists would fain derive from the East ; amber also, which,
it is supposed, was imported from the shores of the Baltic.

At Wangen near Stein, on the Lake of Constance, another
of the most ancient of the lake-dwellings, hatchets of serpentine and greenstone, and arrow-heads of quartz, have been
met with. This settlement belonged exclusively to the stone
period, for Sir J. Lubbock informs us that among 5,000
objects collected there, no trace of metal has been found.
Remains of a kind of cloth, supposed to be of flax, not woven

* Sur les Habitations Lacustres.

but plaited, have been detected. Professor Heer has recognised lumps of carbonized wheat (*Triticum vulgare*), and
grains of another kind (*T. dicoccum*), and barley (*Hordeum
distichon*), and flat round cakes of bread; and at Robben-
hausen and elsewhere *Hordeum hexastichon* in fine ears, the
same kind of barley which is found associated with Egyptian
mummies, showing clearly that in the stone period, the
lake-dwellers cultivated all these cereals, besides having
domesticated the dog, the ox, the sheep and the goat.

Carbonized apples and pears of small size, such as still
grow in the Swiss forests, stones of the wild plum, seeds of
the raspberry and blackberry, and beech-nuts, also occur in
the mud, and hazel-nuts in great plenty.

Near Morges, on the lake of Geneva, a settlement of the
bronze period, no less than forty hatchets of that metal have
been dredged up, and in many other localities the number
and variety of weapons and utensils discovered, in a fine state
of preservation, are truly astonishing.

It is remarkable that as yet nearly all the settlements
of the bronze period are confined to Western and Central
Switzerland. In the more eastern lakes, such as Lakes
Constance and Zurich, they belong almost exclusively to the
stone period; and it has been suggested that the use of stone
may have existed in the east of Switzerland, while bronze
had already been introduced in the west. But one settlement
of the bronze age has been found on Lake Constance;[*] and
this occurrence of dwellings of two distinct periods in the
same lake, the one affording weapons exclusively of stone,
and the other of bronze, is of great use in establishing the
fact that they must have flourished in succession; for had
the lake been contemporaneously inhabited by races using
severally weapons of the two periods, a mixture of the two
kinds of implements would have been inevitable.

* Lubbock, Prehistoric Times, 2nd ed. p. 207.

The tools, ornaments, and pottery of the bronze period in
Switzerland bear a close resemblance to those of the same
period in Denmark, attesting the wide spread of a uniform
civilization over Central Europe at that era. In some few
of the Swiss aquatic stations, as in the lakes of Bienne and
Neufchatel, a mixture of bronze and iron implements has been
observed, but no coins. At Tiefenau, near Berne, in ground
supposed to have been a battle-field, coins and medals of
bronze and silver, struck at Marseilles, and of Greek manufacture, and iron swords, have been found, all belonging to
the first and pre-Roman division of the age of iron. But no
Roman coins, pottery, or other relics have ever been found
associated with bronze weapons.

In the settlements of the bronze era the wooden piles are
not so much decayed as are those of the stone period; the
latter having wasted down quite to the level of the mud,
whereas the piles of the bronze age (as in the Lake of Bienne,
for example) still project above it. Some of the piles of the
bronze age are found at a depth of fifteen feet below the
present surface of the water, and Sir J. Lubbock observes
that this affords evidence that the Swiss lakes cannot have
stood at a much higher level than now at the time when
the dwellings were erected, for buildings could not have been
constructed over water of much greater depth.[*]

Professor Rütimeyer of Basle, well known to paleontologists
as the author of several important memoirs on fossil vertebrata, has published a scientific description of great interest
of the animal remains dredged up at various stations where
they had been embedded for ages in the mud into which the
piles were driven.[†] These bones bear the same relation to
the primitive inhabitants of Switzerland and some of their
immediate successors as do the contents of the Danish

* Prehistoric Times, 2nd ed. p. 177.
† Die Fauna der Pfahlbauten in der Schweiz. Basel, 1861.

'refuse-heaps' to the ancient fishing and hunting tribes who lived on the shores of the Baltic.

The list of wild mammalia enumerated in this excellent treatise contains no less than twenty-four species, exclusive of several domesticated ones: besides which there are eighteen species of birds, the wild swan, goose, and two species of ducks being among them; also three reptiles, including the eatable frog and fresh-water tortoise; and lastly, nine species of fresh-water fish.   All these (amounting to fifty-four species) are with one exception still living in Europe. The exception is the wild bull (*Bos primigenius*), which, as before stated, survived in historical times. The following are the mammalia alluded to:—The bear (*Ursus Arctos*), badger, common marten, polecat, ermine, weasel, otter, wolf, fox, wild cat, hedgehog, squirrel, field-mouse (*Mus sylvaticus*), hare, beaver, hog (comprising two races, namely, the wild boar and swamp hog), the stag (*Cervus Elaphus*), roe-deer, fallow-deer, elk, steinbock (*Capra Ibex*), chamois, Lithuanian bison, and wild bull. The domesticated species comprise the dog, horse, ass, pig, goat, sheep, and several bovine races.

The greater number, if not all, of these animals served for food, and all the bones which contained marrow have been split open in the same way as the corresponding ones found in the shell-mounds of Denmark before mentioned. The bones both of the wild bull and the bison are invariably split in this manner. Mr. Pengelly, however, does not think that this was done solely for the purpose of obtaining the marrow, which could easily be extracted by breaking off the end of the bone. He believes that the bones were split up as material for making bone tools, such as have been found in Kent's cavern and elsewhere.* As a rule, the lower jaws with teeth occur in greater abundance than any other parts

* Condition of Bones in Kent's Cavern, Trans. Devon. Assoc., 1869, p. 466.

of the skeleton,—a circumstance which geologists know holds good in regard to fossil mammalia of all periods. As yet the reindeer is missing in the Swiss lake-settlements as in the Danish 'refuse-heaps,' although this animal in more ancient times ranged over France, together with the mammoth, as far south as the Pyrenees.

A careful comparison of the bones from different sites has shown that in settlements such as Wangen and Moosseedorf, belonging to the earliest lake-dwellings, when stone implements were in use, and when the habits of the hunter predominated over the pastoral, venison, or the flesh of the stag and roe, was more eaten than the flesh of the domestic cattle and sheep. This was afterwards reversed in the later stone period and in the age of bronze. At that later period also the tame pig, which is wanting in some of the oldest stations, had replaced the wild boar as a common article of food. In the beginning of the age of stone in Switzerland, the goats outnumbered the sheep, but towards the close of the same period the sheep were more abundant than the goats.

The fox in the first era was very common, but it nearly disappears in the bronze age, during which period a large hunting-dog, supposed to have been imported into Switzerland from some foreign country, becomes the chief representative of the canine genus.

A single fragment of the bone of a hare (*Lepus timidus*) has been found at Moosseedorf. The almost universal absence of this quadruped is supposed to imply that the Swiss lake-dwellers were prevented from eating that animal by the same superstition which now prevails among the Laplanders, and which Julius Cæsar found in full force amongst the ancient Britons.* That the lake-dwellers should have fed so largely on the fox, while they abstained from touching

* Commentaries, lib. v. ch. 12.

the hare, establishes, says Rütimeyer, a singular contrast
between their tastes and ours.

Even in the earliest settlements, as already hinted,
several domesticated animals occur, namely, the ox, sheep,
goat, and dog. Of the last three, each was represented by
one race only; but there were two races of cattle, the most
common being of small size, and called by Rütimeyer *Bos
brachyceros* (*Bos longifrons* Owen), or the marsh cow, the
other derived from the wild bull (*Bos primigenius*);
though, as no skull has yet been discovered, this identification is not so certain as could be wished. It is, however,
beyond question that at a later era, namely, towards the
close of the stone and beginning of the bronze period, the
lake-dwellers had succeeded in taming that formidable
brute, the *Bos primigenius*, the Urus of Cæsar, which he
described as fierce, swift, and strong, and scarcely inferior to
the elephant in size. In a tame state its bones were somewhat
less massive and heavy, and its horns were somewhat smaller
than in wild individuals. Still, in its domesticated form, it
rivalled in dimensions the largest living cattle, those of
Friesland, in North Holland, for example. When most
abundant, as at Concise on the Lake of Neufchatel, it had
nearly superseded the smaller race, *Bos brachyceros*, and
was accompanied there for a short time by a third bovine
variety, called *Bos trochoceros*, an Italian race, supposed to
have been imported from the southern side of the Alps.[*]
The last-mentioned race, however, seems only to have lasted
a short time in Switzerland. Remains of a fourth race,
Bos frontosus, now common in the north of Switzerland,
and believed by Rütimeyer to be a variety of the Urus,
occur sparingly in the lake-dwellings, more particularly
in those belonging to the bronze period.[†]

* Cæsar's Commentaries, lib. v. ch. Ingis, 1868, p. 219; and Lubbock,
12, p. 181.   Prehistoric Times, p. 197.
† Rütimeyer, Archiv für Anthropo-

The wild bull (*Bos primigenius*) is supposed to have flourished for a while both in a wild and tame state, just as now in Europe the domestic pig co-exists with the wild boar; but some of the most eminent living zoologists are still at issue as to whether our larger domestic cattle of Northern Europe are the descendants of this wild bull.[*]

In the later division of the stone period there were two tame races of the pig, according to Rütimeyer; one large, and derived from the wild boar, the other smaller, called the 'marsh-hog,' or *Sus Scrofa palustris*. It may be asked how the osteologist can distinguish the tame from wild races of the same species by their skeletons alone. Among other characters, the diminished thickness of the bones and the comparative smallness of the ridges which afford attachment to the muscles, are relied on; also the smaller dimensions of the tusks in the boar, and of the whole jaw and skull; and in like manner, the diminished size of the horns of the bull and other modifications, which are the effects of a regular supply of food, and the absence of all necessity of exerting their activity and strength to obtain subsistence and defend themselves against their enemies.

A middle-sized race of dogs continued unaltered throughout the whole of the stone period: but the people of the bronze age possessed a larger hunting-dog, and with it a small horse, of which genus very few traces have been detected in the earlier settlements,—a single tooth, for example, at Wangen, and only a few bones at two or three other places.

In passing from the oldest to the most modern sites, the extirpation of the elk and beaver, and the gradual reduction in numbers of the bear, stag, roe, and fresh-water tortoise are distinctly perceptible. The aurochs, or Lithuanian bison,

[*] Bell, British Quadrupeds. p. 410; and Owen, British Fossil Mammal. p. 500.

appears to have died out in Switzerland about the time when
weapons of bronze came into use. It is only in a few of the
most modern lake-dwellings, such as Noville and Chavannes,
in the Canton de Vaud, (which the antiquaries refer to the
sixth century,) that some traces are observable of the domestic cat, as well as of a sheep with crooked horns, and
with them bones of the domestic fowl.

After the sixth century, no extinction of any wild quadruped, nor introduction of any tame one, appears to have
taken place, but the fauna was still modified by the wild
species continuing to diminish in number, and the tame
ones to become more diversified by breeding and crossing,
especially in the case of the dog, horse, and sheep. On the
whole, however, the divergence of the domestic races from
their aboriginal wild types, as exemplified at Wangen and
Moosseedorf, is confined, according to Professor Rütimeyer,
within narrow limits. As to the goat, it has remained nearly
constant and true to its pristine form, and the small race of
goat-horned sheep still lingers in some Alpine valleys in the
Upper Rhine ; and in the same region a race of pigs, corresponding to the domesticated variety of *Sus Scrofa palustris*, may still be seen.

Amidst all this profusion of animal remains extremely
few bones of Man have been discovered ; and only one skull,
dredged up from Meilen, on the Lake of Zurich, of the early
stone period, seems as yet to have been carefully examined.
Respecting this specimen, Professor His observes that it exhibits, instead of the small and rounded form proper to the
Danish peat-mosses, a type much more like that now prevailing in Switzerland, which is intermediate between the
long-headed and short-headed form.[*]

It is still a question whether any of these subaqueous
repositories of ancient relics in Switzerland go back so far

* Rütimeyer, Die Fauna der Pfahlbauten in der Schweiz, p. 151.

in time as the shell-mounds of Denmark, for in these last
there are no domesticated animals, except the dog, and no
signs of the cultivation of wheat or barley; whereas we have
seen that, in one of the oldest of the Swiss settlements, at
Wangen, no less than three cereals make their appearance,
with four kinds of domestic animals. Yet there is no small
risk of error in speculating on the relative claims to antiquity of such ancient tribes, for some of them may have
remained isolated for ages, and stationary in their habits,
while others advanced and improved.

We know that nations, both before and after the introduction of metals, may continue in very different stages
of civilisation, even after commercial intercourse has been
established between them, and where they are separated by
a less distance than that which divides the Alps from the
Baltic.

The attempts of the Swiss geologists and archæologists
to estimate definitely in years the antiquity of the bronze
and stone periods, although as yet confessedly imperfect,
deserve some notice. The most elaborate calculation is that
made by the late M. Morlot, respecting the delta of the
Tinière, a torrent which flows into the Lake of Geneva
near Villeneuve. This small delta, to which the stream is
annually making additions, is composed of gravel and sand.
Its shape is that of a flattened cone, and its internal
structure has been laid open to view in a railway cutting
one thousand feet long, and thirty-two feet deep. The
regularity of its structure throughout implies, according to
M. Morlot, that it has been formed very gradually, and by
the uniform action of the same causes. Three layers of
vegetable soil, each of which must at one time have formed
the surface of the cone, have been cut through at different
depths. The first of these was traced over a surface of
15,000 square feet, having an average thickness of five

inches, and being about four feet below the present surface
of the cone. This upper layer contained tiles and a coin,
supposed by M. Morlot to belong to the Roman period.
The second layer, followed over a surface of 25,000 square
feet, was six inches thick, and lay at a depth of ten feet.
In it were found fragments of unvarnished pottery, and a
pair of tweezers in bronze, indicating the bronze epoch.
The third layer, followed for 35,000 square feet, was six or
seven inches thick, and nineteen feet from the surface. In
it were fragments of rude pottery, pieces of charcoal, broken
bones, and a human skeleton having a small, round, and
very thick skull. M. Morlot, assuming the Roman period
to represent an antiquity of from thirteen to eighteen
centuries, assigns to the bronze age a date of between 3,000
and 4,000 years, and to the oldest layer, that of the stone
period, an age of from 5,000 to 7,000 years.

Another calculation has been made by M. Troyon to
obtain the approximate date of the remains of an ancient
settlement, built on piles and preserved in a peat-bog at
Chamblon, near Yverdun, on the Lake of Neufchatel. The
site of the ancient Roman town of Eburodunum (Yverdon),
once on the borders of the lake, and between which and the
shore there now intervenes a zone of newly-gained dry land,
2,500 feet in breadth, shows the rate at which the bed of
the lake has been filled up with river sediment in fifteen
centuries. Assuming the lake to have retreated at the same
rate before the Roman period, the pile-works of Chamblon,
which are of the bronze period, must be at the least 3,300
years old.

For the third calculation, communicated to me by M.
Morlot, we are indebted to M. Victor Gilliéron, of Neuve-
ville, on the Lake of Bienne. It relates to the age of a
pile-dwelling, the mammalian bones of which are considered
by M. Rütimeyer to indicate the earliest portion of the

stone period of Switzerland, and to correspond in age with
the settlements of Moosseedorf.

The piles in question occur at the Pont de Thièle,
between the Lakes of Bienne and Neufchatel. The old
convent of St. Jean, founded 750 years ago, and built
originally on the margin of the Lake of Bienne, is now at
a considerable distance from the shore, and affords a measure of the rate of the gain of land in seven centuries and
a half. Assuming that a similar rate of the conversion of
water into marshy land prevailed antecedently, we should
require an addition of sixty centuries for the growth of the
morass intervening between the convent and the aquatic
dwelling of Pont de Thièle, in all 6,750 years. M. Morlot,
after examining the ground, thinks it highly probable that
the shape of the bottom on which the morass rests is
uniform; but this important point has not yet been tested
by boring. The result, if confirmed, would agree exceedingly well with the chronological computation before
mentioned of the age of the stone period of Tinière.

Post-glacial Lake-dwelling in the North of Italy.

We learn from M. de Mortillet that in the peat which
has filled up one of the 'morainic lakes' formed by the
ancient glacier of the Ticino, M. Moro has discovered at
Mercurago the piles of a lake-dwelling like those of Switzerland, together with various utensils, and a canoe hollowed
out of the trunk of a tree. From this fact we learn that
south of the Alps, as well as north of them, a primitive
people having similar habits flourished after the retreat of
the great glaciers.

Irish Lake-dwellings, or Crannoges.

The lake-dwellings of the British Isles, although not
explored as yet with scientific zeal, as those of Switzerland

have been in the last ten years, are yet known to be very
numerous, and when carefully examined will not fail to
throw great light on the history of the bronze and stone
periods.

In the lakes of Ireland alone, no less than forty-six
examples of artificial islands, called *crannoges*, have been
discovered. They occur in Leitrim, Roscommon, Cavan,
Down, Monaghan, Limerick, Meath, King's County, and
Tyrone.* One class of these 'stockaded islands,' as they have
been sometimes called, was formed, according to Mr. Digby
Wyatt, by placing horizontal oak beams at the bottom of
the lake, into which oak posts, from six to eight feet high,
were mortised, and held together by cross beams, till a
circular enclosure was obtained.

A space of 520 feet diameter, thus inclosed at Lagore,
was divided into sundry timbered compartments, which
were found filled up with mud or earth, from which were
taken 'vast quantities of the bones of oxen, swine, deer,
goats, sheep, dogs, foxes, horses, and asses.' All these were
discovered beneath sixteen feet of bog, and were used for
manure; but specimens of them are said to be preserved in
the museum of the Royal Irish Academy. From the same
spot were obtained a great collection of antiquities, which,
according to Lord Talbot de Malahide and Mr. Wylie, were
referable to the ages of stone, bronze, and iron.†

In Ardekillin Lake, in Roscommon, an islet of an oval
form was observed, made of a layer of stones resting on logs
of timber. Round the artificial islet or *crannoge* thus
formed, was a stone wall raised on oak piles. A careful
description has been put on record by Captain Mudge, R.N.,
of a curious log-cabin discovered by him in 1833 in Drum-
kellin bog, in Donegal, at a depth of fourteen feet from the

* W. M. Wylie, Archæologia, vol. xxxviii. p. 8. 1859.
† W. M. Wylie, ibid.

surface. It was twelve feet square, and nine feet high, being divided into two stories, each four feet high. The planking was of oak, split with wedges of stone, one of which was found in the building. The roof was flat. A staked inclosure had been raised round the cabin, and remains of other similar huts adjoining were seen but not explored. A stone celt, found in the interior of the hut, and a piece of leather sandal, also an arrow-head of flint, and in the bog close at hand a wooden sword, give evidence of the remote antiquity of this building, which may be taken as a type of the early dwellings of the *crannoge* islands.

'The whole structure,' says Captain Mudge, 'was wrought with the rudest kind of implements, and the labour bestowed on it must have been immense. The wood of the mortises was more bruised than cut, as if by a blunt stone chisel.' [*] Such a chisel lay on the floor of the hut, and by comparing it with the marks of the tool used in forming the mortises, they were found 'to correspond exactly, even to the slight curved exterior of the chisel; but the logs had been hewn by a larger instrument, in the shape of an axe. On the floor of the dwelling lay a slab of freestone, three feet long and fourteen inches thick, in the centre of which was a small pit three quarters of an inch deep, which had been chiselled out. This is presumed to have been used for holding nuts to be cracked by means of one of the round shingle stones, also found there, which had served as a hammer. Some entire hazel-nuts and a great quantity of broken shells were strewed about the floor.'

The foundations of the house were made of fine sand, such as is found with shingle on the sea-shore about two miles distant. Below the layer of sand the bog or peat was ascertained, on probing it with an instrument, to be at least

[*] Mudge, Archæologia, vol. xxvi.

fifteen feet thick. Although the interior of the building
when discovered was full of 'bog' or peaty matter, it seems
when inhabited to have been surrounded by growing trees,
some of the trunks and roots of which are still preserved in
their natural position. The depth of overlying peat affords
no safe criterion for calculating the age of the cabin or
village, for I have shown elsewhere* that both in England
and Ireland, within historical times, bogs have burst and
sent forth great volumes of black mud, which has been
known to creep over the country at a slow pace, flowing
somewhat at the rate of ordinary lava-currents, and sometimes overwhelming woods and cottages, and leaving a
deposit upon them of bog-earth fifteen feet thick.

None of these Irish lake-dwellings were built, like those
of Helvetia, on platforms supported by piles deeply driven
into the mud. 'The *crannoge* system of Ireland seems,'
says Mr. Wylie, 'well nigh without a parallel in Swiss
waters.'

* Principles of Geology, 11th ed. vol. ii. p. 811.

CHAPTER III.

FOSSIL HUMAN REMAINS AND WORKS OF ART OF THE
RECENT PERIOD,

Continued.

DELTA AND ALLUVIAL PLAIN OF THE NILE—BURNT BRICKS IN EGYPT
BEFORE THE ROMAN ERA—BORINGS IN 1851-54—ANCIENT MOUNDS
OF THE VALLEY OF THE OHIO—THEIR ANTIQUITY—DELTA OF THE
MISSISSIPPI—ANCIENT HUMAN REMAINS IN CORAL REEFS OF FLORIDA
—CHANGES IN PHYSICAL GEOGRAPHY IN THE HUMAN PERIOD—BURIED
CANOES IN MARINE STRATA NEAR GLASGOW—NO UPHEAVAL OF THE
SHORES OF THE FIRTH OF FORTH SINCE THE ROMAN OCCUPATION—
ROMAN SLAB MARKING THE HEIGHT ABOVE THE SEA-LEVEL OF THE
WALL OF ANTONINE—FOSSIL WHALES NEAR STIRLING—UPRAISED MARINE
STRATA OF SWEDEN ON SHORES OF THE BALTIC AND THE OCEAN—
ATTEMPTS TO COMPUTE THEIR AGE.

Delta and Alluvial Plain of the Nile.

SOME new facts of high interest illustrating the geology of
the alluvial land of Egypt were brought to light between
the years 1851 and 1854, in consequence of investigations
suggested to the Royal Society by Mr. Leonard Horner, and
which were carried out partly at the expense of the Society.
The practical part of the undertaking was entrusted by Mr.
Horner to an Armenian officer of engineers, Hekekyan Bey,
who had for many years pursued his scientific studies in
England, and was in every way highly qualified for the task.

It was soon found that to obtain the required information
respecting the nature, depth, and contents of the Nile mud
in various parts of the valley, a larger outlay was called for
than had been originally contemplated. This expense the
viceroy, Abbas Pacha, munificently undertook to defray out

D 2

of his treasury, and his successor, after his death, continued
the operations with the same princely liberality.

Several engineers and a body of sixty workmen were
employed under the superintendence of Hekekyan Bey, men
inured to the climate, and able to carry on the sinking of
shafts and borings during the hot months, after the waters
of the Nile had subsided, and in a season which would have
been fatal to Europeans.

The results of chief importance arising out of this enquiry
were obtained from two sets of shafts and borings sunk at
intervals in lines crossing the great valley from east to west.
One of these consisted of no fewer than fifty-one pits and
artesian borings, made where the valley is sixteen miles
wide from side to side between the Arabian and Libyan
deserts, in the latitude of Heliopolis, about eight miles above
the apex of the delta. The other line of borings and pits,
twenty-seven in number, was in the parallel of Memphis,
where the valley is only five miles broad.

Everywhere in these sections the sediment passed through
was similar in composition to the ordinary Nile mud of the
present day, except near the margin of the valley, where thin
layers of quartzose sand, such as is sometimes blown from the
adjacent desert by violent winds, were observed to alternate
with the loam.

A remarkable absence of lamination and stratification was
observed almost universally in the sediment brought up from
all points except where the sandy layers above alluded to occurred, the mud agreeing closely in character with the ancient
loam of the Rhine, called loess. Mr. Horner attributed this
want of all indication of successive deposition to the extreme thinness of the film of matter which is thrown down
annually on the great alluvial plain during the season of inundation.   The tenuity of this layer must indeed be extreme,
if the French engineers are tolerably correct in their estimate

of the amount of sediment formed in a century, which they suppose not to exceed on the average five inches. When the waters subside, this thin layer of new soil, exposed to a hot sun, dries rapidly, and clouds of dust are raised by the winds. The superficial deposit, moreover, is disturbed almost everywhere by agricultural labours, and even were this not the case, the action of worms, insects, and the roots of plants would suffice to confound together the deposits of many successive years.

All the remains of organic bodies, such as land-shells, and the bones of quadrupeds, found during the excavations belonged to living species. Bones of the ox, hog, dog, dromedary, and ass were not uncommon, but no vestiges of extinct mammalia. No marine shells were anywhere detected; but this was to be expected, as the borings, though they sometimes reached as low as the level of the Mediterranean, were never carried down below it,—a circumstance much to be regretted, since where artesian perforations have been made in deltas, as in those of the Po and Ganges, to the depth of several hundred feet below the sea level, it has been found, contrary to expectation, that the deposits passed through were fluviatile throughout, implying, probably, that a general subsidence of those deltas and alluvial formations has taken place.[*] Whether there has been in like manner a sinking of the land in Egypt, we have as yet no means of proving; but Sir Gardner Wilkinson infers it from the position in the delta on the shore near Alexandria of the tombs commonly called Cleopatra's Baths, which cannot, he says, have been originally built so as to be exposed to the sea which now fills them, but must have stood on land above the level of the Mediterranean. The same author adduces, as additional signs of subsidence, some ruined towns, now half under water, in the Lake Menzaleh, and channels of ancient arms of the Nile submerged with their banks beneath the waters of that same lagoon.

[*] Principles of Geology, 11th ed. vol. I. p. 463.

In some instances, the excavations made under the superintendence of Hekekyan Bey were on a large scale for the
first sixteen or twenty-four feet, in which cases jars, vases,
pots, and a small human figure in burnt clay, a copper knife,
and other entire articles were dug up; but when water soaking
through from the Nile was reached, the boring instrument
used was too small to allow of more than fragments of works of
art being brought up. Pieces of burnt brick and pottery were
extracted almost everywhere, and from all depths, even where
they sank sixty feet below the surface towards the central parts
of the valley. In none of these cases did they get to the bottom
of the alluvial soil.

It has been objected, among other criticisms, that the
Arabs can always find whatever their employers desire to
obtain. Even those who are too well acquainted with the
sagacity and energy of Hekekyan Bey to suspect him of
having been deceived, have suggested that the artificial
objects might have fallen into old wells which had been filled
up. This notion is inadmissible for many reasons. Of the
ninety-five shafts and borings, seventy or more were made
far from the sites of towns or villages; and allowing that
every field may once have had its well, there would be but
small chance of the borings striking upon the site even of a
small number of them in seventy experiments.

Others have suggested that the Nile may have wandered
over the whole valley, undermining its banks on one side
and filling up old channels on the other. It has also been
asked whether the delta with the numerous shifting arms of
the river may not once have been at every point where
the auger pierced.[*] To all these objections there are two
obvious answers:—First, in historical times the Nile has on
the whole been very stationary, and has not shifted its position

[*] For a detailed account of these
sections see Mr. Horner's paper in the
Philosophical Transactions for 1855-
1858.

in the valley ; secondly, if the mud pierced through had been thrown down by the river in ancient channels, it would have been stratified, and would not have corresponded so closely with inundation mud. We learn from Captain Newbold that he observed in some excavations in the great plain alternations of sand and clay, such as are seen in the modern banks of the Nile; but in the borings made by Hekekyan Bey, such stratification seems scarcely in any case to have been detected.

The great aim of the criticisms above enumerated has been to get rid of the supposed anomaly of finding burnt brick and pottery at depths and places which would give them claim to an antiquity far exceeding that of the Roman domination in Egypt. For until the time of the Romans, it is said, no clay was burnt into bricks in the valley of the Nile. But a distinguished antiquary, Mr. S. Birch, assures me that this notion is altogether erroneous, and that he has under his charge in the British Museum, first, a small rectangular baked brick, which came from a Theban tomb, which bears the name of Thothmes, a superintendent of the granaries of the god Amen Ra, the style of art, inscription, and name, showing that it is as old as the 18th dynasty (about 1450 B.C.); secondly, a brick bearing an inscription, partly obliterated, but ending with the words ' of the temple of Amen Ra.' This brick, decidedly long anterior to the Roman dominion, is referred conjecturally, by Mr. Birch, to the 19th dynasty, or 1300 B.C. Sir Gardner Wilkinson has also in his possession pieces of mortar which he took from each of the three great pyramids, in which bits of broken pottery and of burnt clay or brick are imbedded.

M. Girard, of the French expedition to Egypt, supposed the average rate of the increase of Nile mud on the plain between Assouan and Cairo, a distance of more than 400 miles, to be five English inches in a century. This con-

clusion, according to Mr. Horner, is very vague, and founded
on insufficient data; the amount of matter thrown down by
the waters in different parts of the plain varying so much,
that to strike an average with any approach to accuracy
must be most difficult. Were we to assume six inches in a
century, the burnt brick met with at a depth of sixty feet
would be 12,000 years old.

Another fragment of red brick was found by Linant Bey,
in a boring seventy-two feet deep, being two or three feet
below the level of the Mediterranean, in the parallel of the
apex of the delta, 200 metres distant from the river, on the
Libyan side of the Rosetta branch.* M. Rosière, in the
great French work on Egypt, has estimated the mean rate
of deposit of sediment in the delta at two inches and three
lines in a century; † were we to take two and a half inches,
a work of art seventy-two feet deep must have been buried
more than 30,000 years ago. But if the boring of Linant
Bey was made where an arm of the river had been silted up
at a time when the apex of the delta was somewhat farther
south, or more distant from the sea than now, the brick in
question might be comparatively very modern.

I have given at length in my 'Principles of Geology,'‡
the experiments by means of which Mr. Horner endeavoured
to obtain an accurate chronometric scale for testing the age of
a given thickness of Nile sediment. The most important result
was obtained from an excavation and boring made near the
base of the pedestal of the colossal statue of Rameses, the middle
of whose reign, according to Lepsius, was 1,361 years B.C. Assuming with Mr. Horner that the foundation of this pedestal
was 14¾ inches below the ground at the time that it was laid,
there had been formed in 3,211 years, (the interval between

* Horner, Philosophical Transac-   Naturelle, tom. ii. p. 494).
tions, 1858.   ‡ 11th ed. vol. i. p. 430.
† Description de l'Égypte (Histoire

1361 B.C. and A.D. 1850,) a deposit of 9 feet 4 inches, which gives a mean increase of 3½ inches in a hundred years. To this mode of computation Mr. Samuel Sharpe objected that the Egyptians were in the habit of enclosing with embankments the areas on which they erected temples, statues, and obelisks, so as to exclude the waters of the Nile. Herodotus tells us that in his time those spots from which the Nile waters had been shut out for centuries appeared sunk, and could be looked down into from the surrounding grounds, which had been raised by the gradual accumulation over them of sediment annually thrown down. The thickness therefore of 9 feet 4 inches accumulated round the pedestal, instead of indicating 3,211 years, would be simply the produce of the much shorter period which has elapsed since Memphis fell into decay, or since the enclosing mounds gave way and allowed the river to inundate the site of the statue.

But Sir John Lubbock, in reply to this objection, has truly remarked that what we are in search of is the extent to which the flat plain of Memphis has been raised by the accumulation of Nile sediment since the statue was erected, and although the river when it broke through the embankments and washed mud from them into the enclosure might perhaps in a few years raise the enclosed area up to the level of the great plain outside, yet it could never heighten that area above the general level. The exceptional rapidity of accumulation would only be the complement of the exceptional want of deposition which had preceded.* Therefore, although immediately round the base of the pedestal, the deposit might be of comparatively modern date, the 9 feet 4 inches by which the whole plain had been raised would represent the whole of the deposition effected by the Nile since 1361 B.C., when the statue was erected.

* Sir J. Lubbock, The Reader, March 26, 1864.

Ancient Mounds of the Valley of the Ohio.

As I have already given several European examples of monuments of pre-historic date belonging to the recent period, I will now turn to the American continent. Before the scientific investigation by Messrs. Squier and Davis of the 'Ancient Monuments of the Mississippi Valley,'* no one suspected that the plains of that river had been occupied, for ages before the French and British colonists settled there, by a nation of older date and more advanced in the arts than the Red Indians whom the Europeans found there. But we now learn that there are hundreds of large mounds in the basin of the Mississippi, and especially in the valleys of the Ohio and its tributaries, some of which are supposed to have served for temples, others for outlook or defence, and others for sepulture. The unknown people by whom they were constructed, judging by the form of several skulls dug out of the burial-places, were of the Mexican or Toltecan race. Some of the earthworks are on so grand a scale as to embrace areas of fifty or a hundred acres within a simple enclosure, and the solid contents of one pyramidal temple mound are estimated at twenty millions of cubic feet, so that four of them would be more than equal in bulk to the Great Pyramid of Egypt, which comprises seventy-five millions. From several of these repositories pottery and ornamental sculpture have been taken, and various articles in silver and copper, also stone weapons, some composed of hornstone unpolished, and much resembling in shape some ancient flint instruments found near Amiens and other places in Europe, to be alluded to in the sequel.

It is clear that the Ohio mound-builders had commercial intercourse with the natives of distant regions, for among

* Smithsonian Contributions, vol. L. 1847.

the buried articles some are made of native copper from
Lake Superior, and there are also found mica from the
Alleghanies, sea-shells from the Gulf of Mexico, and obsidian
from the Mexican mountains.

The extraordinary number of the mounds implies a long
period, during which a settled agricultural population had
made considerable progress in civilisation, so as to require
large temples for their religious rites, and extensive fortifications to protect them from their enemies. The mounds
were almost all confined to fertile valleys or alluvial plains,
and some at least are so ancient that rivers have had time
since their construction to encroach on the lower terraces
which support them, and again to recede for the distance of
nearly a mile, after having undermined and destroyed a part
of the works. When the first European settlers entered the
valley of the Ohio, they found the whole region covered with
an uninterrupted forest, and tenanted by the Red Indian
hunter, who roamed over it without any fixed abode, or any
traditionary connection with his more civilised predecessors.
The only positive data as yet obtained for calculating the
minimum of time which must have elapsed since the mounds
were abandoned, have been derived from the age and nature
of the trees found growing on some of these earthworks.
When I visited Marietta in 1842, Dr. Hildreth took me to
one of the mounds, and showed me where he had seen a tree
growing on it, the trunk of which when cut down displayed
eight hundred rings of annual growth.* But the late
General Harrison, President in 1841 of the United States,
who was well skilled in woodcraft, has remarked, in a memoir
on this subject, that several generations of trees must have
lived and died before the mounds could have been overspread
with that variety of species which they supported when the
white man first beheld them, for the number and kinds of

* Lyell's Travels in North America, vol. ii. p. 29.

trees were precisely the same as those which distinguished
the surrounding forest. 'We may be sure,' observed Harrison, 'that no trees were allowed to grow so long as the earthworks were in use; and when they were forsaken, the ground,
like all newly-cleared land in Ohio, would for a time be
monopolised by one or two species of trees, such as the yellow
locust and the white or black walnut. When the individuals
which were the first to get possession of the ground had died
out one after the other, they would in many cases, instead
of being replaced by the same species, be succeeded by other
kinds, till at last, after a great number of centuries (several
thousand years, perhaps), that remarkable diversity of species
characteristic of North America, and far exceeding what is
seen in European forests, would be established.'

Delta of the Mississippi.

I have shown elsewhere * that the deposits forming
the delta and alluvial plain of the Mississippi consist
of sedimentary matter, extending over an area of 30,000
square miles, and known in some parts to be several
hundred feet deep. Although we cannot estimate correctly how many years it may have required for the river
to bring down from the upper country so large a quantity
of earthy matter—the data for such a computation being
as yet incomplete—we may still approximate to a minimum of the time which such an operation must have
taken, by ascertaining experimentally the annual discharge
of water by the Mississippi, and the mean annual amount of
solid matter contained in its waters. Dr. Riddell, previous
to 1846, and Messrs. Humphreys and Abbot in 1861, made
a series of careful measurements to ascertain this point,

* Second Visit to United States. 1846, vol. II. p. 250.

and their estimates agree very nearly, Dr. Riddell giving the
quantity of sediment contained in the water as $\frac{1}{1617}$ in
weight, and the later surveyors estimating it at $\frac{1}{1311}$. At
this rate of deposit, I made the calculation in 1846 that
it would require 67,000 years to form the delta, which
extends over an area of about 13,000 square miles, and
which I estimated roughly to have a thickness of 528 feet;
and another 33,500 years for the accumulation of the
alluvial plain above, supposing it to be only half the depth
of the delta. But in the course of their survey, Messrs.
Humphreys and Abbot came to the conclusion that the
quantity of water annually discharged by the Mississippi
into the gulf had been greatly underrated. They also
remarked that the river pushes along the bottom of its
channel, even to its mouth, a certain quantity of sand and
gravel, equal, according to them, to about $\frac{1}{5}$ of the mud
held in suspension by the river, and of which I had taken
no account. Allowing therefore for this addition, and for
the larger discharge of muddy water, they estimate the
annual discharge of earthy matter as being twice as great as
I made it, which would reduce the number of years required
for the growth of the delta and alluvial plain to one-half,
or to 50,000 years, assuming the average thickness of
alluvium to be 528 feet, as I conjectured. But there are,
on the other hand, many facts which lead to the conclusion
that I had greatly underrated this thickness. In 1854,
an artesian well, bored at New Orleans, through strata
containing shells of recent species, reached the depth of
630 feet, without any signs of the foundations of the modern
deposit having been reached; and the depth of the Gulf
of Mexico, within twelve miles of the mouth of the South
Pass, has been ascertained to be 570 feet, increasing rapidly
to 6,000 feet before we arrive at the Florida Straits. More-

over, although no doubt the heavy sand and coarser material
brought down by the river would be deposited at the mouths
of the delta, yet large quantities of impalpable sand, held
in suspension in the waters, would be swept away to indefinite distances, on reaching the sea, by the rapid currents
which set for many months every year across the mouth of
the Mississippi. On the whole, therefore, I am not disposed
to regard the estimate which I made in 1846, of the
time required for the accumulation of the delta, as extravagant. The rate at which the river accomplishes a given
amount of work is no doubt nearly double what I supposed;
but, on the other hand, the quantity of work done, or of
mud and sand carried down into the gulf, is far greater
than I assumed as the basis of my calculation.

It was from one part of the modern delta of New
Orleans, where a large excavation had been made for gas-
works, that Dr. B. Dowler described, in 1852, the human
skeleton to which he is inclined to attribute an antiquity
of fifty thousand years. The formation passed through
was composed of a succession of beds, almost wholly made
up of vegetable matter, such as we now see forming in the
cypress swamps of the neighbourhood, where the deciduous
cypress (*Taxodium distichum*), with its strong and spreading
roots, plays a conspicuous part. In one of these beds, at
the depth of sixteen feet from the surface, beneath four
buried forests, superimposed one upon the other, the workmen are stated by Dr. B. Dowler to have found some charcoal
and a human skeleton, the cranium of which is said to
belong to the type of the aboriginal American race. As
the discovery in question had not been made when I saw
the excavation in progress at the gas-works in 1846, I
cannot form an opinion as to the value of the chronological
calculations which have led Dr. Dowler to ascribe to this

skeleton such a high antiquity as 50,000 years. In several sections, both natural in the banks of the Mississippi and its numerous arms, and where artificial canals had been cut, I observed erect stumps of trees, with their roots attached, buried in strata at different heights, one over the other. I also remarked, that many cypresses which had been cut through, exhibited many hundreds of rings of annual growth, and it then struck me that nowhere in the world could the geologist enjoy a more favourable opportunity for estimating in years the duration of certain portions of the recent epoch.[*]

Coral Reefs of Florida.

Professor Agassiz has described a low portion of the peninsula of Florida as consisting of numerous reefs of coral, which have grown in succession, so as to give rise to a continual annexation of land, gained gradually from the sea in a southerly direction. This growth is still in full activity, and assuming the rate of advance of the land to be one foot in a century, the reefs being built up from a depth of seventy-five feet, and that each reef has in its turn added ten miles to the coast, Professor Agassiz calculates that it has taken 135,000 years to form the southern half of this peninsula. Yet the whole is of post-tertiary origin, the fossil zoophytes and shells being all of the same species as those now inhabiting the neighbouring sea.[†]

In a bluff on the shores of Lake Monroe, in Florida, forming part of the above-mentioned series of reefs, Count Pourtalès found human remains, consisting of jaws and teeth, with some bones of the foot. The remains were contained

[*] Dowler, cited by Dr. W. Usher, in Nott and Glidden's Types of Mankind, p. 336.

[†] Agassiz, in Nott and Glidden, ibid. p. 352.

in a lacustrine formation composed of rotten coral-reef, limestone, and shells, chiefly of the same species of Paludina and
Ampullaria as are now found in the St. John river which
drains Lake Monroe. 'It is certain,' says Professor Agassiz,
'that the whole of the southern extremity of Florida, with
the everglades, has been added to that part of the continent
since the basin has been in existence, in which the conglomerate, with human bones, has been accumulating,' * and
from a careful calculation of the time required for such an
extension of the reefs, he considers the antiquity of the
human remains to be at least 10,000 years.

Recent Deposits of Seas and Lakes.

I have shown, in the 'Principles of Geology,' where the
recent changes of the earth illustrative of geology are described at length, that the deposits accumulated at the
bottom of lakes and seas within the last 4,000 or 5,000 years
cannot be insignificant either in volume or in extent. They
lie hidden, for the most part, from our sight; but we have
opportunities of examining them at certain points where
newly-gained land in the deltas of rivers has been cut through
during floods, or where coral reefs are growing rapidly, or
where the bed of a sea or lake has been heaved up by subterranean movements and laid dry.

As examples of such changes of level by which marine
deposits of the recent period have become accessible to
human observation, I have adduced the strata near Naples
in which the Temple of Serapis at Pozzuoli was entombed.†
These upraised strata, the highest of which are about twenty-
five feet above the level of the sea, form a terrace skirting
the eastern shore of the Bay of Baiæ. They consist partly
of clay, partly of volcanic matter, and contain fragments of

* Agassiz, in Nott and Gliddon's   † Principles of Geology. Index
Types of Mankind, p. 362.   'Serapis.'

sculpture, pottery, and the remains of buildings, together with great numbers of shells, retaining in part their colour, and of the same species as those now inhabiting the neighbouring sea. Their emergence can be proved to have taken place since the beginning of the sixteenth century.

In the same work, as an example of a fresh-water deposit of the recent period, I have described certain strata in Cashmere, a country where violent earthquakes, attended by alterations in the level of the ground, are frequent. In these beds fresh-water shells of species now inhabiting the lakes and rivers of that region are embedded, together with the remains of pottery, often at the depth of fifty feet, and in which a splendid Hindoo temple has lately been discovered, and laid open to view by the removal of the lacustrine silt which had enveloped it for four or five centuries.

In the same treatise (ch. xxviii.) it is stated, that the west coast of South America, between the Andes and the Pacific, is a great theatre of earthquake movements, and that permanent upheavals of the land of several feet at a time have been experienced since the discovery of America. In various parts of the littoral region of Chili and Peru, strata have been observed enclosing shells in abundance, all agreeing specifically with those now swarming in the Pacific. In one bed of this kind, in the island of San Lorenzo, near Lima, Mr. Darwin found, at the altitude of eighty-feet above the sea, pieces of cotton-thread, plaited rush, and the head of a stalk of Indian corn, the whole of which had evidently been embedded with the shells. At the same height, on the neighbouring mainland, he found other signs corroborating the opinion that the ancient bed of the sea had there also been uplifted eighty-five feet since the region was first peopled by the Peruvian race. But similar shelly masses are also met with at much higher elevations, at innumerable points between the Chilian and Peruvian Andes

E

and the sea-coast in which no human remains have as yet
been observed. The preservation for an indefinite period of
such perishable substances as thread is explained by the
entire absence of rain in Peru. The same articles, had they
been enclosed in the permeable sands of a European raised
beach, or in any country where rain falls even for a small
part of the year, would probably have disappeared entirely.

In the literature of the last century, we find frequent
allusion to the ' era of existing continents,' a period supposed
to have coincided in date with the first appearance of Man
upon the earth, since which event it was imagined that the
relative level of the sea and land had remained stationary,
no important geographical changes having occurred, except
some slight additions to the deltas of rivers, or the loss of
narrow strips of land where the sea had encroached upon its
shores. But modern observations have tended continually
to dispel this delusion, and the geologist is now convinced
that at no given era of the past have the boundaries of land
and sea, or the height of the one and depth of the other, or
the geographical range of the species inhabiting them,
whether of animals or plants, become fixed and unchange-
able. Of the extent to which fluctuations have been going
on since the globe had already become the dwelling-place of
Man, some idea may be formed from the examples which I
shall give in this and the next nine chapters.

*Upheaval since the Human Period of the Central
District of Scotland.*

It has long been a fact familiar to geologists, that, both on
the east and west coasts of the central part of Scotland, there
are lines of raised beaches, containing marine shells of the
same species as those now inhabiting the neighbouring sea.[*]

* R. Chambers, 'Sea Margins,'   Jordanhill, Mem. Wern. Soc., vol. viii.,
1848; and papers by Mr. Smith, of   and by Mr. C. Maclaren.

The two most marked of these littoral deposits occur at heights of about forty and twenty-five feet above high-water mark, that of forty feet being considered as the more ancient, and owing its superior elevation to a longer continuance of the upheaving movement. They are seen in some places to rest on the arctic shell-beds and boulder clay of the glacial period, which will be described in future chapters.

In those districts where large rivers, such as the Clyde, Forth, and Tay, enter the sea, the lower of the two deposits, or that of twenty-five feet, expands into a terrace, fringing the estuaries, and varying in breadth from a few yards to several miles. Of this nature are the flat lands which occur along the margin of the Clyde at Glasgow, which consist of finely laminated sand, silt, clay, and gravel. Mr. John Buchanan, a zealous antiquary, writing in 1855, informs us, that in the course of the eighty years preceding that date, no less than seventeen canoes had been dug out of this estuarine silt, and that he had personally inspected a large number of them before they were exhumed. Five of them lay buried in silt under the streets of Glasgow, one in a vertical position with the prow uppermost as if it had sunk in a storm. In the inside of it were a number of marine shells. Twelve other canoes were found about a hundred yards back from the river, at the average depth of about nineteen feet from the surface of the soil, or seven feet above high-water mark ; but a few of them were only four or five feet deep, and consequently more than twenty feet above the sea-level. One was sticking in the sand at an angle of 45°, another had been capsized, and lay bottom uppermost : all the rest were in a horizontal position, as if they had sunk in smooth water.[*] Within the last few years (1869) three other canoes have been found in the silts of the Clyde, between Bowling and

---

[*] J. Buchanan, Brit. Ass. Rep. 1855, p. 80; also, 'Glasgow Past and Present,' 1866.

Dumbarton, and are preserved for inspection in the adjacent
grounds of Auchentorlie. Two of these had been exhumed
from the bed of the river near Dunglass. They were found
lying abreast of each other, embedded in tenacious clay,
containing water-worn boulders, overlaid by a deposit of
alluvial mud.*

Almost every one of these ancient boats was formed out of
a single oak-stem, hollowed out by blunt tools, probably stone
axes, aided by the action of fire; a few were cut beautifully
smooth, evidently with metallic tools. Hence a gradation
could be traced from a pattern of extreme rudeness to one
showing great mechanical ingenuity. Two of them were built
of planks, one of the two, dug up on the property of Bankton
in 1853, being eighteen feet in length, and very elaborately
constructed. Its prow was not unlike the beak of an antique
galley; its stern, formed of a triangular-shaped piece of oak,
fitted in exactly like those of our day. The planks were
fastened to the ribs, partly by singularly shaped oaken pins,
and partly by what must have been square nails of some
kind of metal; these had entirely disappeared, but some of
the oaken pins remained. This boat had been upset, and
was lying keel uppermost, with the prow pointing straight up
the river. In one of the canoes, a beautifully polished celt
or axe of greenstone was found, in the bottom of another a
plug of cork, which, as Professor Geikie remarks, 'could
only have come from the latitudes of Spain, Southern
France, or Italy.'†

There can be no doubt that some of these buried vessels
are of far more ancient date than others. Those most
roughly hewn, may be relics of the stone period; those more
smoothly cut, of the bronze age; and the regularly built boat
of Bankton may perhaps come within the age of iron. The

* A. Currie, Trans. Geol. Soc. Glasgow, vol. iii. p. 370. 1869.

† Geikie, Geol. Quart. Journ., vol. xviii. p. 224.

occurrence of all of them in one and the same upraised
marine formation by no means implies that they belong
to the same era, for in the beds of all great rivers and
estuaries, there are changes continually in progress brought
about by the deposition, removal, and redeposition of gravel,
sand, and fine sediment, and by the shifting of the channel
of the main currents from year to year, and from century to
century. All these it behoves the geologist and antiquary
to bear in mind, so as to be always on their guard, when
they are endeavouring to settle the relative date, whether of
objects of art or of organic remains embedded in any set
of alluvial strata.   Some judicious observations on this head
occur in Professor Geikie's memoir above cited, which are so
much in point that I shall give them in full, and in his own
words.

' The relative position in the silt, from which the canoes
were exhumed, could help us little in any attempt to ascertain their relative ages, unless they had been found vertically
above each other.   The varying depths of an estuary, its
banks of silt and sand, the set of its currents, and the influence of its tides in scouring out alluvium from some parts
of its bottom and redepositing it in others, are circumstances
which require to be taken into account in all such calculations.
Mere coincidence of depth from the present surface of the
ground, which is tolerably uniform in level, by no means
necessarily proves contemporaneous deposition.   Nor would
such an inference follow even from the occurrence of the
remains in distant parts of the very same stratum.   A canoe
might be capsized and sent to the bottom just beneath low-
water mark ; another might experience a similar fate on the
following day, but in the middle of the channel.   Both
would become silted up on the floor of the estuary ; but as
that floor would be perhaps twenty feet deeper in the centre
than towards the margin of the river, the one canoe might

actually be twenty feet deeper in the alluvium than the other ;
and on the upheaval of the alluvial deposits, if we were to
argue merely from the depth at which the remains were
embedded, we should pronounce the canoe found at the one
locality to be immensely older than the other, seeing that the
fine mud of the estuary is deposited very slowly and that it
must therefore have taken a long period to form so great a
thickness as twenty feet. Again, the tides and currents of
the estuary, by changing their direction, might sweep away
a considerable mass of alluvium from the bottom, laying bare
a canoe that may have foundered many centuries before.
After the lapse of so long an interval, another vessel might go
to the bottom in the same locality, and be there covered up
with the older one, on the same general plane. These two
vessels, found in such a position, would naturally be classed
together as of the same age, and yet it is demonstrable that a
very long period may have elapsed between the date of the
one and that of the other. Such an association of these
canoes, therefore, cannot be regarded as proving synchronous
depositions; nor, on the other hand, can we affirm any
difference of age from mere relative position, unless we see
one canoe actually buried beneath another.' *

At the time when the ancient vessels, above described,
were navigating the waters, where the city of Glasgow now
stands, the whole of the low lands which bordered the present
estuary of the Clyde formed the bed of a shallow sea. The
emergence appears to have taken place gradually and by
intermittent movements, for Mr. Buchanan describes several
narrow terraces one above the other on the site of the city
itself, with steep intervening slopes composed of the laminated estuary formation. Each terrace and steep slope
probably mark pauses in the process of upheaval, during

* J. Buchanan. Brit. Ass. Rep. 1855. p. 80; also Glasgow Past and Present,
1856.

which low cliffs were formed, with beaches at their base.
Five of the canoes were found within the precincts of the
city at different heights on or near such terraces.

As to the date of the upheaval, the greater part of it
cannot be assigned to the stone period, but must have
taken place after tools of metal had come into use. In
a former edition of this work I attempted to show, on
the authority of some eminent geographers and geologists, that it might even have been post-Roman, but I
am now convinced that the balance of evidence is strongly
in favour of no alteration having occurred in the relative
level of land and sea, in the central district of Scotland,
since the construction of the Roman or Pictish wall (the
'Wall of Antonine'), which reached from the Firth of Forth
to that of the Clyde. The late Mr. Smith, of Jordanhill,
had always held this opinion, and he stated in 1862 that
'the Roman wall in Scotland, which crosses the island from
sea to sea, has evidently been formed at both ends with
reference to the present level; and the same observation
applies to British tumuli and vitrified forts, which are
perhaps of still greater antiquity.'* More lately Mr. Milne
Home has also called attention to the fact that the Roman
military road to their station north of the Forth passed
along the low alluvial plain near Stirling, and crossed
the river at a ford called the 'Drip,' to the west of that
town, where the foundations have been dug up.† 'The
river at this place,' says Mr. Milne Home, 'is about eight
feet above the level of the sea at high-water mark. Before
the last change in the relative levels of sea and land, the
sea prevailed over the whole carse to the west of Stirling,
rendering the use or even formation of a road across it at
that time utterly impossible, for neither ford nor river could

* Post-Pliocene Geology, p. 15, 1862.   † Nimmo, History of Stirlingshire, 1777. p. 23.

then have existed at the Drip; the sea must have been
more than twenty feet deep there at high water.' He also
shows that lower down the Forth below Stirling there was a
road across the river, and a fortlet or castellum on a spot
where the depth of the sea before the last change of level
must have been many feet greater than even at the Drip
Ford. Therefore, if the Romans really had a road across
the Forth at this place, and also a fortress, it is impossible
that the sea-level was then twenty-five feet higher than now,
because in that case both road and fortress would have been
covered by the sea at all times of the tide.[*]

A most important discovery bearing on this question was
made in the year 1869, on the estate of Mr. Henry Cadell,
at Bridgeness, near Bo'ness, Linlithgowshire, to which my
attention was first called by Mr. Milne Home.   On a rocky
promontory above a harbour, a sculptured slab of sandstone,
8 inches thick and 9 feet in length, by 3 in width, was found
lying under 1 to 2½ feet of garden soil, with its face downwards.   It is broken in three pieces, the pieces lying so near
to each other as to suggest the idea that in falling forward
it had been shattered by its concussion on the ground.
This tablet is divided into three panels, the centre one
bearing a Roman inscription, by which we learn that it
was erected in the reign of Antoninus Pius, by the Second
Legion, in commemoration of the completion by them of
four thousand six hundred and fifty-two paces of the Roman
wall.   This would give the tablet a date of about A.D. 139 to
161.   In the left panel (see fig. 2), is a representation of a
Roman soldier trampling down the naked Caledonians, and in
the right a priest is offering up on the altar a bull, a sheep,
and a pig.   Mr. Cadell presented the tablet to the Society of
Antiquaries in Edinburgh, of whose museum it now forms one

* Estuary of the Forth, 1871, p. 116.

Fig. 2

From a Block engraved for the Society of Antiquaries of Scotland.

Roman sculptured tablet discovered at Bridgness, Linlithgowshire, 1868, 19 feet above high-water mark, at the eastern end of the wall of Antonine.

Reading of the centre panel—Imperatori Caesari Tito Ælio Hadriano Antonino Augusto Pio Patriæ Legio Secunda Augusta, per mille passuum IIII. D.C.LII., fecit.

of the most valuable relics, and I am indebted to their liberality for permission to make use of a fac-simile of their block.
Similar tablets have been found at various times indicating
the number of paces made by different legions, one of which
is very near the western termination of the wall on the Clyde.

Mr. Milne Home has recently visited the ground at
Bo'ness, and has ascertained that the spot where the tablet
was found is about nineteen feet above high-water mark,
and that remains of an old sea-wall exist at the shore at a
distance of about thirty-six yards. A beach of washed
shells and sand also shows that the sea has flowed round the
promontory within ten yards of the stone. It is clear from
the above statement that if an upheaval of land to the
amount of twenty-five feet had taken place in the last
seventeen centuries, the tablet must have been erected
in such a position as to be six feet under the sea at every
high tide, and so exposed to the beating of the waves that
neither the tablet nor the wall would have stood many
weeks.[*]

With regard to the west coast of Scotland, Mr. John
Young, curator of the museum at Glasgow, informs me that
the evidence against a rise of land on the shores of the Clyde
since the Roman occupation, is of the same nature as that
obtained from the Forth. Although the Roman wall termi-
nated, in all probability, on Chapel Hill, West Kilpatrick,
yet there was a line of detached forts extending westwards
along the low marshy shores of the Clyde to the town of
Dumbarton—the Theodosia of the Romans—which could not
have been built there had the land been much lower than it
is now. Vestiges of these forts were in existence down to
the latter part of the seventeenth century. It appears therefore that we must date back the last elevation of land in
Scotland to a period prior to the Roman occupation, and

[*] Milne Home 'On Supposed Upheaval of Land, &c.,' Royal Soc. of Edinburgh, Jan. 20, 1873.

ascribe the recent gain of land, observed both on the Forth
and Clyde, to a large amount of sediment carried down by
rivers during floods from the cultivated lands. The older
inhabitants of the Forth affirm that the clay shores or sleeks
upon its western termination are at present perceptibly
higher, and not so often overflowed by the tide, as they were
in their younger years;* and on the Clyde, in the lower
reaches above Dumbarton, the gain of land by the accumulation of sediment has been considerable. This gain has been
increased by drainage and tillage, and by the deepening and
straightening of the course of the river, and elevation of the
banks along various low tracts during recent years.

But though the last rise of land, denoted by the old
sea-margins of Scotland, must apparently have been pre-
Roman, we have nevertheless evidence that it took place
within the human period. In the Carse of Stirling, a low
tract of land about twenty-five feet above high-water mark,
several skeletons of whales have been found in loamy and
peaty beds. One of these was dug up at Airthrie,† near
Stirling, about a mile from the river, and seven miles from
the sea. Mr. Bald mentions, that near it were found two
pieces of stag's horn, artificially cut, through one of which
a hole, about an inch in diameter, had been perforated.
Another whale, eighty-five feet long, was found at Dunmore,
a few miles below Stirling,‡ which, like that of Airthrie, lay
about twenty feet above high-water mark. Three other
skeletons of whales were found at Blair Drummond, between
the years 1819 and 1824, seven miles up the estuary above
Stirling,§ also at an elevation between twenty and thirty feet
above the sea. Near two of these whales, pointed instruments of deer's horn were found, one of which retained part

* Nimmo, History of Stirlingshire,
p. 432.
† Bald, Edinburgh Philosophical
Journal, i. p. 393; and Memoirs,
Wernerian Society, iii. p. 327.

‡ Edinburgh Philosophical Journal, xi. pp. 220, 413.
§ Memoirs, Wernerian Society, v.
p. 440.

of a wooden handle, probably preserved by having been
enclosed in peat. This weapon is now in the museum at
Edinburgh.

The position of these whales and their association with
human implements, imply that at the time when they were
cast ashore by a tide rising twenty or thirty feet beyond the
present high-water mark, man was already an inhabitant of
Scotland; and their great size, indicating that they belonged
to the Greenland whale, which only frequents seas of floating
ice, would point to an arctic climate in these regions before
the last change of level occurred. This inference, says Mr.
Milne Home, 'agrees with a conclusion come to by the late
James Smith of Jordanhill, who, on the lowest ancient
beach on the west of Scotland, found a large ancient boulder
which could not, in his opinion, have come there except on
floating ice.' *

The same upward movement which laid dry the raised
beaches of the Forth, and the marine strata with canoes on
the Clyde, was also felt as far north as the estuary of the
Tay. This may be inferred from the Celtic name of *Inch*
being attached to many hillocks, which rise above the
general level of the alluvial plains, implying that these
eminences were once surrounded by water or marshy ground.
At various localities also in the silt of the Carse of Gowrie
iron implements have been found.

The raised beach, also containing a great number of
marine shells of recent species, traced up to a height of
fourteen feet above the sea by Mr. W. J. Hamilton at Elie,
on the southern coast of Fife, is doubtless another effect of
the same extensive upheaval.† A similar movement would
also account for some changes which antiquaries have recorded much farther south, on the borders of the Solway

* Estuary of the Forth, p. 116.
† Proceedings of Geological Society, 1838, vol. ii. p. 280.

Firth; though in this case, as in that of the estuary of the
Forth, the conversion of sea into land has always been
referred to the silting up of estuaries, and not to upheaval.
Thus Horsley insists on the difficulty of explaining the
position of certain Roman stations, on the Solway, the Forth,
and the Clyde, without assuming that the sea has been
excluded from certain areas which it formerly occupied.*

On a review of the whole evidence, geological and ar-
chæological, afforded by the Scottish coast-line, we may
conclude that the last upheaval of twenty-five feet took
place not only since the first human population settled in
the island, but probably long after metallic implements had
come into use.

But the twenty-five feet rise is only the last stage of a
long antecedent process of elevation, for examples of recent
marine shells have been observed forty feet and upwards
above the sea in Ayrshire, and near Paisley in Renfrewshire.
At one of these localities, Mr. Smith of Jordanhill informs
me that a rude ornament made of cannel coal has been found
on the coast in the parish of Dundonald, lying fifty feet
above the sea-level, on the surface of the boulder clay or till,
and covered with gravel containing marine shells.

Coast of Cornwall.

Sir H. De la Beche has adduced several proofs of changes
of level, in the course of the human period, in his ' Report
on the Geology of Cornwall and Devon for 1839.' He there
mentions (p. 406) the discovery in 1829, by Mr. Colenso, of
several human skulls and works of art buried in an estuary
deposit, which were found in mining gravel for tin, at Pen-
tuan near St. Austell. The skulls were lying at a depth of forty
feet from the surface, and the overlying strata were marine,

* Britannia. p. 157. 1860.

containing sea-shells of living species, and bones of whales,
besides the remains of several existing species of mammalia.
The whale remains, which are now in the Penzance museum,
have lately been identified by Mr. Flower as belonging to
the *Eschrichtius robustus* Lilljeborg, a whale supposed to be
now extinct, and differing in some of its characters from all the
three principal existing genera, and of which remains have
also been found in the Swedish island of Gräso, in a bed of
sandy clay, ten or fifteen feet above the sea-level, together
with shells of *Mytilus edulis* and *Tellina balthica.*[*]

Human skulls were also found by Mr. Henwood in 1829
at a depth of fifty-three feet in ground covering the tin
works at Carnon, near Falmouth, some miles south of St.
Austell.   The overlying strata were composed of three feet
of river sand and mud, about forty-eight feet of sand and
silt containing marine shells, and a vegetable bed 1·6 feet
in thickness, containing mammalian remains.   As the top of
the section was itself twelve or fifteen feet below the level
of spring tide, the top of the tin ground was at least sixty-
seven feet below this level.[†]

Other examples of works of art, such as stone hatchets,
canoes, and ships, buried in ancient river-beds in England,
and in peat and shell-marl, are mentioned in my work before
cited.[‡]

Sweden and Norway.

In the same work I have shown that near Stockholm, in
Sweden, there occur, at slight elevations above the sea-level,
horizontal beds of sand, loam, and marl, containing the same
peculiar assemblage of testacea which now live in the brackish
waters of the Baltic.   Mingled with these, at different depths,

[*] Evans, Sub-fossil Whale in Cornwall, Ann. and Mag. of Nat. Hist.
June, 1872.
[†] Trans. Royal Geol. Soc. of Cornwall, vol. iv. p. 57, 1829 ; and Pengelly, Trans. Devon. Association, 1867.
[‡] Principles of Geology.

have been detected various works of art implying a rude state
of civilisation, and some vessels built before the introduction
of iron, and even the remains of an ancient hut, the marine
strata containing it, which had been formed during a previous
depression, having been upraised, so that the upper beds
are now sixty feet higher than the surface of the Baltic.   In
the neighbourhood of these recent strata, both to the north-
west and south of Stockholm, other deposits similar in
mineral composition occur, which ascend to greater heights,
in which precisely the same assemblage of fossil shells is met
with, but without any intermixture, so far as is yet known,
of human bones or fabricated articles.

On the opposite or western coast of Sweden, at Uddevalla,
post-tertiary strata, containing recent shells, not of that
brackish-water character peculiar to the Baltic, but such as
now live in the Northern Ocean, ascend to the height of
200 feet ; and beds of clay and sand of the same age attain
elevations of 300 and even 600 feet in Norway, where they
have been usually described as 'raised beaches.'   They are,
however, thick deposits of submarine origin, spreading far
and wide, and filling valleys in the granite and gneiss, just
as the tertiary formations, in different parts of Europe, cover
or fill depressions in the older rocks.

Although the fossil fauna characterising these upraised
sands and clays consists exclusively of existing northern
species of testacea, it is more than probable that they may
not all belong to that division of the post-tertiary strata
which we are now considering.   If the contemporary mammalia were known, they would, in all likelihood, be found to
be referable, at least in part, to extinct species ; for, according
to Lovén (an able living naturalist of Norway), the species
do not constitute such an assemblage as now inhabits corresponding latitudes in the German Ocean.   On the contrary,
they decidedly represent a more arctic fauna.   In order to

find the same species flourishing in equal abundance, or in many cases to find them at all, we must go northwards to higher latitudes than Uddevalla in Sweden, or even nearer the pole than Central Norway.

Judging by the uniformity of climate now prevailing from century to century, and the insensible rate of variation in the geographical distribution of organic beings in our own times, we may presume that an entirely lengthened period was required, even for so slight a modification in the range of the molluscous fauna, as that of which the evidence is here brought to light. There are also other independent reasons for suspecting that the antiquity of these deposits may be indefinitely great as compared with the historical period. I allude to their present elevation above the sea, some of them rising, in Norway, to the height of 600 feet or more. The upward movement now in progress in parts of Norway and Sweden, extends, as I have elsewhere shown,* throughout an area about 1,000 miles north and south, and for an unknown distance east and west, the amount of elevation always increasing as we proceed towards the North Cape, where it is said to equal five feet in a century. If we could assume that there had been an average rise of two and a half feet in each hundred years for the last fifty centuries, this would give an elevation of 125 feet in that period. In other words, it would follow that the shores, and a considerable area of the former bed of the North Sea, had been uplifted vertically to that amount, and converted into land in the course of the last 5,000 years. A mean rate of continuous vertical elevation of two and a half feet in a century would, I conceive, be a high average; yet, even if this be assumed, it would require 24,000 years for parts of the sea-coast of Norway, where the post-tertiary marine strata occur, to attain the height of 600 feet.

* Principles, 11th ed. ch. xxxi.

CHAPTER IV.

PLEISTOCENE PERIOD—BONES OF MAN AND EXTINCT MAMMALIA IN BELGIAN CAVERNS.

RESEARCHES IN 1833 OF DR. SCHMERLING IN THE LIÉGE CAVERNS—
SCATTERED PORTIONS OF HUMAN SKELETONS ASSOCIATED WITH BONES
OF ELEPHANT AND RHINOCEROS—DISTRIBUTION AND PROBABLE MODE
OF INTRODUCTION OF THE BONES — IMPLEMENTS OF FLINT AND BONE
—SCHMERLING'S CONCLUSIONS AS TO THE ANTIQUITY OF MAN IGNORED
—PRESENT STATE OF THE BELGIAN CAVES—HUMAN REMAINS FOUND
IN CAVE OF ENGIHOUL — ENGULFED RIVERS — STALAGMITIC CRUST
—ANTIQUITY OF THE HUMAN REMAINS IN BELGIUM NOW PROVED—
RESEARCHES OF M. DUPONT IN THE CAVERNS OF THE LESSE—DURA-
TION OF ANCIENT RACES SHOWN BY THE OBJECTS IN THE CAVERNS.

HAVING hitherto considered those formations in which both the fossil shells and the mammalia are of living species, we may now turn our attention to those of older date, in which the shells being all recent, some of the accompanying mammalia are extinct, or belong to species not known to have lived within the times of history or tradition.

Researches, in 1833–1834, of Dr. Schmerling in the Caverns near Liége.

The late Dr. Schmerling of Liége, a skilful anatomist and palæontologist, after devoting several years to the exploring of the numerous ossiferous caverns which border the valleys of the Meuse and its tributaries, published two volumes, descriptive of the contents of more than forty

F

caverns.   One of these volumes consisted of an atlas of plates, illustrative of the fossil bones.[*]

Many of the caverns had never before been entered by scientific observers, and their floors were encrusted with unbroken stalagmite.   At a very early stage of his investigations, Dr. Schmerling found the bones of Man so rolled and scattered, as to preclude all idea of their having been intentionally buried on the spot.   He also remarked that they were of the same colour, and in the same condition as to the amount of animal matter contained in them, as those of the accompanying animals, some of which, like the cave-bear, hyæna, elephant, and rhinoceros, were extinct; others, like the wild cat, beaver, wild boar, roe-deer, wolf, and hedgehog, still extant.   The fossils were lighter than fresh bones, except such as had their pores filled with carbonate of lime, in which case they were often much heavier.   The human remains of most frequent occurrence were teeth detached from the jaw, and the carpal, metacarpal, tarsal, metatarsal, and phalangial bones separated from the rest of the skeleton.   The corresponding bones of the cave-bear, the most abundant of the accompanying mammalia, were also found in the Liége caverns more commonly than any others, and in the same scattered condition.   Occasionally, some of the long bones of mammalia were observed to have been first broken across, and then reunited or cemented again by stalagmite, as they lay on the floor of the cave.

No gnawed bones nor any coprolites were found by Schmerling.   He therefore inferred that the caverns of the province of Liége had not been the dens of wild beasts, but that their organic and inorganic contents had been swept into them by streams communicating with the surface

---

[*] Recherches sur les Ossements fossiles découverts dans les Cavernes de   la Province de Liége.   Liége, 1833 –1834.

·

of the country. The bones, he suggested, may often have
been rolled in the beds of such streams before they reached
their underground destination. To the same agency the
introduction of many land-shells dispersed through the cave-
mud was ascribed, such as *Helix nemoralis*, *H. lapicida*,
H. pomatia, and others of living species. Mingled with
such shells, in some rare instances, the bones of fresh-
water fish, and of a snake (*Coluber*), as well as of several
birds, were detected.

The occurrence here and there of bones in a very perfect
state, or of several bones belonging to the same skeleton in
natural juxtaposition, and having all their most delicate
apophyses uninjured, while many accompanying bones in
the same breccia were rolled, broken, or decayed, was
accounted for by supposing that portions of carcasses were
sometimes floated in during floods while still clothed with
their flesh. No example was discovered of an entire skeleton, not even of one of the smaller mammalia, the bones of
which are usually the least injured.

The incompleteness of each skeleton was especially ascertained in regard to the human subjects, Dr. Schmerling being
careful, whenever a fragment of such presented itself, to
explore the cavern himself, and see whether any other bones
of the same skeleton could be found. In the Engis cavern,
distant about eight miles to the south-west of Liége, on the
left bank of the Meuse, the remains of at least three human
individuals were disinterred. The skull of one of these, that
of a young person, was embedded by the side of a mammoth's tooth. It was entire, but so fragile, that nearly all of
it fell to pieces during its extraction. Another skull, that
of an adult individual, and the only one preserved by Dr.
Schmerling in a sufficient state of integrity to enable the
anatomist to speculate on the race to which it belonged, was
buried five feet deep in a breccia, in which the tooth of a

rhinoceros, several bones of a horse, and some of the reindeer, together with some ruminants, occurred. This skull,
now in the museum of the University of Liége, is figured in
Chap. V. (fig. 4, p. 86), where further observations will be
offered on its anatomical character, after a fuller account
of the contents of the Liége caverns has been laid before the
reader.

On the right bank of the Meuse, on the opposite side of
the river to Engis, is the cavern of Engihoul. Bones of
extinct animals mingled with those of Mau, were observed
to abound in both caverns; but with this difference, that
whereas in the Engis cave there were several human crania
and very few other bones, in Engihoul there occurred
numerous bones of the extremities belonging to at least
three human individuals, and only two small fragments of a
cranium. The like capricious distribution held good in
other caverns, especially with reference to the cave-bear, the
most frequent of the extinct mammalia. Thus, for example,
in the cave of Chokier, skulls of the bear were few, and
other parts of the skeleton abundant, whereas in several
other caverns these proportions were exactly reversed, while
at Goffontaine skulls of the bear and other parts of the
skeleton were found in their natural numerical proportions.
Speaking generally, it may be said that human bones, where
any were met with, occurred at all depths in the cave-mud
and gravel, sometimes above and sometimes below those of
the bear, elephant, rhinoceros, hyæna, &c.

Some rude flint implements of the kind commonly called
flint knives or flakes, of a triangular form in the cross section
(as in fig. 21, p. 163), were found by Schmerling dispersed
generally through the cave-mud, but he was too much en-
grossed with his osteological enquiries to collect them
diligently. He preserved some few of them, however, which
I have seen in the museum at Liége. He also discovered in

the cave of Chokier, two and a half miles south-west from
Liége, a polished and jointed needle-shaped bone, with a
hole pierced obliquely through it at the base; such a cavity,
he observed, as had never given passage to an artery. This
instrument was embedded in the same matrix with the
remains of a rhinoceros.*

Another cut bone and several artificially shaped flints
were found in the Engis cave, near the human skulls before
alluded to. Schmerling observed, and we shall have to refer
to the fact in the sequel (Chap. VIII.), that although in some
forty fossiliferous caves explored by him human bones were
the exception, yet these flint implements were universal, and
he added that ' none of them could have been subsequently
introduced, being precisely in the same position as the
remains of the accompanying animals.' ' I therefore,' he
continues, ' attach great importance to their presence; for
even if I had not found the human bones under conditions
entirely favourable to their being considered as belonging to
the antediluvian epoch, proofs of Man's existence would still
have been supplied by the cut bones and worked flints.'†

Dr. Schmerling, therefore, had no hesitation in concluding from the various facts ascertained by him, that Man
once lived in the Liége district contemporaneously with the
cave-bear and several other extinct species of quadrupeds.
But he was much at a loss when he attempted to invent a
theory to explain the former state of the fauna of the region
now drained by the Meuse; for he shared the notion, then
very prevalent among naturalists, that the mammoth and the
hyæna‡ were beasts of a warmer climate than that now
proper to Western Europe. In order to account for the
presence of such ' tropical species,' he was half-inclined to
imagine that they had been transported by a flood from some

* Schmerling, part ii. p. 177.   † Ibid. part ii. p. 179.
‡ Ibid. part ii. pp. 70, 96.

distant region; then again he raised the question whether
they might not have been washed out of an older alluvium,
which may have pre-existed in the neighbourhood. This
last hypothesis was directly at variance with his own statements, that the remains of the mammoth and hyæna were
identical in appearance, colour, and chemical condition with
those of the bear and other associated fossil animals, none
of which exhibited signs of having been previously enveloped
in any dissimilar matrix. Another enigma which led
Schmerling astray in some of his geological speculations was
the supposed presence of the agouti, a South-American
rodent, ' proper to the torrid zone.' My friend M. Lartet,
guided by Schmerling's figures of the teeth of this species,
suggests, and I have little doubt with good reason, that they
appertain to the porcupine, a genus found fossil in Pleistocene deposits of certain caverns in the south of France.

In the year 1833, I passed through Liége, on my way to
the Rhine, and conversed with Dr. Schmerling, who showed
me his splendid collection, and when I expressed some incredulity respecting the alleged antiquity of the fossil
human bones, he pointedly remarked, that if I doubted their
having been contemporaneous with the bear or rhinoceros, on
the ground of Man being a species of more modern date, I
ought equally to doubt the coexistence of all the other living
species, such as the red-deer, roe, wild cat, wild boar, wolf,
fox, weasel, beaver, hare, rabbit, hedgehog, mole, dormouse,
field-mouse, water-rat, shrew, and others, the bones of which
he had found scattered everywhere indiscriminately through
the same mud with the extinct quadrupeds. The year after
this conversation I cited Schmerling's opinions, and the
facts bearing on the antiquity of Man, in the 3rd edition of
my ' Principles of Geology' (p. 161, 1834), and in succeeding
editions, without pretending to call in question their trust-
worthiness, but at the same time without giving them the

weight to which I now consider they were entitled. He had
accumulated ample evidence to prove that Man had been
introduced into the earth at an earlier period than geologists
were then willing to believe.

One positive fact, it will be said, attested by so com-
petent a witness, ought to have outweighed any amount of
negative testimony, previously accumulated, respecting the
non-occurrence elsewhere of human remains in formations
of the like antiquity. In reply, I can only plead that a
discovery which seems to contradict the general tenor of
previous investigations is naturally received with much
hesitation. To have undertaken in 1832, with a view of
testing its truth, to follow the Belgian philosopher through
every stage of his observations and proofs, would have been
no easy task even for one well-skilled in geology and osteo-
logy. To be let down, as Schmerling was, day after day,
by a rope tied to a tree, so as to slide to the foot of the first
opening of the Engis cave,* where the best-preserved human
skulls were found; and, after thus gaining access to the
first subterranean gallery, to creep on all fours through a
contracted passage to larger chambers, there to superintend
by torchlight, week after week and year after year, the
workmen who were breaking through the stalagmitic crust
as hard as marble, in order to remove piece by piece the
underlying bone-breccia nearly as hard; to stand for hours
with one's feet in the mud, and with water dripping from
the roof on one's head, in order to mark the position and
guard against the loss of each single bone of a skeleton; and
at length, after finding leisure, strength, and courage for all
these operations, to look forward, as the fruits of one's
labour, to the publication of unwelcome intelligence, opposed
to the prepossessions of the scientific as well as of the unscientific public;—when these circumstances are taken into

* Schmerling. part i. p. 30.

account, we need scarcely wonder, not only that a passing
traveller failed to stop and scrutinise the evidence, but that
a quarter of a century should have elapsed before even the
neighbouring professors of the University of Liége came
forth to vindicate the truthfulness of their indefatigable and
clear-sighted countryman.

In 1860, when I revisited Liége, twenty-six years after
my interview with Schmerling, I found that several of the
caverns described by him had in the interval been annihi-
lated. Not a vestige, for example, of the caves of Engis,
Chokier, and Goffontaine remained. The calcareous stone,
in the heart of which the cavities once existed, had been
quarried away, and removed bodily for building and lime-
making. Fortunately, a great part of the Engihoul cavern,
situated on the right bank of the Meuse, was still in the
same state as when Schmerling delved into it in 1831, and
drew from it the bones of three human skeletons. I determined, therefore, to examine it, and was so fortunate as to
obtain the assistance of a zealous naturalist of Liége, Professor
Malaise, who accompanied me to the cavern, where we
engaged some workmen to break through the crust of
stalagmite, so that we could search for bones in the undisturbed earth beneath. Bones and teeth of the cave-bear
were soon found, and several other extinct quadrupeds which
Schmerling has enumerated. My companion, continuing
the work perseveringly for weeks after my departure, succeeded at length in extracting from the same deposit, at the
depth of two feet below the crust of stalagmite, three
fragments of a human skull, and two perfect lower jaws with
teeth, all associated in such a manner with the bones of
bears, large pachyderms, and ruminants, and so precisely
resembling these in colour and state of preservation, as to
leave no doubt in his mind that Man was contemporary with
the extinct animals. Professor Malaise has given figures of

the human remains in the Bulletin of the Royal Academy of
Belgium for 1860.*

The rock in which the Liége caverns occur belongs
generally to the carboniferous or mountain limestone, in
some few cases only to the older Devonian formation.
Whenever the work of destruction has not gone too far,
magnificent sections, sometimes 200 and 300 feet in height,
are exposed to view. They confirm Schmerling's doctrine,
that most of the materials, organic and inorganic, now filling
the caverns, have been washed into them through narrow
vertical or oblique fissures, the upper extremities of which
are choked up with soil and gravel, and would scarcely ever
be discoverable at the surface, especially in so wooded a
country. Among the sections obtained by quarrying, one of
the finest which I saw was in the beautiful valley of Fond
du Forêt, above Chaudefontaine, not far from the village of
Magnée, where one of the rents communicating with the
surface has been filled up to the brim with rounded and
half-rounded stones, angular pieces of limestone and shale,
besides sand and mud, together with bones, chiefly of the
cave-bear. Connected with this main duct, which is from
one to two feet in width, are several minor ones, each from
one to three inches wide, also extending to the upper country
or table-land, and choked up with similar materials. They
are inclined at angles of 30° and 40°, their walls being generally coated with stalactite, pieces of which have here and there
been broken off and mingled with the contents of the rents,
thus helping to explain why we so often meet with detached
pieces of that substance in the mud and breccia of the
Belgian caves. It is not easy to understand how a solid
horizontal floor of hard stalagmite should, after its formation, be broken up by running water; but when the walls of
steep and tortuous rents, serving as feeders to the principal

* Tom. x. p. 346.

fissures and to inferior vaults and galleries, are encrusted
with stalagmite, some of the incrustation may readily be
torn up when heavy fragments of rock are hurried by a flood
through passages inclined at angles of 30° or 40°.

The decay and decomposition of the fossil bones seem to
have been arrested in most of the caves by a constant supply of water charged with carbonate of lime, which dripped
from the roofs while the caves were becoming gradually
filled up. By similar agency the mud, sand, and pebbles
were usually consolidated.

The following explanation of this phenomenon has been
suggested by the eminent chemist Liebig. On the surface
of Franconia, where the limestone abounds in caverns, is a
fertile soil in which vegetable matter is continually decaying.
This mould or humus, being acted on by moisture and air,
evolves carbonic acid, which is dissolved by rain. The rain-
water, thus impregnated, permeates the fissured limestone,
dissolves a portion of it, and afterwards, when the excess of
carbonic acid evaporates in the caverns, parts with the calcareous matter and forms stalactite. So long as a stream of
water flows through a cavern no layer of pure stalagmite can
be produced ; hence the formation of such a layer is generally
an event posterior in date to the cessation of the old system
of drainage, an event which might be brought about by an
earthquake causing new fissures, or by the river wearing its
way down to a lower level, and thenceforth running in a
new channel.

In all the subterranean cavities, more than forty in
number, explored by Schmerling, he only observed one cave,
namely, that of Chokier, where there were two regular layers
of stalagmite, divided by fossiliferous cave-mud. Two similar
layers have since been found to occur in the caverns of Brixham and Kent's Hole in Devonshire. In such cases it would
appear that the alternate formation of cave-earth and stalag-

mite should be referred to the small changes produced by
the ordinary processes of denudation and deposition always
going on in caves. We may suppose that the stream, after
flowing for a long period at one level, cut its way down to
an inferior suite of caverns, and, flowing through them for
centuries, choked them up with débris; after which it rose
once more to its original higher level: just as in the mountain limestone district of Yorkshire some subterranean rivers
are occasionally unable to discharge all their water through
the cave in which they habitually run; in which case they
rise and rush through a higher subterranean passage, which
was at some former period in the regular line of drainage.

There are now in the basin of the Meuse, not far from
Liége, several examples of engulfed brooks and rivers; some
of them like that of St. Hadelin, east of Chaudefontaine,
which reappears after an underground course of a mile or
two; others like the Vesdre, which is lost near Goffontaine,
and after a time re-emerges: some again, like the torrent near
Magnée, which, after entering a cave, never again comes to
the day. In the season of floods such streams are turbid at
their entrance, but clear as a mountain-spring where they
issue again, so that they must be slowly filling up cavities in
the interior with mud, sand, pebbles, snail-shells, and the
bones of animals which may be carried away during floods.

The manner in which some of the large thigh and shank
bones of the rhinoceros and other pachyderms are rounded,
while some of the smaller bones of the same creatures, and
of the hyæna, bear, and horse, are reduced to pebbles, shows
that they were often transported for some distance in the
channels of torrents before they found a resting-place.

When we desire to reason or speculate on the probable
antiquity of human bones found fossil in such situations as
the caverns near Liége, there are two classes of evidence to
which we may appeal for our guidance. First, considerations

of the time required to allow of many species of carnivorous
and herbivorous animals, which flourished in the cave period,
becoming first scarce, and then so entirely extinct as we
have seen that they had become before the era of the Danish
peat and Swiss lake-dwellings: secondly, the great number
of centuries necessary for the conversion of the physical
geography of the Liége district from its ancient to its present
configuration; so many old underground channels, through
which brooks and rivers flowed in the cave period, being now
laid dry and choked up.

The great alterations which have taken place in the
shape of the valley of the Meuse and some of its tributaries
are often demonstrated by the abrupt manner in which the
mouths of fossiliferous caverns open in the face of perpendicular precipices 200 feet or more in height above the
present streams. There appears also, in many cases, to be
such a correspondence in the openings of caverns on opposite
sides of some of the valleys, both large and small, as to
incline one to suspect that they originally belonged to a
series of tunnels and galleries which were continuous before
the present system of drainage came into play, or before the
existing valleys were scooped out. Other signs of subsequent
fluctuations are afforded by gravel containing elephant's
bones at slight elevations above the Meuse and several of its
tributaries. The loess also, in the suburbs and neighbourhood of Liége, occurring at various heights in patches lying
at between 20 and 200 feet above the river, cannot be explained without supposing the filling up and re-excavation
of the valleys at a period posterior to the washing in of the
animal remains into most of the old caverns. It may be
objected that, according to the present rate of change, no
lapse of ages would suffice to bring about such revolutions
in physical geography as we are here contemplating. This
may be true. It is more than probable that the rate of

change was once far more active than it is now in the basin
of the Meuse. Some of the nearest volcanoes, namely, those
of the Lower Eifel, about sixty miles to the eastward, seem
to have been in eruption in Pleistocene times, and may
perhaps have been connected and coeval with repeated risings
or sinkings of the land in the Liége district. It might be
said, with equal truth, that according to the present course
of events, no series of ages would suffice to reproduce such
an assemblage of cones and craters as those of the Eifel (near
Andernach, for example); and yet some of them may be of
sufficiently modern date to belong to the era when Man was
contemporary with the mammoth and rhinoceros in the basin
of the Meuse.

But, although we may be unable to estimate the minimum of time required for the changes in physical geography
above alluded to, we cannot fail to perceive that the duration
of the period must have been very protracted, and that other
ages of comparative inaction may have followed, separating
the Pleistocene from the historical periods, and constituting
an interval no less indefinite in its duration.

Caverns of the Lesse.

In the year 1864 M. E. Dupont, at the request of the
Belgian Government, began a series of extensive researches
in the caverns of the valley of the Lesse, where this river
joins the Meuse. These caverns are chiefly distinguished
from those of Perigord by the thick deposits of loess or river-
mud in which the bones and implements are embedded. Out
of forty-three caves, twenty-five furnished traces of men.
M. Dupont has divided these into three periods: first, the
Mammoth period, which he considers to be the equivalent
of the rolled pebbles and stratified clay of the river-valley;
second, the Reindeer period, corresponding to the brick-
earths and angular pebbles; and thirdly, the Neolithic or

polished stone period, which is, however, more feebly represented than the other two.[*]

The deposits of the Mammoth period contain the bones of *Elephas primigenius* and *Rhinoceros tichorhinus*, together with other extinct animals, and the bones of the reindeer. With these are associated flint instruments of the rudest type, and instruments of reindeer horn; and in one cave, the Trou de la Naulette, was found a remarkable human lower jaw, which in its characters 'exaggerates,' says M. Dupont, ' those points in which the most inferior of the living races are distinguished from ourselves.' In another cave, the Trou Magrite, rude carvings on reindeer bone occur similar to, but more rude than, those found in the caves of Dordogne presently to be mentioned (p.138).

The celebrated Trou du Frontal is the one which perhaps best exhibits the superposition of remains of men of the Reindeer period upon the deposits of the age of the Mammoth. It was evidently, according to M. Dupont, a place of sepulture. The deposits of the Mammoth period, which sloped down gently to the mouth of the cave, were covered by a mass of yellow clay mixed with fragments of the adjacent rock. Above this were found in great confusion the bones of sixteen human skeletons, the cave being closed by a large block of stone. The associated animals were of living species, but many, such as the reindeer and chamois, now belong to more northern regions, or to the snowy summits of the centre of Europe. Well cut, but unpolished, flint implements, ornaments of artificially pierced fossil shells, ornaments of pierced fluorine and the fragments of an urn, were also found. Some of the shells have a peculiar interest, being of species, such as the enormous *Cerithium giganteum*, which must have been brought from the Paris basin in the environs of Versailles and Rheims. Indeed, M. Dupont

points out that we can trace the route northwards followed
by these ancient races by means of the objects which they
have transported into the Lesse caverns of the Reindeer
period.   The material of many of the worked flints can only
have come from Vertus in the department of Marne, whilst
we find also corals from Vouziers, liassic jet from Jamoigne,
Silurian slate from Fumay, and Devonian fossils from Givet.

Remains of the Neolithic period are, as I have said, comparatively rare in these caverns, occurring only in five cases,
the most important of which is the Trou de Gendron, a
sepulchre containing sixteen skeletons.   It is worthy, however, of remark, that the flint implements in these Neolithic
deposits are all made of stone from the immediate province of Hainault.

CHAPTER V.

PLEISTOCENE PERIOD—FOSSIL HUMAN SKULLS OF THE
NEANDERTHAL AND ENGIS CAVES.

HUMAN SKELETON FOUND IN CAVE NEAR DÜSSELDORF—ITS GEOLOGICAL
POSITION AND PROBABLE AGE—ITS ABNORMAL AND APE-LIKE CHA-
RACTER—FOSSIL HUMAN SKULL OF THE ENGIS CAVE NEAR LIÉGE—
PROFESSOR HUXLEY'S DESCRIPTION OF THESE SKULLS—COMPARISON
OF EACH, WITH EXTREME VARIETIES OF THE NATIVE AUSTRALIAN
RACE—RANGE OF CAPACITY IN THE HUMAN AND SIMIAN BRAINS—
SKULLS FROM BORREBY IN DENMARK—CONCLUSIONS OF PROFESSOR
HUXLEY — BEARING OF THE PECULIAR CHARACTERS OF THE NEAN-
DERTHAL SKULL ON THE HYPOTHESIS OF TRANSMUTATION.

Fossil Human Skeleton of the Neanderthal Cave near Düsseldorf.

BEFORE I speak more particularly of the opinions which
anatomists have expressed respecting the osteological
characters of the human skull from Engis, near Liége,
mentioned in the last chapter and described by Dr. Schmerling, it will be desirable to say something of the geological
position of another skull, or rather skeleton, which, on
account of its peculiar conformation, has excited no small
interest. I allude to the skull found in 1857, in a cave
situated in that part of the valley of the Düssel, near
Düsseldorf, which is called the Neanderthal. The spot
is a deep and narrow ravine about seventy English miles
north-east of the region of the Liége caverns treated of
in the last chapter, and close to the village and railway
station of Hochdal between Düsseldorf and Elberfeld. The
cave occurs in the precipitous southern or left side of the
winding ravine, about sixty feet above the stream, and a

hundred feet below the top of the cliff.   The accompanying
section will give the reader an idea of its position.

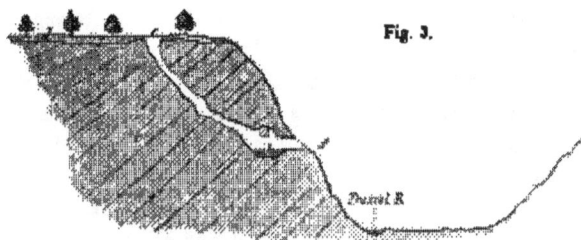

Fig. 3.

Section of the Neanderthal Cave near Düsseldorf.

a Cavern 60 feet above the Düssel, and 100 feet below the surface
of the country at e.
b Loam covering the floor of the cave near the bottom of which the
human skeleton was found.
b, c Real connecting the cave with the upper surface of the country.
d Superficial sandy loam.
e Devonian limestone.
f Terrace, or ledge of rock.

When Dr. Fuhlrott of Elberfeld first examined the cave,
he found it to be high enough to allow a man to enter.
The width was seven or eight feet, and the length or depth
fifteen.   I visited the spot in 1860, in company with Dr.
Fuhlrott, who had the kindness to come expressly from
Elberfeld to be my guide, and who brought with him the
original fossil skull, and a cast of the same, which he presented to me.   In the interval of three years, between 1857
and 1860, the ledge of rock, f, on which the cave opened,
and which was originally twenty feet wide, had been almost
entirely quarried away, and, at the rate at which the work
of dilapidation was proceeding, its complete destruction
seemed near at hand.

In the limestone are many fissures, one of which, still

G

partially filled with mud and stones, is represented in the
section at *a c* as continuous from the cave to the upper
surface of the country. Through this passage the loam,
and possibly the human body to which the bones belonged,
may have been washed into the cave below. The loam,
which covered the uneven bottom of the cave, was sparingly
mixed with rounded fragments of chert, and was very similar
in composition to that covering the general surface of that
region.

There was no crust of stalagmite overlying the mud in
which the human skeleton was found, and no bones of other
animals in the mud with the skeleton; but just before our
visit in 1860 the tusk of a bear had been met with in some
mud in a lateral embranchment of the cave, in a situation
precisely similar to *b*, fig. 3, and on a level corresponding
with that of the human skeleton. This tusk, shown us by
the proprietor of the cave, was two and a half inches long and
quite perfect; but whether it was referable to a recent or
extinct species of bear, I could not determine.

From a printed letter of Dr. Fuhlrott we learn that on
removing the loam, which was five feet thick, from the cave,
the human skull was first noticed near the entrance, and,
further in, the other bones lying in the same horizontal
plane. It is supposed that the skeleton was complete, but
the workmen, ignorant of its value, scattered and lost most
of the bones, preserving only the larger ones.*

The cranium, which Dr. Fuhlrott showed me, was covered,
both on its outer and inner surface, and especially on the
latter, with a profusion of dendritical crystallisations, and
some other bones of the skeleton were ornamented in the
same way. These markings, as Dr. Hermann von Meyer
observes, afford no sure criterion of antiquity, for they have

---

* Fuhlrott, Letter to Professor   Review. No. 2. p. 156. See also
Schaaffhausen, cited Natural History   Naturhistorisch Vereins Bonn, 1859.

been observed on Roman bones. Nevertheless, they are more common in bones that have been long embedded in the earth. The skull and bones, moreover, of the Neanderthal skeleton had lost so much of their animal matter as to adhere strongly to the tongue, agreeing in this respect with the ordinary condition of fossil remains of the Pleistocene period. On the whole, I think it probable that this fossil may be of about the same age as those found by Schmerling in the Liége caverns, and Professor Huxley, who visited the cave in 1870, and found there bones of the rhinoceros at the bottom of the loess, inclines to the same opinion. Its position lends no countenance whatever to the supposition of its being more ancient.

When the skull and other parts of the skeleton were first exhibited at a German scientific meeting at Bonn, in 1857, some doubts were expressed by several naturalists, whether it was truly human. Professor Schaaffhausen, who, with the other experienced zoologists, did not share these doubts, observed that the cranium, which included the frontal bone, both parietals, part of the squamous, and the upper third of the occipital, was of unusual size and thickness, the forehead narrow and very low, and the projection of the supra-orbital ridges enormously great. He also stated that the absolute and relative length of the thigh-bone, humerus, radius, and ulna, agreed well with the dimensions of a European individual of like stature at the present day; but that the thickness of the bones was very extraordinary, and the elevation and depression for the attachment of muscles were developed in an unusual degree. Some of the ribs, also, were of a singularly rounded shape and abrupt curvature, which was supposed to indicate great power in the thoracic muscles.[*]

[*] Professor Schaaffhausen's Memoir, translated, Natural History Review, No. 2, April 1861.

In the same memoir, the Prussian anatomist remarks that the depression of the forehead (see fig. 5, p. 87) is not due to any artificial flattening, such as is practised in various modes by barbarous nations in the Old and New World, the skull being quite symmetrical, and showing no indication of counter-pressure at the occiput; whereas, according to Morton, in the Flat-heads of the Columbia, the frontal and parietal bones are always unsymmetrical.* On the whole, Professor Schaaffhausen concluded that the individual to whom the Neanderthal skull belonged must have been distinguished by small cerebral development, and uncommon strength of corporeal frame.

When on my return to England I showed the cast of the cranium to Professor Huxley, he remarked at once that it was the most ape-like skull he had ever beheld. Mr. Busk, after giving a translation of Professor Schaaffhausen's memoir in the 'Natural History Review,'† added some valuable comments of his own on the characters in which this skull approached that of the gorilla and chimpanzee.

Professor Huxley afterwards studied the cast with the object of assisting me to give illustrations of it in this work, and in doing so discovered what had not previously been observed, that it was quite as abnormal in the shape of its occipital as in that of its frontal or superciliary region. Before citing his words on the subject, I will offer a few remarks on the Engis skull, which the same anatomist has compared with that of the Neanderthal.

Fossil Skull of the Engis Cave near Liége.

Among six or seven human skeletons, portions of which were collected by Dr. Schmerling from three or four caverns

* Natural History Review, No. 2, p. 160.   † No. 2, 1861.

near Liége, embedded in the same matrix with the remains
of the elephant, rhinoceros, bear, hyæna, and other extinct
quadrupeds, the most perfect skull, as I have before stated
(p. 67), was that of an adult individual found in the cavern of
Engis. This skull, Dr. Schmerling figured in his work,
observing that it was too imperfect to enable the anatomist
to determine the facial angle, but that one might infer, from
the narrowness of the frontal portion, that it belonged to an
individual of small intellectual development. He speculated on its Ethiopian affinities, but not confidently, observing
truly that it would require many more specimens to enable
an anatomist to arrive at sound conclusions on such a point.
M. Geoffroy St. Hilaire and other osteologists, who examined
the specimen, denied that it resembled a negro's skull. When
I saw the original in the museum at Liége, I invited Dr.
Spring, one of the professors of the university, to whom we
are indebted for a valuable memoir on the human bones
found in the cavern of Chauvaux near Namur, to have a
cast made of this Engis skull. He not only had the kindness to comply with my request, but rendered a service to
the scientific world by adding to the original cranium
several detached fragments which Dr. Schmerling had
obtained from Engis, and which were found to fit in exactly,
so that the cast represented in fig. 4 is more complete than
that given in the first plate of Schmerling's work.   It
exhibits on the right side the position of the auditory
foramen (see fig. 8, p. 93), which was not included in
Schmerling's figure. Mr. Busk, when he saw this cast,
remarked to me that, although the forehead was, as Schmerling had truly stated, somewhat narrow, it might nevertheless
be matched by the skulls of individuals of European race,
an observation since fully borne out by measurements, as
will be seen in the sequel.

OBSERVATIONS BY PROFESSOR HUXLEY ON THE HUMAN SKULLS
OF ENGIS AND NEANDERTHAL.

'The Engis skull, as originally figured by Professor Schmerling,
was in a very imperfect state; but other fragments have since been
added to it by the care of Dr. Spring, and the cast upon which my
observations are based (fig. 4) exhibits the frontal, parietal, and
occipital regions, as far as the middle of the occipital foramen, with
the squamous and mastoid portions of the right temporal bone
entire, or nearly so, while the left temporal bone is wanting. From
the middle of the occipital foramen to the middle of the roof of each
orbit, the base of the skull is destroyed, and the facial bones are
entirely absent.

Fig. 4.

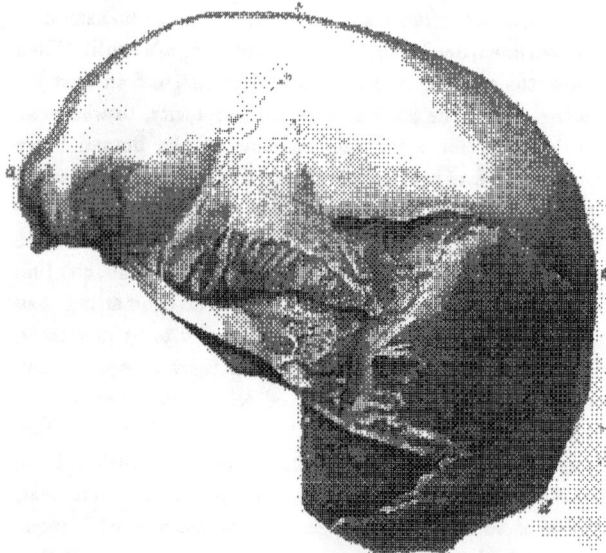

Side view of the cast of part of a human skull found by Dr. Schmerling
embedded amongst the remains of extinct mammalia in the cave of Engis, near
Liège.

a Superciliary ridge and glabella.   c The apex of the lambdoidal suture.
b Coronal suture.   d The occipital protuberance.

'The extreme length of the skull is 7·7 inches, and as its extreme
breadth is not more than 5·25, its form is decidedly dolichocephalic.
At the same time its height (4¾ inches from the plane of the
glabello-occipital line (a d) to the vertex) is good, and the forehead
is well arched ; so that while the horizontal circumference of the
skull is about 20½ inches, the longitudinal arc from the nasal spine of
the frontal bone to the occipital protuberance (d) measures about 13¾
inches. The transverse arc from one auditory foramen to the other
across the middle of the sagittal suture measures about 13 inches.
The sagittal suture (b c) is 5½ inches in length. The superciliary
prominences are well, but not excessively, developed, and are separated by a median depression in the region of the glabella. They
indicate large frontal sinuses. If a line joining the glabella and
the occipital protuberance (a d) be made horizontal, no part of the
occiput projects more than ⅒th of an inch behind the posterior extremity of that line; and the upper edge of the auditory foramen
is almost in contact with the same line, or rather with one drawn
parallel to it on the outer surface of the skull.

Fig. 4.

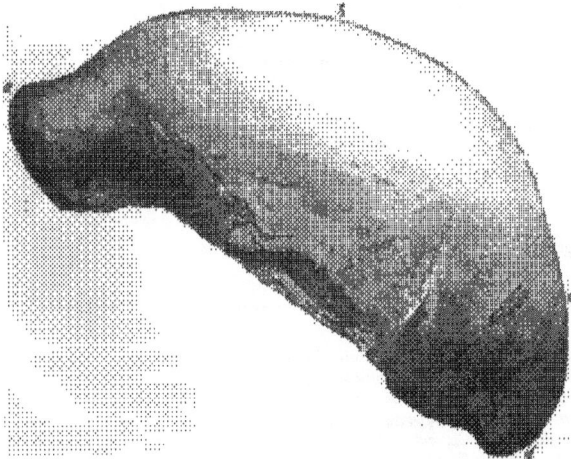

Side view of the cast of a part of a human skull from a cave in the Neanderthal
near Düsseldorf.

a The superciliary ridge and glabella.   c The apex of the lambdoidal suture.
b The coronal suture.   d The occipital protuberance.

' The Neanderthal skull, with which also I am acquainted only by means of Professor Schaaffhausen's drawings, of an excellent cast, and of photographs, is so extremely different in appearance from the Engis cranium, that it might well be supposed to belong to a distinct race of mankind. It is 8 inches in extreme length and 5·75 inches in extreme breadth, but only measures 3·4 inches from the glabello-occipital line to the vertex. The longitudinal arc, measured as above, is 12 inches; the transverse arc cannot be exactly ascertained, in consequence of the absence of the temporal bones, but was probably about the same, and certainly exceeded 10¼ inches. The horizontal circumference is 23 inches. This great circumference arises largely from the vast development of the superciliary ridges, which are occupied by great frontal sinuses whose inferior apertures are displayed exceedingly well in one of Dr. Fuhlrott's photographs, and form a continuous transverse prominence, somewhat excavated in the middle line, across the lower part of the

Fig. 6.

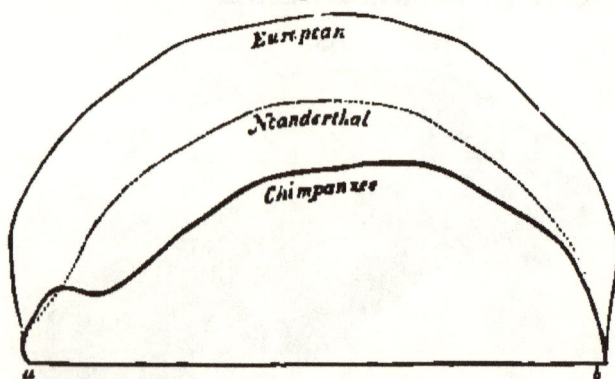

Outline of the skull of an adult Chimpanzee, of that from the Neanderthal, and of that of a modern European, drawn to the same absolute size, in order better to exhibit their relative differences. The superciliary region of the Neanderthal skull appears less prominent than in fig. 5, as the contours are all taken along the middle line where the superciliary projection of the Neanderthal skull is least marked.

    a The glabella.
    b The occipital protuberance, or the point on the exterior of each skull which corresponds roughly with the attachment of the tentorium, or with the inferior boundary of the posterior cerebral lobes.

brows. In consequence of this structure, the forehead appears still lower and more retreating than it really is. To an anatomical eye the posterior part of the skull is even more striking than the anterior. The occipital protuberance occupies the extreme posterior end of the skull when the glabello-occipital line is made horizontal, and so far from any part of the occipital region extending beyond it, this region of the skull slopes obliquely upward and forward, so that the lambdoidal suture is situated well upon the upper surface of the cranium. At the same time, notwithstanding the great length of the skull, the sagittal suture is remarkably short (4¼ inches), and the squamosal suture is very straight.

' In human skulls, the superior curved ridge of the occipital bone and the occipital protuberance correspond, approximatively, with the level of the tentorium and with the lateral sinuses, and consequently with the inferior limit of the posterior lobes of the brain. At first, I found some difficulty in believing that a human brain could have its posterior lobes so flattened and diminished as must have been the case in the Neanderthal man, supposing the ordinary relation to obtain between the superior occipital ridges and the tentorium; but on my application, through Sir Charles Lyell, Dr. Fuhlrott, the possessor of the skull, was good enough not only to ascertain the existence of the lateral sinuses in their ordinary position, but to send convincing proofs of the fact, in excellent photographic views of the interior of the skull, exhibiting clear indications of these sinuses.

' There can be no doubt that, as Professor Schaaffhausen and Mr. Busk have stated, this skull is the most brutal of all known human skulls, resembling those of the apes not only in the prodigious development of the superciliary prominences and the forward extension of the orbits, but still more in the depressed form of the brain-case, in the straightness of the squamosal suture, and in the complete retreat of the occiput forward and upward, from the superior occipital ridges.

' But the cranium, in its present condition, is stated by Professor Schaaffhausen to contain 1033·24 cubic centimetres of water, or, in other words, about 63 English cubic inches. As the entire skull could hardly have held less than 12 cubic inches more, its minimum capacity may be estimated at 75 cubic inches. The most capacious healthy European skull yet measured had a capacity of 114 cubic inches, the smallest (as estimated by weight of brain) about 55 cubic inches, while, according to Professor Schaaffhausen, some Hindoo skulls have as small a capacity as about 46 cubic inches

(27 oz. of water). The largest cranium of any gorilla yet measured contained 34·5 cubic inches. The Neanderthal cranium
stands, therefore, in capacity, very nearly on a level with the mean
of the two human extremes, and very far above the pithecoid
maximum.

' Hence, even in the absence of the bones of the arm and thigh,
which, according to Professor Schaaffhausen, had the precise proportions found in Man, although they were very much stouter than
ordinary human bones, there could be no reason for ascribing this
cranium to anything but a man; while the strength and development of the muscular ridges of the limb-bones are characters in
perfect accordance with those exhibited, in a minor degree, by the
bones of such hardy savages, exposed to a rigorous climate, as the
Patagonians.

' The Neanderthal cranium has certainly not undergone com-
pression, and, in reply to the suggestion that the skull is that of an
idiot, it may be urged that the onus probandi lies with those who
adopt the hypothesis. Idiocy is compatible with very various forms
and capacities of the cranium, but I know of none which present
the least resemblance to the Neanderthal skull; and, furthermore, I
shall proceed to show that the latter manifests but an extreme degree
of a stage of degradation exhibited, as a natural condition, by the
crania of certain races of mankind.

' Mr. Busk drew my attention, some time ago, to the resemblance
between some of the skulls taken from tumuli of the stone period at
Borreby in Denmark, of which Mr. Busk possesses numerous accurate
figures, and the Neanderthal cranium. One of the Borreby skulls
in particular (fig. 7, p. 91) has remarkably projecting superciliary
ridges, a retreating forehead, a low flattened vertex, and an occiput
which shelves upward and forward. But the skull is relatively
higher and broader, or more brachycephalic, the sagittal suture
longer, and the superciliary ridges less projecting, than in the
Neanderthal skull. Nevertheless, there is, without doubt, much
resemblance in character between the two skulls,—a circumstance
which is the more interesting, since the other Borreby skulls have
better foreheads and less prominent superciliary ridges, and exhibit
altogether a higher conformation.

' The Borreby skulls belong to the stone period of Denmark, and
the people to whom they appertained were probably either contemporaneous with, or later than, the makers of the "refuse-heaps" of
that country. In other words, they were subsequent to the last great
physical changes of Europe, and were contemporaries of the urus

and bison, not of the *Elephas primigenius*, *Rhinoceros tichorhinus*, and *Hyæna spelæa*.

'Supposing for a moment, what is not proven, that the Neanderthal skull belonged to a race allied to the Borreby people, and was as

Fig. 7.

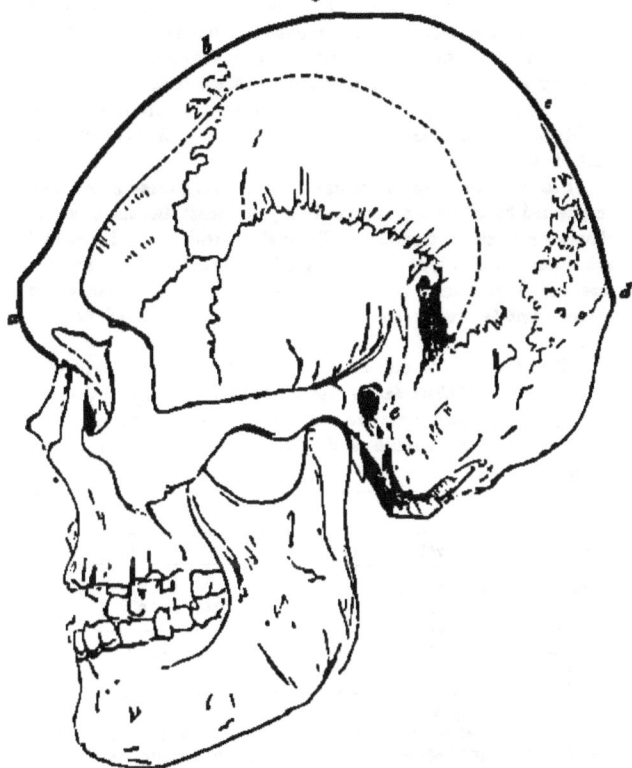

Skull associated with ground flint implements, from a tumulus at Borreby in Denmark, after a camera lucida drawing by Mr. G. Busk, F.R.S. The thick dark line indicates so much of the skull as corresponds with the fragment from the Neanderthal.

    a Superciliary ridge.   c The apex of the lambdoidal suture.
    b Coronal suture.   d The occipital protuberance.
                 e The auditory foramen.

modern as they, it would be separated by as great a distance of time as of anatomical character from the Engis skull, and the possibility of its belonging to a distinct race from the latter might reasonably appear to be greatly heightened.

'To prevent the possibility of reasoning in a vicious circle, however, I thought it would be well to endeavour to ascertain what amount of cranial variation is to be found in a pure race at the present day; and as the natives of Southern and Western Australia are probably as pure and homogeneous in blood, customs, and language, as any race of savages in existence, I turned to them, the more readily as the Hunterian museum contains a very fine collection of such skulls.

'I soon found it possible to select from among these crania two (connected by all sorts of intermediate gradations), the one of which should very nearly resemble the Engis skull, while the other should somewhat less closely approximate the Neanderthal cranium in form, size, and proportions. And at the same time others of these skulls presented no less remarkable affinities with the low type of Borreby skull.

'That the resemblances to which I allude are by no means of a merely superficial character, is shown by the accompanying diagram (fig. 8, p. 93), which gives the contours of the two ancient and of one of the Australian skulls, and by the following table of measurements:—

	A	B	C	D	E	F
Engis . . .	20½	13½	12½	4½	7½	5½
Australian, No. 1	20½	13	12	4½	7	5¾
Australian, No. 2	22	12½	10½	3½	7½	6½
Neanderthal .	23	12	10	3½	8	6

A The horizontal circumference in the plane of a line joining the glabella with the occipital protuberance.

B The longitudinal arc from the nasal depression along the middle line of the skull to the occipital tuberosity.

C From the level of the glabello-occipital line on each side, across the middle of the sagittal suture to the same point on the opposite side.

D The vertical height from the glabello-occipital line.

E The extreme longitudinal measurement.

F The extreme transverse measurement.*

* I have taken the glabello-occipital line as a base in these measurements, simply because it enables me to compare all the skulls, whether fragments or entire, together. The greatest circumference of the English skull lies in a plane considerably above that of the glabello-occipital line, and amounts to twenty-two inches.

'The question whether the Engis skull has rather the character of one of the high races or of one of the lower has been much disputed, but the following measurements of an English skull, noted in the catalogue of the Hunterian museum as typically Caucasian (see fig. 6), will serve to show that both sides may be right, and that cranial measurements alone afford no safe indication of race.

English . .	A 2½	B 13¾	C 12½	D 4¹⁄₁₀	E 7¾	F 5¼

'In making the preceding statement, it must be clearly understood that I neither desire to affirm that the Engis and Neanderthal skulls belong to the Australian race, nor to assert even that the ancient

Fig. 6.

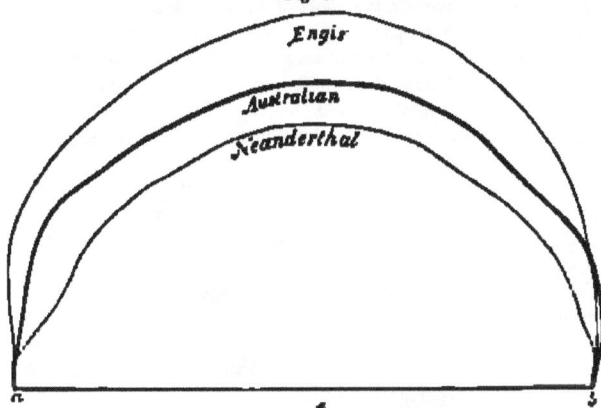

Outlines of the skull from the Neanderthal, of an Australian skull from Port Adelaide, and of the skull from the Cave of Engis, drawn to the same absolute length, in order the better to contrast their proportions.

a, b As in fig. 6, p. 88.
c The position of the auditory foramen of the Engis skull.

skulls belong to one and the same race, so far as race is measured by language, colour of skin, or character of hair.   Against the conclusion that they are of the same race as the Australians, various minor anatomical differences of the ancient skulls, such as the great development of the frontal sinuses, might be urged; while against the supposition of either the identity, or the diversity, of race of the two arises the known independence of the variation of cranium on the one hand, and of hair, colour, and language on the other.

'But the amount of variation of the Borreby skulls, and the fact that the skulls of one of the purest and most homogeneous of existing races of men can be proved to differ from one another in the same characters, though perhaps not quite to the same extent, as the Engis and Neanderthal skulls, seem to me to prohibit any cautious reasoner from affirming the latter to have been necessarily of distinct races.

'The marked resemblances between the ancient skulls and their modern Australian analogues, however, have a profound interest, when it is recollected that the stone axe is as much the weapon and the implement of the modern as of the ancient savage; that the former turns the bones of the kangaroo and of the emu to the same account as the latter did the bones of the deer and the urus; that the Australian heaps up the shells of devoured shellfish in mounds which represent the "refuse-heaps" or "Kjökkenmöddings" of Denmark; and, finally, that, on the other side of Torres Straits, a race akin to the Australians are among the few people who now build their houses on pile-works, like those of the ancient Swiss lakes.

'That this amount of resemblance in habit and in the conditions of existence is accompanied by as close a resemblance in cranial configuration, illustrates on a great scale that what Cuvier demonstrated of the animals of the Nile valley is no less true of men; circumstances remaining similar, the savage varies little more, it would seem, than the ibis or the crocodile, especially if we take into account the enormous extent of the time over which our knowledge of man now extends, as compared with that measured by the duration of the sepulchres of Egypt.

'Finally, the comparatively large cranial capacity of the Neanderthal skull, overlaid though it may be by pithecoid bony walls, and the completely human proportions of the accompanying limb-bones, together with the very fair development of the Engis skull, clearly indicate that the first traces of the primordial stock whence Man has proceeded need no longer be sought, by those who entertain any form of the doctrine of progressive development, in the newest tertiaries; but that they may be looked for in an epoch more distant from the age of the *Elephas primigenius* than that is from us.'

The two skulls which form the subject of the preceding comments and illustrations have given rise to nearly an equal amount of surprise for opposite reasons; that of Engis, because, being so unequivocally ancient, it approaches so

near to the highest or Caucasian type; that of the Neanderthal, because, having no such decided claims to antiquity, it
departs so widely from the normal standard of humanity.
Professor Huxley's observation regarding the wide range of
variation, both as to shape and capacity, in the skulls of so
pure a race as the native Australian, removes to no small
extent this supposed anomaly, by showing the probability
that both varieties may have coexisted in the Pleistocene
period in Western Europe.

As to the Engis skull, we must remember that although
associated with the elephant, rhinoceros, bear, tiger, and
hyena, all of extinct species, it nevertheless is also accompanied by a bear, stag, wolf, fox, beaver, and many other
quadrupeds of species still living. Indeed many eminent
palæontologists, and among them Professor Pictet, think that,
numerically considered, the larger portion of the mammalian
fauna agrees specifically with that of our own period, so that
we are scarcely entitled to feel surprised if we find human
races of the Pleistocene epoch undistinguishable from some
living ones. It would merely tend to show that Man has
been as constant in his osteological characters as many other
mammalia now his contemporaries. The expectation of
always meeting with a lower type of human skull, the older
the formation in which it occurs, is based on the theory of
progressive development, and it may prove to be sound;
nevertheless we must remember that as yet we have no distinct geological evidence that the appearance of what are
called the inferior races of mankind has always preceded in
chronological order that of the higher races.

It is now admitted that the differences in character
between the brain of the highest races of Man and that of
the lowest, though less in degree, are of the same order as
those which separate the human from the simian brain;"
and the same rule holds good in regard to the shape of the

* Natural History Review, 1861, p. 81.

skull. The average Negro skull differs from that of the
European in having a more receding forehead, more prominent superciliary ridges, and more largely developed
prominences and furrows for the attachment of muscles; the
face also, and its lines, are larger proportionally. The brain
is somewhat less voluminous on the average in the lower
races of mankind, its convolutions rather less complicated,
and those of the two hemispheres more symmetrical, in all
which points an approach is made to the simian type. It
will also be seen, by reference to the late Dr. Morton's works,
and by the foregoing statements of Professor Huxley, that
the range of size or capacity between the highest and lowest
human brain is greater than that between the highest simian
and lowest human brain; but the Neanderthal skull, although
in several respects it is more ape-like than any human skull
previously discovered, is, in regard to volume, by no means
contemptible.

Eminent anatomists have shown that in the average proportions of some of the bones the Negro differs from the
European, and that in most of these characters he makes a
slightly nearer approach to the anthropoid quadrumana;*

* 'The inferior races of mankind
exhibit proportions which are in many
respects intermediate between the
higher, or European, orders, and the
monkeys. In the Negro, for instance,
the stature is less than in the European. The cranium, as is well known,
bears a small proportion to the face.
Of the extremities the upper are pro-
portionately longer, and there is, in
both upper and lower, a less marked
preponderance of the proximal over the
distal segments. For instance, in the
Negro, the thigh and arm are rather
shorter than in the European; the leg
is actually of equal length in both
races, and is therefore, relatively, a
little longer in the Negro; the fore-arm
in the latter is actually, as well as
relatively, a little longer: the foot is
an eighth, and the hand a twelfth
longer than in the European. It is
well known that the foot is less well
formed in the Negro than in the
European. The arch of the instep,
the perfect conformation of which is
essential to steadiness and ease of
gait, is less elevated in the former
than in the latter. The foot is
thereby rendered flatter as well as
longer, more nearly resembling the
monkey's, between which and the
European there is a marked difference in this particular.'—From 'A
Treatise on the Human Skeleton,' by
Dr. Humphry, Lecturer on Surgery
and Anatomy in the Cambridge University Medical School, p. 91.

but Professor Schaaffhausen has pointed out that in these
proportions the Neanderthal skeleton does not differ from
the ordinary standard, so that the skeleton by no means
indicates a transition between Homo and Pithecus.

There is doubtless, as shown in the diagram fig. 6, a
nearer resemblance in the outline of the Neanderthal skull
to that of a chimpanzee than had ever been observed before
in any human cranium; and Professor Huxley's description
of the occipital region shows that the resemblance is not
confined to the mere excessive prominence of the superciliary
ridges.

The direct bearing of the ape-like character of the Neanderthal skull on Lamarck's doctrine of progressive development and transmutation, or on that modification of it
which has been so ably advocated by Mr. Darwin, consists
in this, that the newly-observed deviation from a normal
standard of human structure is not a casual or random mon-
strosity, but just what might have been anticipated if the
laws of variation were such as the transmutationists require.
For if we conceive the cranium to be very ancient, it exem-
plifies a less advanced stage of progressive development
and improvement. If it be a comparatively modern race,
owing its peculiarities of conformation to degeneracy, it is an
illustration of what botanists call ' atavism,' or the tendency
of varieties to revert to an ancestral type, which type, in
proportion to its antiquity, would be of lower grade.　To
this hypothesis, of a genealogical connection between Man
and the lower animals, I shall again allude in the concluding
chapters.

CHAPTER VI.

FLINT IMPLEMENTS IN ENGLISH CAVE-DEPOSITS.

DISCOVERY OF FLINT IMPLEMENTS IN KENT'S CAVERN IN 1825—SYSTE-
MATIC EXPLORATION OF BRIXHAM CAVERN IN 1858—SUPERPOSITION
OF DEPOSITS IN THE CAVE—PALÆOLITHIC FLINT IMPLEMENTS WITH
BONES OF EXTINCT MAMMALIA OCCURRING IN THE CAVE-EARTH UNDER
A THICK FLOOR OF STALAGMITE—RECENT EXPLORATIONS OF KENT'S
CAVERN—FINDING OF A TOOTH OF MACHAIRODUS LATIDENS CONFIRMING
THE PREVIOUS DISCOVERIES OF MAC ENERY—FLINT IMPLEMENTS IN A
SOMERSETSHIRE CAVE ALSO CONTAINING HYÆNA AND OTHER EXTINCT
MAMMALIA—CAVES OF THE GOWER PENINSULA IN WALES—RHINOCEROS
HEMITŒCHUS—GRIMALDI'S CAVES NEAR PALERMO—SICILY ONCE PART
OF AFRICA—RISE OF BED OF THE MEDITERRANEAN TO THE HEIGHT
OF THREE HUNDRED FEET IN THE HUMAN PERIOD IN SARDINIA.

ABOUT the time that Schmerling was exploring the Liége caves, the Rev. Mr. Mac Enery, a Roman Catholic priest, residing near Torquay, had found in a cave one mile east of that town, called 'Kent's Hole,' in red loam covered with stalagmite, not only bones of the mammoth, tichorhine rhinoceros, cave-bear, and other mammalia, but several remarkable flint tools, some of which he supposed to be of great antiquity, and which are now known to be of a distinctly Palæolithic type,* while there were also remains of Man in the same cave of a later date.† To make known his discovery he prepared, in common with Dr. Buckland, a joint memoir on the contents of 'Kent's Hole,' but he died before it could be completed, and it remained in MS. until 1859, when an abridgment of it was published by Mr. Vivian, of Torquay, and more lately the whole has been printed in full by Mr. Pengelly.

* Evans's Stone Implements of Great Britain, p. 443.

† Trans. Devon. Assoc. vol. iii. p. 221, 1869.

Dr. Buckland, in his celebrated work, entitled Reliquiæ
Diluvianæ, published in 1823, in which he treated of the
organic remains contained in caves, fissures, and 'diluvial
gravel' in England, had given a clear statement of the results
of his own original observations, and had declared that none
of the human bones or stone implements met with by him
in any of the caverns could be considered to be as old as the
mammoth and other extinct quadrupeds. Opinions in harmony with these conclusions continued in vogue for many years
after Mr. Mac Enery's discoveries, although it would appear
from the memoir, that Mr. Mac Enery himself only refrained,
out of deference to Dr. Buckland, from declaring his belief
in the contemporaneity of the ancient flint implements *
and the bones of the extinct animals.

About ten years afterwards, in a memoir on the Geology
of South Devon, which was published in 1842 by the Geological Society of London, Mr. Godwin-Austen declared
that he had obtained in the same cave (Kent's Hole) works
of Man from undisturbed loam or clay, under stalagmite,
mingled with the remains of extinct animals, and that all
these must have been introduced 'before the stalagmite
flooring had been formed.' He maintained that such facts
could not be explained away by the hypothesis of sepulture,
because in the Devon cave the flint implements were widely
distributed through the loam, and lay beneath the stalagmite.

Brixham Cave near Torquay.

Before alluding to renewed explorations made twenty-
two years afterwards in 'Kent's Hole,' mention must be
made of a discovery which effected a sudden change of
opinion in England respecting the probable coexistence of
Man with many extinct mammalia. A new and intact bone

* Given in plate T of the memoir.

h 2

cave was brought to light in 1858, at Brixham, about four miles south of Torquay. At the instance of the late Dr. Falconer, prompt measures were taken to have this cave thoroughly and systematically examined. The Royal Society made two grants towards defraying the expenses, and the Baroness Burdett Coutts, Sir James Kay-Shuttleworth, and the late Mr. R. Arthington, of Leeds, contributed liberally towards the same object. A committee of geologists was charged with the investigations, among whom Dr. Falconer and Mr. Prestwich took a prominent part, visiting Torquay while the excavations were in progress. Mr. Pengelly, another member of the committee, well qualified for the task by nearly twenty years' previous experience in cave explorations, zealously directed and superintended the work. By him, in 1859, I was conducted through the subterranean galleries after they had been cleared out; and Dr. Falconer, who was also at Torquay, showed me the numerous fossils which had been discovered, and which he was then studying, all numbered and labelled, with reference to a journal in which the geological position of each specimen was recorded with scrupulous care.*

The discovery of the existence of this suite of caverns near the sea at Brixham was made accidentally by the roof of one of them being broken through in quarrying. None of the four external openings now exposed to view in steep cliffs or in the valley, were visible before the breccia and earthy matter which blocked them up were removed during the late exploration. According to a ground plan drawn up by Professor Ramsay, the cavern consists of two sets of passages, one running nearly north and south, and the other almost east and west—the directions in fact of the joints of the rocks in the district. The two principal passages, known

* An abstract of the whole of the explorations is now published in the   Royal Society's Proceedings, vol. xx., No. 137, p. 514.

as the 'Reindeer' and 'Flint-Knife' galleries, open into one another, and possess characters which leave no doubt of their having been excavated by the long-continued· action of running water.

In its western portion, the 'Flint-Knife' gallery was filled to the roof with mud; but in the remainder of it, as well as throughout the entire length of the 'Reindeer' gallery, there was a considerable space between the roof and the deposit, and the latter was covered with a continuous floor of stalagmite. The eastern entrance, opening into the 'Reindeer' gallery, was one hundred feet above the level of mean tide, and seventy-eight above the bottom of the adjoining valley. The following was the general succession of the deposits :—

1st. At the top, a layer of stalagmite, varying in thickness from one to fifteen inches, and containing bones, amongst which were the entire humerus of bear, and an antler of reindeer, the latter being firmly cemented to the upper surface of the floor, but rising above it in strong relief. The 'Reindeer' gallery took its name from this specimen.

2nd. Next below, loam or cave-earth, of an ochreous red colour, with angular pieces of limestone, and some pebbles of hematite of iron, quartz, and greenstone, from two to thirteen feet thick. This was the 'bone bed.'

3rd. At the bottom of all, gravel, consisting mainly of rounded pebbles. This was everywhere removed so long as the passages, which narrowed downwards, were wide enough to be worked. It proved to be almost entirely barren of fossils.

The mammalia obtained from the cave-earth and the stalagmite were as follows :—*

* See 'The Distribution of the   Mr. Boyd Dawkins, Quart. Journ. British Postglacial Mammals.' By   Geol. Soc. vol. xxv. p. 192, 1869.

1. *Ursus spelæus*, Gold.	10. *Cervus megacerus*, Hart.
2. *Elephas primigenius*, Blum.	11. *Rhinoceros tichorhinus*, Cuv.
3. *Lagomys spelæus*, Old.	12. *Cervus Tarandus*, L.
4. *Hyæna spelæa*, Gold.	13. *Felis Leo* (var. *spelæa*. Gold.)
5. *Ursus arctos*, L.	14. *Ursus ferox*, L.
6. *Canis Vulpes*, L.	15. *Canis Lupus*, L.
7. *Cervus Capreolus*, L.	16. *Cervus Elaphus*, L.
8. *Bos primigenius*, Boj.	17. *Equus Caballus*, L.
9. *Lepus cuniculus*, Pall.	18. *Lepus timidus*, Erxl.

Messrs. Dawkins and Sandford place the first four in the list of ‘extinct species,’ and regard the remainder as either actual, or varieties of, existing forms; some of them, as 5 and 6, being now confined to northern climates, others, as 7 and 8, to low latitudes, whilst the remainder still inhabit temperate Europe,[*] with the exception of the grizzly bear (*Ursus ferox*), which is restricted to North America, over which it ranges from Mexico to 61° N. lat.

No human bones were obtained anywhere during these excavations, but many flint implements and flakes, chiefly from the lowest part of the cave-earth. Neglecting the less perfect specimens, some of which were met with even in the lowest gravel, about fifteen implements, recognised as artificially formed by the most experienced antiquaries, were taken from the cave-earth, and usually from near the bottom. Their forms show that they must be referred to the Palæolithic or Early Stone Age, and this is borne out by the absence of any trace of polish on them, and also by the mode of their association with the remains of extinct animals, deep in a cave-earth sealed up with a floor of stalagmite, on which lay the reindeer's antler already mentioned. If they were not all of contemporary date, it is clear from this cave that the reindeer lived in Devon after the tools were buried, or in other words, that Man in this district preceded that animal.

A glance at the position of Windmill Hill, in which

---

[*] See ‘Pleist. Mam.’ Pal. Soc. Part I. pp. xxxviii.-xliii.

the caverns are situated, and a brief survey of the valleys which bound it on three sides, are enough to satisfy a geologist that the drainage and geographical features of this region have undergone great changes since the gravel and bone-earth were carried by streams into the subterranean cavities above described. The pebbles of hematite, quartz, and greenstone, can only have come from their nearest parent rocks at a period when the valleys immediately adjoining the caves were much shallower than they now are. The reddish loam in which the bones are embedded is such as may be seen on the surface of limestone in the neighbourhood, but the currents which were formerly charged with such mud must have run at a level seventy-eight feet above that of the stream now flowing in the same valley. It was remarked by Mr. Pengelly, that the stones and bones in the loam had their longest axes parallel to the direction of the tunnels and fissures, showing that they were deposited by the action of a stream.*

It appears that so long as the flowing water had force enough to propel stony fragments, no layer of fine mud could accumulate, and so long as there was a regular current capable of carrying in fine mud and bones, no superficial crust of stalagmite. In some passages, as before stated, stalagmite was wanting, while in one place five alternations of stalagmite and sand were observed, seeming to indicate a prevalence of more rainy seasons, succeeded by others, when the water was for a time too low to flood the area where the calcareous incrustation accumulated.

If the regular sequence of the three deposits of pebbles, mud, and stalagmite was the result of the causes above explained, the order of superposition would be constant, yet we could not be sure that the gravel in one passage

* Pengelly, Geologist, vol. iv. p. 133, 1861.

might not sometimes be coeval with the bone-earth or
stalagmite in another. If, therefore, the flint knives had
not been very widely dispersed, and if they had not all
been beneath the stalagmite, itself covered with the reindeer antler above described, their antiquity relative to the
extinct mammalia might have been questioned.

No coprolites were found in the Brixham excavations, and
very few gnawed bones. These few may have been brought
from some distance, before they reached their place of rest.
Upon the whole, the same conclusion which Dr. Schmerling
came to respecting the filling up of the caverns near Liège
seems applicable to the caves of Brixham.

Recent Explorations in Kent's Cavern.

While the Brixham explorations were producing a revolution in public opinion, Kent's Cave remained undisturbed
from 1846 until 1864. But one point of great interest had
been mooted by Mr. Mac Enery, upon which the Brixham
researches failed to throw any light. He had stated that
among the extinct mammalia in Kent's Hole, he had found
several teeth of *Ursus cultridens*, (fig. 9) a large carnivore
since called *Muchairodus latidens* by Owen. This genus is
one of considerable antiquity in the tertiary formations of
Europe, having been first known in Miocene deposits in
Touraine and Eppelsheim, and though occuring in Pliocene
formations in Auvergne and the Val d'Arno, it had never
been met with in cavern or fluviatile deposits of Pleistocene
date. In the hope of clearing up all doubts on this subject,
and of verifying, if possible, Mac Enery's important discovery,
the British Association, in 1864, gave a grant, which has
been renewed each subsequent year, for further explorations
of Kent's Cavern.* The work has been carried on ever since

* See Brit. Assoc. Reports, 1865 to 1872, inclusive,

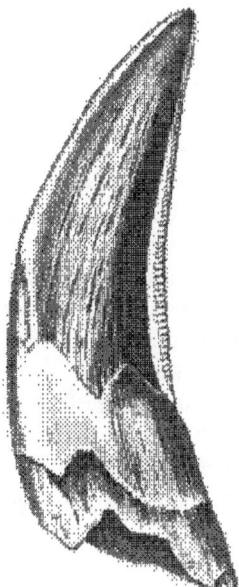

Fig. 9.

Canine of *Machairodus latidens* (*Ursus cultridens*) found by Mr. Mac Enery in
the cave-earth of Kent's Hole, Torquay, Jan. 1826.

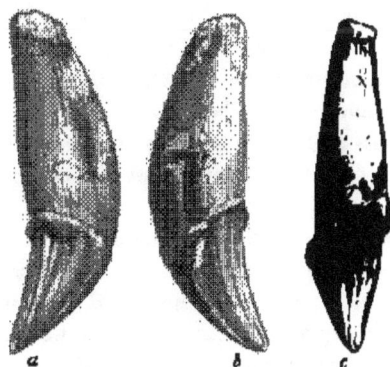

Fig. 10.

a   *b*   *c*

Incisor of *Machairodus latidens* from the cave-earth of Kent's Hole, Torquay,
July 1872.—*a b* Side views of the incisor. *c* Front view of the incisor.

this date, under the superintendence of Mr. Vivian and Mr.
Pengelly, of Torquay, but it was not till the end of nearly
eight years' labour that they were rewarded last July (1872)
by the discovery of a fine incisor of *Machairodus latidens*
in the uppermost part of the cave-earth (see fig. 10), thus
confirming the accuracy of the labours of Mac Enery, and establishing the fact that Man had been the contemporary of
the Machairodus in England.

The succession of deposits in Kent's Cavern may be briefly
enumerated in descending order, as follows:—

1. Large blocks of limestone fallen from the roof and
sometimes cemented by stalagmite.

2. Black mould, containing articles of various kinds of
medieval, Romano-British, and pre-Roman date.

3. A stalagmite floor, from sixteen to twenty inches
thick, sometimes attaining five feet, and containing large
fragments of limestone, a human jaw, and the remains of
extinct animals.

4. A local band of black earth, containing charcoal and
other evidence of fire, and also bone and flint implements.

5. The red cave-earth containing Palæolithic implements, and bones and teeth of extinct animals, including the
tooth of *Machairodus latidens.*

6. A second stalagmite floor, from three to twelve feet
thick, containing bones of bears only.

7. A mass of dark red sandy loam, also containing only
bears' bones. Three flint implements and one flint chip have
been found in this lowest layer.

The black mould, No. 2, contains the bones of living
species only; but immediately on arriving at the stalagmite
floor, bones of extinct species are found intermingled, and
below the stalagmite, in the cave-earth, we come upon a
large series of unpolished flint implements of Palæolithic
type, together with the bones of the cave-lion, cave-bear,

mammoth, tichorhine rhinoceros, 'Irish elk,' reindeer, hyæna, and other Pleistocene species, and finally, the long-sought-for tooth of *Machairodus latidens.*

We must however bear in mind that although the *genus* to which this tooth belongs is very ancient in Europe, the *species* differs from the *Machairodus cultridens* of the Continent. It is therefore impossible to fix, with any certainty, the exact age implied by its presence. Mr. Boyd Dawkins, on account of the Pliocene affinities of the genus, and its general absence in Pleistocene caves, is inclined to believe that the teeth found by Mac Enery were derived from the lowest deposit No. 7, which in the form of a breccia has since become partially mixed in places with the cave-earth No. 5, and he has therefore thrown out a suggestion that the Machairodus may have lived in England[*] in a very early stage of the Pleistocene period, before the maximum point of cold of the Glacial period was reached. Mr. Pengelly, however, points out some very forcible objections to this hypothesis. The teeth of Machairodus, he says, are not mineralised as are the bears' bones which have been derived from the breccia; on the contrary, they resemble the specimens of other animals found in the cave-earth, and three of them bear decided marks of having been gnawed, which points to their having been contemporary with the hyæna; moreover, though the teeth are delicate and finely serrated, they are quite uninjured, a fact scarcely consistent with the hypothesis of their being derivative; moreover, the tooth which has been brought to light since Mr. Boyd Dawkins wrote, was lying in the uppermost part of the cave-earth, having the teeth of hyæna, horse, and bear vertically under it. On the whole therefore, after having seen this specimen and noticed its perfect condition, I scarcely

[*] Dawkins and Sanford's British Pleistocene Mammalia, Part IV., Palæonto-graphical Soc., 1872, 190.

think it possible that it can be a derivative fossil. It may
however be well to remember, that in either case its bearing
on the antiquity of Man is equally significant.   If, as I
believe, it was a contemporary of the mammoth and hyæna,
it still lived on in England after the works of Man had
already been entombed in the red loam No. 7, and sealed
down with a floor of stalagmite. And if it was derivative
from the breccia, Man was still equally its contemporary in
that earlier period.

*Works of Art associated with extinct Mammalia in a
Cavern in Somersetshire.*

The hyæna-den of Wokey Hole, near Wells, in Somersetshire, from which implements resembling those of Amiens
have been obtained, was first discovered in 1849, but did not
attract the attention it deserved until after the investigations
of Brixham Cave had led geologists to attach importance to
the association of such relics with extinct mammalia. It
occurs near the cavern of Wokey Hole, from the mouth of
which the river Axe issues on the southern flanks of the
Mendip Hills. No one had suspected that on the left side
of the ravine, through which the river flows after escaping
from its subterranean channel, there were other caves and
fissures concealed beneath the green sward of the steep
sloping bank.   But in 1849 a canal was made, several
hundred yards in length, for the purpose of leading the
waters of the Axe to a paper-mill, now occupying the
middle of the ravine.   In carrying out this work, about
twelve feet of the left bank was cut away, and a cavernous
fissure, choked up to the roof with ossiferous loam, was then,
for the first time, exposed to view. This great cavity, originally nine feet high and thirty-six wide, traversed the
dolomitic conglomerate; and fragments of that rock, some
angular and others water-worn, were scattered through the

red mud of the cave, in which fossil remains were abundant. For an account of them and the position they occupied we are indebted to Mr. Dawkins, who, in company with Mr. Williamson, explored the cavern in 1859, and obtained from it the bones of the *Hyæna spelæa* in such numbers as to lead him to conclude that the cavern had for a long time been a hyæna's den. Among the accompanying animals found fossil in the same bone-earth, were observed *Elephas primigenius, Rhinoceros tichorhinus, Ursus spelæus, Bos primigenius, Megaceros Hibernicus, Cervus tarandus* (and other species of *Cervus*), *Felis spelæa, Canis Lupus, Canis Vulpes,* and teeth and bones of the genus *Equus* in great numbers.

Intermixed with the above fossil bones were some arrowheads, made of bone, and many chipped flints, and chipped pieces of chert, a white or bleached flint weapon of the spear-head Amiens type, which was taken out of the undisturbed matrix by Mr. Williamson himself, together with a hyæna's tooth, showing that Man had either been contemporaneous with or had preceded the extinct fauna. After penetrating thirty-four feet from the entrance, Mr. Dawkins found the cave bifurcating into two branches, one of which was vertical. By this rent, perhaps, some part of the contents of the cave may have been introduced.[*]

When I examined the spot in 1860, after I had been shown some remains of the hyæna collected there, I felt convinced that the changes which had taken place in the physical geography of the district since the time of the extinct quadrupeds must have been so great that it would be a hopeless task to attempt to restore the ancient topography.

Caves of Gower in Glamorganshire, South Wales.

The ossiferous caves of the peninsula of Gower, in Glamorganshire, have been diligently explored of late years by

[*] W. B. Dawkins, F.G.S., Geological Society's Proceedings, January 1862.

the late Dr. Falconer and Lieutenant-Colonel E. R. Wood, who have thoroughly investigated the contents of many which were previously unknown. Among these Dr. Falconer's skilled eye recognised the remains of almost every quadruped which he had elsewhere found fossil in British caves: in some places the *Elephas primigenius,* accompanied by its usual companion the *Rhinoceros tichorhinus,* in others *Elephas antiquus* associated with *Rhinoceros hemitœchus* Falconer; the extinct animals being often embedded, as in the Belgian caves, in the same matrix with species now living in Europe, such as the common badger (*Meles Taxus*), the common wolf, and the fox.

In a cavernous fissure called the Raven's Cliff, teeth of several individuals of *Hippopotamus major,* both young and old, were found; and this in a district where there is now scarce a rill of running water, much less a river in which such quadrupeds could swim. In one of the caves, called Spritsail Tor, bones of the elephants above named were observed, with a great many other quadrupeds of recent and extinct species.

From one fissure, called Bosco's Den, no less than one thousand antlers of the reindeer, chiefly of the variety called *Cervus Guettardi,* were extracted by the persevering exertions of Colonel Wood, who estimated that several hundred more still remained in the bone-earth of the same rent.

Those which he showed me were mostly shed horns, and of young animals; and had been washed into the rent with other bones, and with angular fragments of limestone, and all enveloped in the same ochreous mud. Among the other bones, which were not numerous, were those of the cave-bear, wolf, fox, ox, stag, and field-mouse.

But the discovery of most importance, as bearing on the subject of the present work, is the occurrence in a newly-discovered cave, called Long Hole, by Colonel Wood, in 1861

of the remains of two species of rhinoceros, *R. tichorhinus*
and *R. hemitœchus* Falconer, in an undisturbed deposit, in
the lower part of which were some well-shaped flint knives,
evidently of human workmanship. It is clear from their
position that Man was coeval with these two species. We
have elsewhere independent proofs of his coexistence with
every other species of the cave fauna of Glamorganshire:
but this is the first well-authenticated example of the
occurrence of *R. hemitœchus* in connection with human implements.

In the fossil fauna of the valley of the Thames, *Rhinoceros
leptorhinus* was mentioned as occurring at Gray's Thurrock
with *Elephas antiquus.* Dr Falconer, in a memoir on the
European Pliocene and Pleistocene species of the genus
Rhinoceros, has shown that, under the above name of *R.
leptorhinus*, three distinct species have been confounded by
Cuvier, Owen, and other palæontologists :—

1. *R. megarhinus* Christol, being the original and typical
R. leptorhinus of Cuvier, founded on Cortesi's Monte Zago
cranium, and the *only* Pliocene, or Pleistocene (Post-Pliocene) European species, that had not a nasal septum.—
Gray's Thurrock, &c.

2. *R. hemitœchus* Falconer, in which the ossification of
the septum dividing the nostrils is incomplete in the middle,
besides other cranial and dental characters distinguishing it
from *R. tichorhinus*, accompanies *Elephas antiquus* in most
of the oldest British bone-caves, such as Kirkdale, Cefn,
Durdham Down, Minchin Hole, and other Gower caverns
—also found at Clacton, in Essex, and in Northamptonshire.

3. *R. etruscus* Falconer, a comparatively slight and
slender form, also with an incomplete bony septum,* occurs
deep in the Val d'Arno deposits, and in the 'Forest Bed,'
and superimposed blue clays, with lignite, of the Norfolk

* See Falconer, Quarterly Geological Journal, vol. xv. p. 602.

coast, but nowhere as yet found in the ossiferous caves in
Britain.

Dr. Falconer announced in 1860 his opinion that the
filling up of the Gower caves in South Wales took place after
the deposition of the marine boulder clay,* an opinion in
harmony with what we have since learnt from the section
of the gravels near Bedford, given at p. 215, where a
fauna corresponding to that of the Welsh caves characterises
the ancient alluvium, and is shown to be clearly post-glacial,
in the sense of being posterior in date to the emergence of
the midland counties from beneath the waters of the glacial
sea. In the same sense the late Edward Forbes declared, in
1846, his conviction that not only the *Cervus megaceros*,
but also the mammoth and other extinct pachyderms and
carnivora, had lived in Britain in post-glacial times.† The
Gower caves in general have their floors strewed over with
sand, containing marine shells, all of living species; and
there are raised beaches on the adjoining coast, and other
geological signs of great alteration in the relative level of
land and sea, since that country was inhabited by the extinct
mammalia, some of which, as we have seen, were certainly
coeval with Man.

The marine shells of recent species in the drift on the
banks of the Severn, one hundred feet or more above the
level of that river, attest the former existence of what Sir
R. Murchison called the Severn Straits, when England and
Wales were separated by a channel of the sea. During the
conversion of this channel from sea to land, there were probably oscillations of level, when among other changes, Ireland
was separated from England. During one part of these
movements, those forests, now submarine, of which we have
the remains in Somersetshire, near Watchet, occupied the

* Geological Quarterly Journal,   † Memoirs of Geol. Survey, pp. 394-
vol. xvi. p. 491. 1860,   397,

present Bristol Channel, and the Severn wandered through a wide low tract, probably fertile, and affording ample feeding-ground for the former inhabitants of the caves of the Gower Peninsula.

Ossiferous Caves in the North of Sicily.

Geologists have long been familiar with the fact that on the northern coast of Sicily, between Termini on the east, and Trapani on the west, there are several caves containing the bones of extinct animals. These caves are situated in rocks of hippurite limestone, a member of the cretaceous series, and some of them may be seen on both sides of the Bay of Palermo. If in the neighbourhood of that city we proceed from the sea inland, ascending a sloping terrace, composed of the marine Newer Pliocene strata, we reach, about a mile from the shore, and at the height of about one hundred and eighty feet above it, a precipice of limestone, at the base of which appear the entrances of several caves. In that of San Ciro, on the east side of the bay, we find at the bottom sand with marine shells, forty species of which have been examined, and found almost all to agree specifically with mollusca now inhabiting the Mediterranean. Higher in position, and resting on the sand, is a breccia, composed of pieces of limestone, quartz, and schist in a matrix of brown marl, through which land-shells are dispersed, together with bones of two species of hippopotamus, as determined by Dr. Falconer. Certain bones of the skeleton were counted in such numbers as to prove that they must have belonged to several hundred individuals. With these were associated the remains of *Elephas antiquus*, and bones of the genera *Bos, Cervus, Sus, Ursus, Canis*, and a large *Felis*. Some of these bones have been rolled as if partially subjected to the action of water, and may have been introduced by streams through rents in the hippurite limestone; but there is now

I

no running water in the neighbourhood, no river such as the
hippopotamus might frequent, not even a small brook, so
that the physical geography of the district must have been
altogether changed since the time when such remains were
swept into fissures, or into the channels of engulfed rivers.

No proofs seem yet to have been found of the existence of
Man at the period when the hippopotamus and *Elephas antiquus* flourished at San Ciro. But there is another cave
called the Grotto di Maccagnone, which much resembles it
in geological position, on the opposite or west side of the
Bay of Palermo, near Carini. In the bottom of this cave a
bone deposit like that of San Ciro occurs, and above it other
materials reaching to the roof, and evidently washed in from
above, through crevices in the limestone. In this upper and
newer breccia, Dr. Falconer discovered flint knives, bone
splinters, bits of charcoal, burnt clay, and other objects indicating human intervention, mingled with entire land-shells,
teeth of horses, coprolites of hyænas, and other bones, the
whole agglutinated to one another and to the roof by the
infiltration of water holding carbonate of lime in solution.
' The perfect condition of the large fragile helices (*Helix
vermiculata*) afforded satisfactory evidence,' says Dr. Falconer,
' that the various articles were carried into the cave by the
tranquil agency of water, and not by any tumultuous action.
At a subsequent period other geographical changes took
place, so that the cave, after it had been filled, was washed
out again, or emptied of its contents with the exception of
those patches of breccia which, being cemented together by
stalactite, still adhere to the roof.' *

Baron Anca, following up these investigations, explored, in
1859, another cave at Mondello, west of Palermo, and north
of Mount Gallo, where he discovered molars of the living
African elephant, and afterwards additional specimens of the

* Quarterly Geological Journal, vol. xvi. p. 105, 1860.

same species in the neighbouring grotto of Olivella. In reference to this elephant, Dr. Falconer has reminded us that
the distance between the nearest part of Sicily and the coast
of Africa, between Marsala and Cape Bon, is not more than
eighty miles, and Admiral Smyth, in his Memoir on the
Mediterranean, states (p. 499) that there is a subaqueous
plateau, named by him Adventure Bank, uniting Sicily to
Africa by a succession of ridges which are not more than
from forty to fifty fathoms under water.* Sicily therefore
might be reunited to Africa by movements of upheaval not
greater than those which are already known to have taken
place within the human period on the borders of the Mediterranean, of which I shall now proceed to cite a well-authenticated example, observed in Sardinia.

*Rise of the Bed of the Sea to the Height of 300 Feet, in the
Human Period, in Sardinia.*

Count Albert de la Marmora, in his description of the geology of Sardinia,† has shown that on the southern coast of
that island, at Cagliari and in the neighbourhood, an ancient
bed of the sea, containing marine shells of living species, and
numerous fragments of antique pottery, has been elevated
to the height of from 230 to 324 feet above the present
level of the Mediterranean. Oysters and other shells, of
which a careful list has been published, including the common
mussel (*Mytilus edulis*), many of them having both valves
united, occur, embedded in a breccia in which fragments of
limestone abound. The mussels are often in such numbers
as to impart, when they have decomposed, a violet colour
to the marine stratum. Besides pieces of coarse pottery,
a flattened ball of baked earthenware, with a hole through

* Cited by Mr. Horner, President   † Partie Géologique, tom. L pp.
of Geological Society. Anniversary   382, 387.
Address, February 1861, p. 42.

I 2

its axis, was found in the midst of the marine shells. It is supposed to have been used for weighting a fishing-net. Of this and of one of the fragments of ancient pottery Count de la Marmora has given figures.

The upraised bed of the sea probably belongs in this instance to the Pleistocene period, for in a bone breccia, filling fissures in the rocks around Cagliari, the remains of extinct mammalia have been detected; among which is a new genus of carnivorous quadruped, named *Cynotherium* by M. Studiati, and figured by Count de la Marmora in his Atlas (pl. vii.), also an extinct species of *Lagomys*, determined by Cuvier in 1825. Embedded in the same bone breccia, and enveloped with red earth like the mammalian remains, were detected shells of the *Mytilus edulis* before mentioned, implying that the marine formation containing shells and pottery had been already upheaved and exposed to denudation before the remains of quadrupeds were washed into these rents and included in the red earth. In the vegetable soil covering the upraised marine stratum, fragments of Roman pottery occur.

If we assume the average rate of upheaval to have been, as before hinted, p. 64, two and a half feet in a century, which, however, is purely conjectural, 300 feet would give an antiquity of 12,000 years to the Cagliari pottery, even if we simply confine our estimate to the upheaval above the sea-level, without allowing for the original depth of water in which the mollusca lived. In this case, moreover, our calculation would merely embrace the period during which the upward movement was going on; and we can form at present no conjecture as to the probable era of its commencement or termination.

I learn from Capt. Spratt, R.N., that the island of Crete or Candia, about 135 miles in length, has been raised at its western extremity about twenty-five feet; so that ancient ports belonging to a late Roman period are now high and dry

above the sea, while at its eastern end it has sunk so much
that the ruins of old towns are seen under water.* Revolutions like these in the physical geography of the countries
bordering the Mediterranean, may well help us to understand
the phenomena of the Palermo caves, and the presence in
Sicily of African species of mammalia.

* Spratt, Travels in Crete, vol. i. p. 141, vol. ii. p. 216.

CHAPTER VII.

PLEISTOCENE PERIOD—BONES OF MAN AND EXTINCT ANIMALS
IN FRENCH CAVES.

EARLIEST DISCOVERIES IN CAVES OF LANGUEDOC OF HUMAN REMAINS
WITH BONES OF EXTINCT MAMMALIA—CAVE OF AURIGNAC, IN THE SOUTH
OF FRANCE—EXTINCT MAMMALIA AND WORKS OF ART FOUND BY M.
LARTET IN THIS CAVE—UNCERTAINTY AS TO THE AGE OF INTERMENT IN
THIS CAVERN—CAVERNS OF THE DORDOGNE WITH RELICS BOTH OF THE
NEOLITHIC AND PALÆOLITHIC PERIODS—CARVING OF A MAMMOTH ON AN
IVORY TUSK IN THE CAVERN OF LA MADELAINE—CAVE OF BRUNIQUEL—
SKELETON LATELY FOUND IN A CAVE AT MENTONE, IN THE SOUTH OF
FRANCE.

Discoveries of MM. Tournal and Christol in 1828, *in the
South of France.*

IN 1832, when treating of the fossil remains found in alluvium, and the mud of caverns, I gave an account of the
investigations made by MM. Tournal and Christol in the
South of France.*

M. Tournal stated in his memoir, that in the cavern of
Bize, in the department of the Aude, he had found human
bones and teeth, together with fragments of rude pottery, in
the same mud and breccia cemented by stalagmite in which
land-shells of living species were embedded, and the bones
of mammalia, some of extinct, others of recent species. The
human bones were declared by his fellow-labourer, M. Marcel
de Serres, to be in the same chemical condition as those of
the accompanying quadrupeds.†

Speaking of these fossils of the Bize cavern five years

* Principles of Geology, 1st ed.   † Annales des Sciences Naturelles,
vol. ii. ch. xiv., 1832.   tom. xv. p. 348: 1828.

later, M. Tournal observed, that they could not be referred,
as some suggested, to a 'diluvial catastrophe,' for they
evidently had not been washed in suddenly by a transient
flood, but must have been introduced gradually, together
with the enveloping mud and pebbles, at successive periods.[*]

M. Christol, who was engaged at the same time in
similar researches in another part of Languedoc, published an
account of them a year later, in which he described some
human bones, as occurring in the cavern of Pondres, near
Nismes, in the same mud with the bones of an extinct hyæna
and rhinoceros.[†] The cavern was in this instance filled up
to the roof with mud and gravel, in which fragments of two
kinds of pottery were detected, the lowest and rudest near
the bottom of the cave, below the level of the extinct
mammalia.

It has never been questioned that the hyæna and rhinoceros
found by M. Christol were of extinct species; but whether
the animals enumerated by M. Tournal might not all of them
be referred to quadrupeds which are known to have been
living in Europe in the historical period seems doubtful.
They were said to consist of a stag, an antelope, and a goat,
all named by M. Marcel de Serres as new ; but the majority
of palæontologists do not agree with this opinion. Still it is
true, as M. Lartet remarks, that the fauna of the cavern of
Bize must be of very high antiquity, as shown by the presence, not only of the Lithuanian aurochs (*Bison europæus*)
but also of the reindeer, which has not been an inhabitant
of the South of France in historical times, and which, in that
country, is almost everywhere associated, whether in ancient
alluvium or in the mud of caverns, with the mammoth.
Before the distinction between the various works of art found

---

[*] Annales de Chimie et de Phy-
sique. p. 161 : 1833.

[†] Christol, Notice sur les Ossements

Humains des Cavernes du Gard. Mont-
pellier, 1829.

associated with stone implements of Palæolithic and of
Neolithic type had been clearly established, considerable
difficulty was often felt in explaining why the older mammalia should never have been found in the sepulchral
monuments.

M. Desnoyers, an observer equally well versed in geology
and archæology, had disputed the conclusion arrived at by
MM. Tournal and Christol, that the fossil rhinoceros, hyæna,
bear, and other lost species, had once been inhabitants of
France contemporaneously with Man. 'The flint hatchets
and arrow-heads,' he said, 'and the pointed bones and coarse
pottery of many French and English caves, agree precisely
in character with those found in the tumuli, and under the
dolmens (rude altars of unhewn stone) of the primitive
inhabitants of Gaul, Britain, and Germany. The human
bones, therefore, in the caves which are associated with such
fabricated objects, must belong not to antediluvian periods,
but to a people in the same stage of civilization as those
who constructed the tumuli and altars.'

'In the Gaulish monuments,' he added, 'we find, together
with the objects of industry above mentioned, the bones of
wild and domestic animals of species now inhabiting Europe,
particularly of deer, sheep, wild boars, dogs, horses, and
oxen. This fact has been ascertained in Quercy,* and other
provinces; and it is supposed by antiquaries that the animals
in question were placed beneath the Celtic altars in memory
of sacrifices offered to the Gaulish divinity Hesus, and in the
tombs to commemorate funeral repasts, and also from a
superstition prevalent among savage nations, which induces
them to lay up provisions for the manes of the dead in a
future life. But in none of these ancient monuments have
any bones been found of the elephant, rhinoceros, hyæna,
tiger, and other quadrupeds, such as are found in caves,

* An old province of France, now the department of the Lot.

which might certainly have been expected, had these species
continued to flourish at the time that this part of Gaul was
inhabited by Man.'[*]

After giving no small weight to the arguments of M. Desnoyers, and the writings of Dr. Buckland on the same subject,
and visiting myself several caves in Germany, I came to the
opinion that the contemporaneity of the human bones mixed
with those of extinct animals, in osseous breccias and cavern
mud, in different parts of Europe, was not conclusive.   The
caverns having been at one period the dens of wild beasts, and
having served at other times as places of human habitation,
worship, sepulture, concealment, or defence, one might easily
conceive that the bones of Man and those of animals, which
were strowed over the floors of subterranean cavities, or
which had fallen into tortuous rents connecting them with
the surface, might, when swept away by floods, be mingled
in one promiscuous heap in the same ossiferous mud or
breccia.[†]

In addition to this we have evidence of interments of
various dates, by which the accumulations of any previous
age might be disturbed and mixed up with the relics of a
later race ; while the agency of burrowing animals in bringing the older remains to the surface, and allowing recent
objects to fall down to almost any depth, must not be lost
sight of.   That such intermixtures have really taken place
in some caverns, and that geologists have occasionally been
deceived, and have assigned to one and the same period
fossils which had really been introduced at successive times,
will readily be conceded.   But of late years we have obtained
convincing proofs, as we shall see in the sequel (Chap. VIII.),
that the mammoth, and many other extinct mammalian
species very common in caves, occur also in undisturbed

* Desnoyers, Bulletin de la Société   Universelle   d'Histoire   Naturelle.
Géologique de France, tom. ii. p. 252;   Paris, 1845.
and article on Caverns, Dictionnaire   † Principles, 6th ed. p. 740.

alluvium, embedded in such a manner as to leave no room
for doubt that Man and the mammoth coexisted.

Cave of Aurignac, in the South of France.

One of the most celebrated caverns containing both
human remains and the bones of extinct quadrupeds is the
cave of Aurignac, so ably described by the late M. Edouard
Lartet; and it seems probable from late researches that this
may be one of the caves above alluded to, in which two or
more periods are represented in the deposits of one vault,
the memorials having become so intermixed at several
successive periods, that it is now almost impossible to
separate them. In 1862 I had the advantage of inspecting
the fossil bones and works of art obtained by M. Lartet
from this grotto, and of conversing with him on the subject;
and more recently, in the spring of 1872, I visited the cave
itself, in company with Mr. T. McKenny Hughes, of the
Geological Survey, lately appointed Woodwardian Professor
at Cambridge.

The town of Aurignac is situated in the department of
the Haute-Garonne, near a spur of the Pyrenees; adjoining
it is the small flat-topped hill of Fajoles, about sixty feet
above the brook called Rodes, which flows at its foot.
It consists of nummulitic limestone, which forms a steep
escarpment to the N.W., but dips away with the surface
slope to the S.E. The ridge does not run equally along
the strike, but sinks with a slight curve to the south-west,
so that the beds which form the highest scar, near the
centre of the ridge, come down to within about twenty feet
of the brook at the south-west end. A softer set of beds
below this scar has yielded to various denuding forces more
readily than the harder mass which forms the scar, and
especially where the rock is fissured, has been eaten out into
irregular caves and grottos.

At the lower or south-west end of the ridge, about fifteen feet above the stream, there is now visible the entrance of one of these grottos, *a*, fig. 11, which opened originally on the terrace, *h, c, k,* which slopes gently towards the valley.

Fig. 11.

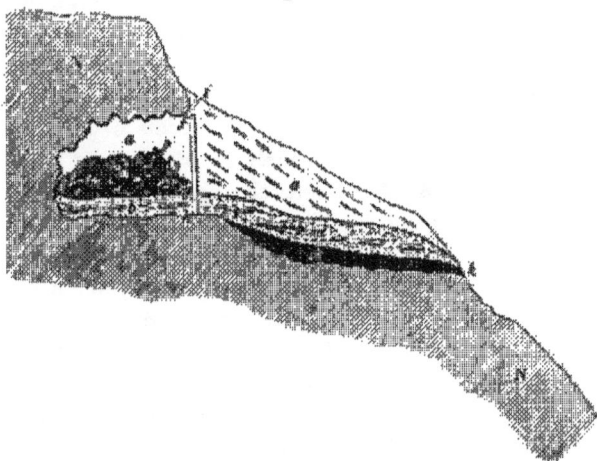

Section of part of the hill of Fajoles passing through the sepulchral grotto of Aurignac (E. Lartet).

a Part of the vault in which the remains of seventeen human skeletons were found.

b Layer of made ground, two feet thick, inside the grotto in which a few human bones, with entire bones of extinct and living species of animals, and many works of art were imbedded.

c Layers of ashes and charcoal, six inches thick, with broken, burnt, and gnawed bones of extinct and recent mammalia; also hearth-stones and works of art; no human bones.

d Deposit with similar contents and a few scattered cinders.

e Talus of rubbish washed down from the hill above.

f, g Slab of rock which closed the vault, not ascertained whether it extended to *h*.

f, i Rabbit burrow which led to the discovery of the grotto.

h, k Original terrace on which the grotto opened.

n Nummulitic limestone of hill of Fajoles.

Until the year 1852, the opening into this grotto was masked by a talus of small fragments of limestone and earthy matter, e, such as the rain may have washed down the slope of the hill. In that year a labourer named Bonnemaison, when seeking the broken rubbly part of the rock, to procure stones for repairing the roads, observed a rabbit-hole in the talus, at i f, fig. 11. On reaching as far into the opening as the length of his arm, he drew out, to his surprise, one of the long bones of a human skeleton; and his curiosity being excited, and having a suspicion that the hole communicated with a subterranean cavity, he commenced digging a trench through the middle of the talus, and in a few hours found himself opposite a large heavy slab of rock, f h, placed vertically against the entrance. Having removed this, he discovered on the other side of it an arched cavity, a, seven or eight feet in its greatest height, ten in width, and seven in horizontal depth. It was almost filled with bones, among which were two entire skulls, which he recognised at once as human. The people of Aurignac, astonished to hear of the occurrence of so many human relics in so lonely a spot, flocked to the cave; and Dr. Amiel, the Mayor, ordered all the bones to be taken out and re-interred in the parish cemetery. But before this was done, having as a medical man a knowledge of anatomy, he ascertained by counting the homologous bones, that they must have formed parts of no fewer than seventeen skeletons of both sexes, and all ages; some so young, that the ossification of the bones was incomplete. Unfortunately, the skulls were injured in the transfer: and what is worse, after the lapse of eight years, when M. Lartet visited Aurignac, the village sexton was unable to tell him in what exact place the trench was dug into which the skeletons had been thrown, so that this rich harvest of anthropological knowledge seems for ever lost to the antiquary and geologist.

M. Lartet having been shown, in 1860, the remains of some extinct animals and works of art, found in digging the original trench made by Bonnemaison through the bed *d* under the talus, and some others brought out from the interior of the grotto, determined to investigate systematically what remained intact of the deposits outside and inside the vault, those inside, underlying the human skeletons, being supposed to consist entirely of made ground. Having obtained the assistance of some intelligent workmen, he personally superintended their labours, and found outside the grotto, resting on the sloping terrace *h k*, the layer of ashes and charcoal, *c*, about six inches thick, extending over an area of six or seven square yards, and going as far as the entrance of the grotto and no farther, there being no cinders or charcoal in the interior. Among the cinders outside the vault were fragments of fissile sandstone, reddened by heat, which were observed to rest on a levelled surface of nummulitic limestone, and seemed to have formed a hearth. The nearest place from whence such slabs of sandstone could have been brought was the opposite side of the valley.

Among the ashes, and in some overlying earthy layers, *d*, separating the ashes from the talus *e*, were a great variety of bones and implements; amongst the latter not fewer than a hundred flint articles — knives, projectiles, slingstones, and chips, and among them one of those siliceous cores or nuclei with numerous facets, from which flint flakes or knives had been struck off, seeming to prove that some instruments were occasionally manufactured on the very spot.

Among other articles outside the entrance was found a stone of a circular form, and flattened on two sides, with a central depression, composed of a tough rock, which does not belong to that region of the Pyrenees. This instrument is supposed by the Danish antiquaries to have been used for striking off flakes, or dressing the edges of flint knives,

the fingers and thumb being placed in the two opposite
depressions during the operation. Among the bone instruments were arrows without barbs, and other tools made of
reindeer horn, and a bodkin formed out of the more compact
horn of the roe-deer. This instrument was well shaped,
and sharply pointed, and in so good a state of preservation
that it might still be used for piercing the tough skins of
animals.

Scattered through the ashes and earth were the bones of
the various species of animals enumerated in the subjoined
lists, with the exception of two, marked with an asterisk,
which only occurred in the interior of the grotto :—

1. CARNIVORA.

		Number of Individuals.
ex. 1.	*Ursus spelæus*, Blum. (cave-bear)	5 — 6
l. 2.	*Ursus arctos?* Linn. (brown-bear)	1
l. 3.	*Meles Taxus*, Owen (badger)	1 — 2
l. 4.	*Putorius vulgaris*, Owen (polecat)	1
ex. 5.*	*Felis spelæa*, Goldf. (cave-lion)	1
l. 6.	*Felis Catus ferus*, Owen (wild cat)	1
ex. 7.	*Hyæna spelæa*, Goldf. (cave-hyæna)	5 — 6
l. 8.	*Canis Lupus*, Linn. (wolf)	2
l. 9.	*Canis Vulpes*, Briss. (fox)	18 — 20

2. HERBIVORA.

ex. 1.	*Elephas primigenius*, Blum. (mammoth, two molars).	
ex. 2.	*Rhinoceros tichorhinus*, Cuv. (Siberian rhinoceros)	1
l. 3.	*Equus Caballus*, Linn. (horse)	12 — 15
l. 4.	*Equus Asinus?* Linn. (ass)	1
l. 5.*	*Sus Scrofa*, Linn. (pig, two incisors).	
l. 6.	*Megaceros Hibernicus*, Owen (gigantic Irish deer)	1
ex. 7.	*Cervus Elaphus*, Linn. (stag)	1
l. 8.	*C. Capreolus*, Linn. (roebuck)	3 — 4
l. 9.	*C. Tarandus*, Linn. (reindeer)	10 — 12
l. 10.	*Bison europæus*, Boj. (aurochs)	12 — 15

ex. extinct.   l. living.

The bones of the herbivora were the most numerous, and
all those on the outside of the grotto which had contained
marrow were invariably split open, as if for its extraction,
many of them being also burnt. The spongy parts, moreover, were wanting, having been eaten off and gnawed after

they were broken, the work, according to M. Lartet, of hyænas, the bones and coprolites of which were mixed with the cinders, and dispersed through the overlying soil *d*.

Among the various proofs that the bones were fresh when brought to the spot, it was remarked that those of the herbivora not only bore the marks of having had the marrow extracted and having afterwards been gnawed and in part devoured as if by carnivorous beasts, but they had also been acted upon by fire (and this was especially noticed in one case of a cave-bear's bone), in such a manner as to show that they retained in them at the time all their animal matter.

Among the bones found in the ashes, were those of a young *Rhinoceros tichorhinus*, which had been, like those of the accompanying herbivora, broken and gnawed by a beast of prey at both extremities.

Outside of the great slab of stone forming the door, not one human bone occurred; inside of it there were found, as before stated, mixed with loose soil, the remains of as many as seventeen human individuals, besides some works of art and bones of animals. We know nothing of the arrangement of these bones when they were first broken into. M. Lartet inferred at first that the bodies were bent down upon themselves in a squatting attitude, a posture known to have been adopted in most of the sepulchres of primitive times; and he has so represented them in his restoration of the cave. But in 1863 he saw reason to retract this opinion, and wrote to me as follows :—

'I was wrong in figuring in the interior of the sepulchral cavern the human skeletons in the bent attitude in which they are represented, and as they so often occur in ancient burying-places. On making a careful examination of the walls of the little grotto, during a third visit in 1862, I found in a recess, or smaller hollow of the south wall, a mass of bones, all bound together by a sort of concretionary

matrix. Among the bones were several of the human foot,
half a radius of the reindeer, a fragment of reddish pottery,
and a little below a calcaneum of an elephant, attached to
the same concretionary and calcareous mass which adhered
to the walls of the grotto. All these were at a very high
level above the floor, and about two feet only below the top
or centre of the arch of the cave. The finding a human foot
at this height contradicts my supposition as to the squatting
attitude given to the bodies, which were probably all buried
horizontally and in successive superposition. It also explains
how Bonnemaison, by putting his hand and arm into the
rabbit's hole, might easily have pulled out a long bone. I
may also add that the elephant's calcaneum, although presenting nearly the same appearance of alteration as the
various other bones taken out of the burying-place, had
evidently been gnawed by large carnivora before it was introduced into its present position. It is the only gnawed bone
that has been found in the interior of the sepulchral vault;
and by the elevated position it occupies, one is led to conclude that it was placed there at the time of some of the
last interments in this grotto.

'On the occasion of my third visit, I determined to search
the refuse heap cast up during the first diggings by Bonnemaison, when they carried away the human bones to the
cemetery of Aurignac. This heap of rubbish, lying to the
left of the grotto, was covered with vegetation, and Bonnemaison assured me that nothing had been left in it. I
found, however, a hundred worked flints in it, some teeth
and bones of carnivora, bones also of reindeer, ox, and
rhinoceros, besides sixty-eight human bones, principally of
the hands and feet, and a human half-jaw containing teeth,
all resemble to adult individuals of small size, except two
pieces, which may have belonged to an individual, if not of
great height, at least relatively pretty large. I have also

found among this rubbish a great number of fragments of pottery, some dried in the sun, others half baked, but all made by hand, and of clay of different fineness. I obtained from the same heap several ornaments made of the hard parts of the bones of the ear of the horse or ox.'

From the above statements we may conclude that the race which made use of the ancient burying-place was one of small stature.

There was no stalagmite in the grotto, and M. Lartet, an experienced investigator of ossiferous caverns in the south of France, came to the conclusion that all the bones and soil found in the inside were artificially introduced. The substratum, b, fig. 11, which remained after the skeletons had been removed, was about two feet thick. In it were found about ten detached human bones, including a molar tooth; and M. Delesse ascertained by careful analysis of one of these, as well as of the bones of a rhinoceros, bear, and some other extinct animals, that they all contained precisely the same proportion of azote, or had lost an equal amount of their animal matter. Mr. Evans has suggested to me that such a fact, taken alone, may not be conclusive in favour of the equal antiquity of the human and other remains. No doubt, had the human skeletons been found to contain more gelatine than those of the extinct mammalia, it would have shown that they were the more modern of the two; but it is possible that after a bone has gone on losing its animal matter up to a certain point, it may then part with no more so long as it continues enveloped in the same matrix. If this be so, it follows, that bones of very different degrees of antiquity, after they have lain for many thousands of years in a particular soil, may all have reached long ago the maximum of decomposition attainable in such a matrix.

Mixed with the human bones inside the grotto which were first removed by Bonnemaison, were eighteen small, round, and

K

flat plates of a white shelly substance, made of some species
of cockle (*Cardium*), pierced through the middle as if for
being strung into a bracelet. In the substratum also in the
interior examined by M. Lartet was found the tusk of a
young bear, probably *Ursus spelæus*, the crown of which had
been stripped of its enamel, and which had been carved
perhaps in imitation of the head of a bird. It was perforated
lengthwise as if for suspension as an ornament or amulet.
A flint knife also was found in the interior which had
evidently never been used, in this respect unlike the
numerous worn specimens found outside; so that it was conjectured that it might, like other associated works of art,
have been placed there as part of the funeral ceremonies.

Thus, on the whole, it was noticed that the bones of
animals inside the vault offered a remarkable contrast to
those of the exterior, being all entire and uninjured, none
of them broken, gnawed, half-eaten, scraped or burnt like
those lying among the ashes on the other side of the great
slab which formed the portal. The bones of the interior
seemed to have been clothed with their flesh when buried in
the layer of loose soil strewed over the floor. In confirmation of this idea, many bones of the skeleton were often
observed to be in juxtaposition, and in one spot all the
bones of the leg of an *Ursus spelæus* were lying together
uninjured, and were supposed by M. Lartet to have been
placed there as provision for the dead on their journey to
the land of spirits. The entire absence in the interior of
cinders and charcoal also confirmed the idea that this might
be an ancient place of sepulture, closed at the opening so
effectually against the hyænas or other carnivora that no
marks of their teeth appeared on any of the bones, whether
human or brute.

John Carver, in his travels in the interior of North
America in 1766–68 (ch. xv.), gave a minute account of the

funeral rites of an Indian tribe, which inhabited the country
now called Iowa, at the junction of the St. Peter's River with
the Mississippi; and Schiller, in his famous 'Nadowessische
Todtenklage,' has faithfully embodied in a poetic dirge all
the characteristic features of the ceremonies so graphically
described by the English traveller, not omitting the many
funeral gifts which, we are told, were placed 'in a cave'
with the bodies of the dead. The lines beginning, ' Bringet
her die letzten Gaben,' have been thus translated, truth-
fully, and with all the spirit of the original, by Sir E. L.
Bulwer [*]—

> ' Here bring the last gifts!—and with these
>   The last lament be said;
> Let all that pleased, and yet may please,
>   Be buried with the dead.

> ' Beneath his head the hatchet hide,
>   That he so stoutly swung:
> And place the bear's fat haunch beside—
>   The journey hence is long!

> ' And let the knife new sharpened be
>   That on the battle-day
> Shore with quick strokes—he took but three—
>   The foeman's scalp away!

> ' The paints that warriors love to use,
>   Place here within his hand,
> That he may shine with ruddy hues
>   Amidst the spirit-land.'

But the question whether Man practised rites of sepulture
such as these in more ancient times is quite distinct from
the question of the co-existence of Man with the great
extinct mammalia; and, even if we were to accept M.
Lartet's interpretation of the ossiferous deposits of Aurignac,
both inside and outside the grotto, they would add nothing
to the palæontological evidence in favour of Man's anti-

---

[*] Poems and Ballads of Schiller.

quity, for we have seen all the same mammalia associated
elsewhere with flint implements, and some species, such as
the *Elephas antiquus*, *Rhinoceros hemitœchus*, and *Hippopotamus major*, missing here, have been met with in other
places. Moreover the views of M. Lartet with regard to the
age of the interment have been much questioned by later
explorers. In 1864, about a year after I published my
account of the cave in the 'Antiquity of Man,' Mr. Evans
expressed his opinion that the sepulture might be referable
to an age long subsequent to that of the mammoth, and
suggested, in explanation of the occurrence of the burnt
bones of rhinoceros among the ashes, that they were thrown
out with the earth and burnt, as were also some stones, on
the disturbed ground where the fire was lighted. I lost no
time in communicating these doubts in the following words
to my friend Mr. King, who was about to visit the South of
France :—'The question is whether a cave at Aurignac,
which had been previously filled with the usual cave-mud
and bone breccia, together with extinct animals of the ancient period of *Ursus spelæus* and *Rhinoceros tichorhinus*,
was not at a much later era appropriated as a place of
sepulture by a certain hunter tribe. If they broke up the
ground composed of the bony deposit with extinct mammalia in order to bury their dead, then, when the whole
was cemented together, no one could distinguish between
relics of the old Amiens period and those of much later
date.'

Mr. King had only time to pay a hurried visit of one day
to Aurignac, but he worked carefully in the cave with
Bonnemaison, and penetrated into a downward fissure which
he found to exist at the further end of the cave. He saw
enough to convince himself how difficult it was to fix the
date of burial, on account of the confused state of the evidence, and Mr. Boyd Dawkins, when arranging the papers

of Mr. King after his death, published in ' Nature ' * a
clear account of the most prominent facts which had been
ascertained and the discussions to which they had given
rise.   More recently MM. Curtaillac and Gontier,† the
former of whom I saw when I was with Mr. Hughes at
Toulouse, have carried on investigations leading to the
conclusion that the interment may, as Mr. Evans suspected,
be of comparatively recent date.   They observed that the
lower deposits of the cavern were of a yellow colour,
covered by a large band of a much lighter tint, and that,
while in the darker ground remains of rhinoceros, reindeer,
and cave-bear were present, together with flint implements,
the higher band contained bones of living wild animals and
man, associated with pierced disks of Cardium and fragments
of pottery ; and as it is the generally received opinion that
pottery did not exist in Palæolithic times, they refer the
sepulchral remains to the Neolithic period.   When I visited
Aurignac last spring (1872) I found that the cave had been
so often searched, and the earth so much disturbed, that it
would be hopeless to seek to determine at present the age
of the interments.   The discoveries of M. Cartaillac
doubtless tend to link the later deposits with the Neolithic
period.   But, on the other hand, we shall see, in the following account of the caverns of the Dordogne that, since I
first gave in 1863 M. Lartet's account of the Aurignac
cave, he has himself brought to light such proofs of the
intelligence of Palæolithic man as render it far from improbable that he should have advanced sufficiently to manufacture rude pottery such as that found associated with unpolished flint implements in the Trou du Frontal (see p. 78),
or to burn or bury his dead, or even to have a belief in a
future state.

* Vol. iv. p. 208: July 13. 1871.
† Matériaux pour l'Histoire de l'Homme, April 1872, p. 20.

Caverns of the Dordogne.   Carving of the Mammoth.

The Dordogne district forms a portion of the great
plateau which extends over a large part of the south of
France, and is drained by rivers flowing into the Bay of
Biscay. It is intersected by deep narrow valleys, in one of
which, that of the Vezère, a tributary of the river Dordogne,
are some most remarkable caves, which Mr. Hughes and I
visited in the spring of 1872, under the guidance of M.
Lagrange. The rocks in which the caves occur is a yellow
or white limestone of Cretaceous age. In general appearance, the cliffs along the Vezère are very similar to those
in mountain limestone districts, especially along the Elwy
in North Wales, where we find the same alternation of hard
massive beds breaking up along the joints, and soft clayey
beds, peeling off in flakes from an inch to $\frac{1}{16}$th of an inch in
thickness, and thus undermining the harder masses above,
which fall down from time to time, burying the accumulated
relics below under a heap of large blocks.* We may well
realize how, under the ordinary operations of nature, such
accumulations would not be continuous; an exceptionally
severe frost, or a slight earthquake shock, would bring down
huge masses of solid rock, while it might take many long
years to undermine this rock sufficiently to produce another
fall. It will be understood that the retreats thus formed
are not like the long tortuous caves which occur near
Torquay, or in Derbyshire, and which are due to the action
of carbonated water on the sides of long fissures, but are
merely rock shelters, produced by the overhanging masses of
harder rock. Under the rock shelters thus formed hunter
tribes took up their abode from time to time, leaving refuse-
heaps and rejected or lost instruments strewn in great

* A. Lagrange—Annales d'Agriculture de Dordogne.

profusion in and around their dwellings. M. Mortillet [*] has grouped the various cave deposits under four heads, classifying them according to slight differences in the forms of the flint tools, which he pointed out to us in the museum at St. Germain, and by the number and character of the bone implements. But the difference in the fashion of the flint tools is not great, and is less apparent on the ground where all the rough specimens are also present, than when a selected series is examined. The absence of the larger forms of the stone implements in what he considers the newer deposits, and the rarity of bone instruments in the older forms the chief distinction.

M. Lartet has also founded a classification upon the prevalence of the remains of certain animals in the débris; the mammoth and cave-bear characterising the earlier, and the reindeer the later deposits. But as the same species occur throughout, and as most of the remains were brought there by Man, the abundance of any particular animal may not indicate the prevalence of that species at the time, but only the success of the hunters, or the sojourn of migratory animals in the neighbourhood.

The caves and rock shelters are very numerous in the valley of the Vezère. More than ten of them have been thoroughly examined and described by MM. Christy and Lartet,[†] and it is from their work that I have copied the accompanying view (fig. 12, p. 136). I visited those of Les Eyzies and Laugerie Haute, Laugerie Basse, Gorgo d'Enfer, and also that of Le Moustier, the greater antiquity of which last has been inferred from the character of its deposits considered in connection with the physical geography of the country. In this cave there is a bed of sand

* Mortillet—Matériaux pour l'Histoire de l'Homme. Mars, 1869; and Evans—Ancient Flint Implements.   1872. p. 436.

† Reliquiæ Aquitanicæ, 1868.

View of the Caves of Les Eyzies, Dordogne.

Two lines, the one vertical, the other horizontal, drawn from e e meet in the centre of the principal cave.

having both above and below it floors similar in character,
containing charcoal, flint implements, and other remains.
This sandy bed, which has an average depth of about 10
inches, has the appearance of a river deposit, and it has
been thought, from the large quantity of mica occurring in
it, that it may have been brought from the central granitic
area. If this be the case, the shelters of the two Laugeries,
which occur at a lower level, running to within 35 or 40
feet of the river, must belong to a subsequent period, as
they could hardly have existed when the river ran at the
higher level of Le Moustier. But though an obscure terrace,
which runs along the hill-side on the level of the upper
cave, makes it probable that the river may have run at
that height during the *formation* of the cave, there is no
evidence that it ran there during its *occupation*, for Mr.
Evans has pointed out that there are sandstone beds on
the hill above containing flakes of mica quite as large as
any found in the cave, and therefore the sand with mica
may belong to any period when the talus at the mouth turned
some little runlet of rain-water into the cave behind it.

No object of worked bone has ever been found in Le
Moustier, and the flint implements are more ancient than
those in most of the other caverns, as many of them belong
to types common in the drift of Amiens and Abbeville.*
In the surface deposits of the Laugeries many objects
belonging to the bronze and polished stone period have been
found, together with various kinds of pottery. Below the
horizon of the pottery, under vast masses of fallen rock and
ancient floors covered with flints of Palæolithic types, and
associated with sculptured bones and antlers of reindeer, a
human skeleton was found lying in such a position under a
great block of stone that it was inferred that the individual

* Lartet and Christy, Cavernes du Périgord, Revue Archéologique, 1864,
p. 23, etc.

must have been killed by the fall of the rock ;* but, as no
geological observer was present when the skeleton was discovered, I should hesitate in offering any opinion as to its age.

By far the most remarkable of the Dordogne caves, on
account of a carving on ivory which has been found in it, is
that of La Madelaine, about halfway between Les Eyzies
and Le Moustier. I was able, so long ago as 1863, already
to mention several rude carvings obtained by M. Lartet in
French caves associated in such a manner with reindeer
remains as to show that the men of that period were capable
of making rough representations of animals. Thus, for example, at Savigné, near Civray, in the department of Vienne,
there is a cave in which there are no extinct mammalia, but
where remains of the reindeer abound. Among these remains
there is a stag's horn, on which figures of two animals, apparently meant for deer, are engraved in outline, as if with a
sharp-pointed instrument. In another cave, that of Massat,
in the department of Arriége, which M. Lartet ascribes to
the period of aurochs (a quadruped which survived the
reindeer in the south of France), there are bone instruments
of a still more advanced state of the arts, as, for example,
barbed arrows with a small canal in each, believed to have
served for the insertion of poison ; also a needle of bird's
bone, finely shaped, with an eye or perforation at one end ;
and a stag's horn, on which is carved a representation of a
bear's head, and a hole at one end as if for suspending it.

But the carved slab which M. Lartet found in the cavern
of La Madelaine surpasses all these in interest on account
of the indisputable evidence which it affords that it was
executed by Palæolithic man. It consists of a fragment of
mammoth tusk on which was rudely carved a representation
of the animal itself. (See fig. 13.) Although the slab was

* Comptes Rendus, Annales des Sciences, tom. lxxiv. 1872.

Fig. 13.

Drawing of *Elephas Primigenius* on a plate of fossil ivory found in the ossiferous cavern of La Madelaine, Perigord.

broken in five fragments, Dr. Falconer, who accompanied
M. Lartet, immediately recognised not only the outline of
the elephant, but a group of falling lines representing the
characteristic long-haired mane of the mammoth. If this
carving had not been brought to light, it might have been
objected that the hunter tribes of the valley of the Vézère
visited the district at a period subsequent to that of the
mammoth, and only made use of the fragments of bone and
ivory which belonged to a previous period, and then, as now,
lay strewn about the caves. Even the evidence from the
state of the bones, that the mammoth formed part of the
food of the cave-men, might not be considered quite satisfactory, as they might have broken and burnt bones of a
previous period if they happened to be lying in the earth on
which their fires were lighted. Or if the representation had
been merely that of an elephant, we might have conjectured
that some African tribe, migrating to the South of France,
had brought with them a drawing of the animal as it still
survives in that country. But the characteristic wavy lines of
the long hair of the mammoth allow of no escape from the
conclusion that the cave-men saw this animal in life, and
that they were sufficiently advanced at that period to make
a tolerably faithful sketch of it.

Carvings, though of less importance, have been found in
some of the other caves. In the Laugerie Basse cavern
MM. Massenat, Lalande, and Curtoilhac have recorded the
discovery of engravings on bone of a young reindeer at full
gallop (*lancé au galop*), a sketch of a hare, and the head
of a horse, carved on reindeer horn, and also a curious animal
with feline characters. The object of these engravings
form an interesting subject of speculation. Are they,
as my friend Mr. Hughes suggests, only the product of
idle hours, when the hunters amused themselves by whit-
tling and carving in bone and other material; or have

we, in these rude figures, the distinguishing badge of a
clan or its chief?   Whatever may have been their use,
they teach us that the hunter tribes of the Dordogne, though
the contemporaries of extinct animals, did not belong to a
more undeveloped stage of humanity than some savage tribes
of the present day.

Cave of Bruniquel.

The cavern of Bruniquel, situated in a limestone rock on
the north side of the valley of the Aveyron, one of the
tributaries of the Tarn and Garonne, which was explored in
1863 by the Viscomte de Lastic St. Jal, deserves notice on
account of the very large number of animal remains and
flint implements which have been exhumed from beneath
its stalagmitic floor.   A fine collection of these is to be
seen in the British Museum.   This cave is one of the most
southern in which the remains of the reindeer have yet been
found : here also the wild horse of Europe no doubt had its
home in summer, or was, like the reindeer, intercepted by
the aborigines during its vernal and autumnal migrations
north and south.   Professor Owen estimates the number of
horse remains at more than 100 individuals, and those
of the reindeer at above 1,000.*   The bones indicate by
their fractured condition that they had all been broken to
obtain the marrow, and the horns cut to form weapons.

Portions of ten human skulls, the jaw of a child about
five years of age, and the remains of an infant, together with
various bones, have been extracted from beneath the breccia
of the cavern.   No trace of pottery has been detected, but
abundance of carbonaceous matter, intermixed with many
thousand flint flakes, implements, and cores.   With these
are also more than a hundred beautifully-executed barbed
harpoons, the largest being serrated on both sides.   Many

* Owen's Phil. Trans., 1869, pp. 1–55.

of these implements are ornamented with patterns composed
of heads of animals, others having the complete figures, such
as the reindeer, horse, ibex, snake, and salmon, carved on
them. Portions of four implements made of the mammoth's
tusk have been obtained, and needles and pins made from
the leg-bone of the horse.

ANIMALS FOUND IN BRUNIQUEL CAVERN.

l. living. ex. extinct.

l. *Cervus Tarandus*, Linn.,	l. *Bos Cerrannorum*
ex. — *palmulatus*, Owen	ex. *Rupicapra Christolii*
ex. — *Guettardi*, Desmar	l. *Antelope saiga*, Pallas
l. — *Elaphus*, Linn.	ex. *Equus spelens*, Owen
ex. *Bos primigenius*, Boj.	l. *Lepus timidus*, Linn.
ex. *Elephas primigenius* (indicated	l. *Canis Lupus*, Linn.
by ivory implements only)	l. *Canis Vulpes*, Briss.

FISH.
Salmo fario.

BIRDS.

Haliætus	*Corvus Corax*
Falco	*Perdix*

From the black carbonized layer were also obtained the
following shells, which may be divided into two classes, the
one characteristic of the Atlantic and the other of the
Mediterranean :—

ATLANTIC ?	MEDITERRANEAN ?
Turritella communis	*Triton olearius*
Littorina obtusata	*Cassis saburon*
Fusus sp.	*Dentalium dentalis*
Pecten maximus	— *novem-costatum*
Cardium echinatum	*Nassa reticulata*
Mytilus edulis	*Pecten Jacobæus*
Cardium rusticum	*Pectunculus pilosus!*
Modiola sp.	*Pecten irradians*

I am therefore inclined to agree with the late Dr. S. P
Woodward, who examined these shells in 1864, that they
imply that the natives of Aveyron had easy access to both
sea-coasts, from whence they returned to mingle the shells
of the Atlantic and Mediterranean in their cave-dwellings.

*Skeleton, probably of Palæolithic Age, found in a Cave at
Mentone, in the South of France.*

One of the most interesting discoveries of late years of
the remains of Man, together with flint implements and
the bones of extinct animals, was made in March 1872, by
Dr. Rivière, in a cave at Mentone, in the south of France.
While Dr. Rivière was engaged in a scientific mission for
the exploration of the well-known quarries and caverns near
Nice and Mentone, he found, *in situ*, in a cave called La
Barma du Cavillon, on the Italian frontier, the bones of a
human foot. He immediately began a careful excavation
around the remains, and, after eight days' labour, disclosed
an entire human skeleton at a depth of 20 feet. The
cavern is a triangular slit in the rock, about 30 feet wide
at the base, and from 60 to 80 feet in height, and extending
from south to north, without any marked curvature, 45 feet
into the cliff. From its form and dimensions, the full
daylight reaches to the inner end, and thus enabled Dr.
Rivière to have an excellent photograph taken of the
skeleton *in situ*. The body lay in the longitudinal direction of the cavern, about 24 feet from the entrance. Surrounding and above it were fifty unpolished flint flakes and
scrapers, and a fragment of a skewer about six inches long.[*]
Mr. Pengelly, whose opinion, from his long experience in
Brixham and Kent's caverns, is of great value, visited
Mentone a few weeks after the skeleton was removed. The
exploration of the cavern was still going on, and he saw
300 flint implements, all unpolished, which had been lately
exhumed. 'There was no metal,' he says, 'found in the
cavern, nor was there any pottery nor any polished flint implements; all were unpolished.'[†] The accompanying shells

* Comptes Rendus de l'Acad. des Sciences, avril 29, 1872.
† Journal of London Institution, vol. iii. 1873. No. 18.

were *Patella ferruginea, Pectunculus glycemeris, Cardium rusticum, Cardium edule, Mytilus edulis, Pecten Jacobæus, P. maximus,* and many others, making a total of 54 marine and 11 terrestrial species. The mammalia found in the soil immediately above the skeleton included the *Bos primigenius, Ursus spelæus, Felis spelæa, Hyæna spelæa, Rhinoceros tichorhinus,* wolf, stag, chamois, and others.*

The skeleton was taken to Paris, to the Jardin des Plantes, where Mr. Hughes and I examined it in company with MM. Desnoyers, and Hamy soon after its arrival. The skeleton is that of a man about 5 feet 9 or 10 inches high, and all but perfect. The skull was of a red colour, and covered with numerous perforated marine shells, of the species *Nassa neritea,* and twenty-two perforated canines of the stag, the whole having probably been a chaplet. A bone instrument, pointed at one end, lay across the forehead. The skull was very dolichocephalic, the occiput much produced, the forehead rather narrow, and the temple flattened; the facial angle measured from 80 to 85 degrees. All the teeth were perfect, but were worn flat, as if by trituration of hard food, as is commonly observed in very ancient skulls, as well as in modern savage races. The thigh-bones were strongly carinate, and the tibia, or shin-bone, was somewhat platycnemic, or flattened; the fibula also was of enormous thickness. The whole attitude indicated that the man had died in his sleep; and from the manner in which his remains were associated with unpolished implements and the bones of extinct animals, it seems not improbable that Dr. Rivière has brought to light a complete human skeleton of Palæolithic age.

* Rivière, Découverte d'un Squelette humain de l'époque paléolithique. Paris, 1873.

CHAPTER VIII.

PLEISTOCENE ALLUVIUM WITH FLINT IMPLEMENTS OF THE VALLEY OF THE SOMME.

GENERAL POSITION OF DRIFT WITH EXTINCT MAMMALIA IN VALLEYS
—DISCOVERIES OF M. BOUCHER DE PERTHES AT ABBEVILLE—FLINT
IMPLEMENTS FOUND ALSO AT ST. ACHEUL, NEAR AMIENS—GEOLOGICAL
STRUCTURE OF THE VALLEY OF THE SOMME AND OF THE SURROUND-
ING COUNTRY—POSITION OF ALLUVIUM OF DIFFERENT AGES—PEAT
NEAR ABBEVILLE—ITS ANIMAL AND VEGETABLE CONTENTS—WORKS OF
ART IN PEAT—PROBABLE ANTIQUITY OF THE PEAT AND CHANGES OF
LEVEL SINCE ITS GROWTH BEGAN—FLINT IMPLEMENTS OF PALÆOLITHIC
TYPE IN OLDER ALLUVIUM—THEIR VARIOUS FORMS AND GREAT
NUMBERS.

*Pleistocene Alluvium containing Flint Implements in the
Valley of the Somme.*

THROUGHOUT a large part of Europe we find at moderate elevations above the present river-channels, usually
at a height of less than forty feet, but sometimes much
higher, beds of gravel, sand, and loam containing bones of
the elephant, rhinoceros, horse, ox, and other quadrupeds,
some of extinct, others of living, species. Many of these
deposits contain fluviatile shells, and have undoubtedly
been accumulated in ancient river-beds. These old channels have long since been dry, the streams which once
flowed in them having shifted their position, deepening
the valleys, and often widening them.

It has naturally been asked, if Man coexisted with the
extinct species of the caves, why were his remains and the
works of his hands never embedded outside the caves in
ancient river-gravel containing the same fossil fauna? Why

L

should it be necessary for the geologist to resort for evidence of the antiquity of our race to the dark recesses of underground vaults and tunnels, which may have served as places of refuge or sepulture to a succession of human beings and wild animals, and where floods may have confounded together in one breccia the memorials of the fauna of more than one epoch? Why do we not meet with a similar assemblage of the relics of Man, and of living and extinct quadrupeds, in places where the strata can be thoroughly scrutinised in the light of day?

Recent researches have at length demonstrated that such memorials, so long sought for in vain, do in fact exist, and their recognition is the chief cause of the more favourable reception now given to the conclusions which MM. Tournal, Christol, Schmerling, and others, arrived at thirty years ago respecting the fossil contents of caverns.

A very important step in this new direction was made thirteen years after the publication of Schmerling's ' Researches,' by M. Boucher de Perthes, who found in ancient alluvium at Abbeville, in Picardy, some flint implements, the relative antiquity of which was attested by their geological position. The antiquarian knowledge of their discoverer enabled him to recognise in their rude and peculiar type a character distinct from that of the polished stone weapons of a later period, usually called ' celts.' In the first volume of his ' Antiquités Celtiques,' published in 1847, M. Boucher de Perthes styled these older tools ' antediluvian,' because they came from the lowest beds of a series of ancient alluvial strata bordering the valley of the Somme, which geologists had termed ' diluvium.' He had begun to collect these implements in 1841. From that time they had been annually dug out of the drift or deposits of gravel and sand, of which fine sections were laid open from twenty to thirty-five feet in depth, whenever excavations were made in

repairing the fortifications of Abbeville; or as often as flints
were wanted for the roads, or loam for making bricks. For
years previously, bones of quadrupeds of the genera elephant,
rhinoceros, bear, hyæna, stag, ox, horse, and others, had been
collected there, and sent from time to time to Paris to be
examined and named by Cuvier, who had described them in
his 'Ossemens Fossiles.' A correct account of the associated
flint tools and of their position was given in 1847 by
M. Boucher de Perthes in his work above cited, and they
were stated to occur at various depths, often twenty or thirty
feet from the surface, in sand and gravel, especially in those
strata which were nearly in contact with the subjacent white
chalk. But the scientific world had no faith in the statement that works of art, however rude, had been met with in
undisturbed beds of such antiquity. Few geologists visited
Abbeville in winter, when the sandpits were open, and when
they might have opportunities of verifying the sections, and
judging whether the instruments had really been embedded
by natural causes in the same strata with the bones of the
mammoth, rhinoceros, and other extinct mammalia. Some
of the tools figured in the 'Antiquités Celtiques' were so
rudely shaped, that many imagined them to have owed their
peculiar forms to accidental fracture in a river's bed ; others
suspected fraud on the part of the workmen, who might
have fabricated them for sale, or that the gravel had been
disturbed, and that the worked flints had got mingled with
the bones of the mammoth long after that animal and its
associates had disappeared from the earth.

No one was more sceptical than the late eminent physician
of Amiens, Dr. Rigollot, who had long before (in the year
1819) written a memoir on the fossil mammalia of the
valley of the Somme. He was at length induced to visit
Abbeville, and, having inspected the collection of M. Boucher
de Perthes, returned home resolved to look for himself for

flint tools in the gravel-pits near Amiens. There, accordingly, at a distance of about thirty miles from Abbeville, he immediately found abundance of similar flint implements, precisely the same in the rudeness of their make, and the same in their geological position; some of them in gravel nearly on a level with the Somme, others in similar deposits resting on chalk at a height of about ninety feet above the river.

Dr. Rigollot having in the course of four years obtained several hundred specimens of these tools, most of them from St. Acheul, in the south-east suburbs of Amiens, communicated an account of them to the scientific world, in a memoir * illustrated by good figures of the worked flints, and careful sections of the beds. These sections were executed by M. Buteux, an engineer well qualified for the task, who had written a good description of the geology of Picardy. Dr. Rigollot, in this memoir, pointed out most clearly that it was not in the vegetable soil, nor in the brick-earth with land and fresh-water shells next below, but in the lower beds of coarse flint-gravel, usually twelve, twenty, or twenty-five feet below the surface, that the implements were met with, just as they had been previously stated by M. Boucher de Perthes to occur at Abbeville. The conclusion, therefore, which was legitimately deduced from all the facts, was that the flint tools and their fabricators were coeval with the extinct mammalia embedded in the same strata.

A further step in advance was made when Dr. Falconer, after aiding in the investigations above alluded to (p. 100), near Torquay, stopped at Abbeville on his way to Sicily, in the autumn of 1858, and saw there the collection of M. Boucher de Perthes. Being at once satisfied that the flints

* Mémoire sur des Instruments en Silex trouvés à St.-Acheul, près Amiens, 1855.

called hatchets had been really fashioned by the hand of
man, he urged Mr. Prestwich, by letter, thoroughly to
explore the geology of the Valley of the Somme. This
Mr. Prestwich accordingly accomplished, in company with
Mr. John Evans, and, before his return that same year,
succeeded in dissipating all doubts from the minds of
his geological friends by extracting, with his own hands,
from a bed of undisturbed gravel, at St. Acheul, a well-
shaped flint hatchet. This implement was buried in the
gravel at a depth of seventeen feet from the surface, and
was lying on its flat side. There were no signs of vertical
rents in the enveloping matrix, nor in the overlying beds of
sand and loam, in which were many land and fresh water
shells; so that it was impossible to imagine that the
tool had gradually worked its way downwards, as some
had suggested, through the incumbent soil, into an older
formation.[*]

There was no one in England whose authority deservedly
had so much weight in overcoming incredulity in regard
to the antiquity of the implements in question. For Mr.
Prestwich, besides having published a series of important
memoirs on the tertiary formations of Europe, had devoted
many years specially to the study of the drift and its
organic remains. His report, therefore, to the Royal
Society, accompanied by a photograph showing the position
of the flint tool *in situ* before it was removed from its
matrix, not only satisfied many inquirers, but induced
others to visit Abbeville and Amiens; and one of these,
Mr. J. W. Flower, who accompanied Mr. Prestwich on his
second excursion to St. Acheul, in June, 1859, succeeded, by
digging into the bank of gravel, in disinterring, at the
depth of twenty-two feet from the surface, a fine, symmetri-

* Prestwich, Proceedings of the Royal Society, 1859; and Philosophical
Transactions, 1860.

cally shaped weapon of an oval form, lying in and beneath
strata which were observed by many witnesses to be perfectly undisturbed.*

Shortly afterwards, in the year 1859, I visited the same
pits, and obtained seventy flint tools, one of which was
taken out while I was present, though I did not see it before
it had fallen from the matrix. I expressed my opinion in
favour of the antiquity of the flint tools to the meeting of
the British Association at Aberdeen, in the same year.† On
my way through Rouen, I stated my convictions on this
subject to M. George Pouchet, who immediately betook
himself to St. Acheul, commissioned by the municipality
of Rouen, and did not quit the pits till he had seen
one of the hatchets extracted from gravel in its natural
position.‡

M. Gaudry also gave the following account of his
researches in the same year to the Royal Academy of
Sciences at Paris. 'The great point was not to leave the
workmen for a single instant, and to satisfy oneself by
actual inspection, whether the hatchets were found *in situ*.
I caused a deep excavation to be made, and found nine
hatchets, most distinctly *in situ* in the diluvium, associated
with teeth of *Equus fossilis* and a species of *Bos*, different
from any now living, and similar to that of the diluvium
and of caverns.' §   In 1859, M. Hébert, an original observer
of the highest authority, declared to the Geological Society
of France that he had, in 1854, or four years before Mr.
Prestwich's visit to St. Acheul, seen the sections at Abbeville and Amiens, and had come to the opinion that the
hatchets were imbedded in the 'lower diluvium,' and that
they were as ancient as the mammoth and the rhinoceros.

* Geological Quarterly Journal,
vol. xvi. p. 190.
† See Proceedings of British Association for 1859.

‡ Actes du Musée d'Histoire naturelle de Rouen, 1860, p. 33.
§ Comptes Rendus, September 26.
and October 3, 1859.

M. Desnoyers also made excavations after M. Gaudry, at
St. Acheul, in 1859, with the same results.*

After a lively discussion on the subject in England and
France, it was remembered, not only that there were numerous recorded cases leading to similar conclusions in regard
to cavern deposits, but, also, that Mr. Frere had, so long
ago as 1797, found flint weapons, of the same type as those
of Amiens, in a fresh-water formation in Suffolk, in conjunc-
tion with elephant remains; and nearly a hundred years
earlier (1715), another tool of the same kind had been
exhumed from the gravel of London, together with bones
of an elephant; to all which examples I shall allude more
fully in the sequel.

I may conclude this summary by quoting a saying of
Professor Agassiz, ' that whenever a new and startling fact is
brought to light in science, people first say, " it is not
true," then that " it is contrary to religion," and lastly,
" that everybody knew it before." '

If I were considering merely the cultivators of geology,
I should say that the doctrine of the former coexistence of
Man with extinct mammalia had already gone through
these three phases in the progress of every scientific truth
towards acceptance. But the grounds of this belief cannot
be too often or too fully laid before the general public, so as
to enable them fairly to weigh and appreciate the evidence.
I shall therefore do my best in this and the next three
chapters to accomplish this task.

Geological Structure of the Somme Valley.

The Valley of the Somme in Picardy is situated geologically in a region of white chalk with flints, the strata of
which are nearly horizontal. The chalk hills which bound the
valley are almost everywhere between 200 and 300 feet in

* Bulletin, vol. xvii. p. 18.

height. On ascending to that elevation, we find ourselves on an extensive table-land, in which there are slight elevations and depressions. The white chalk itself is scarcely ever exposed at the surface on this plateau, although seen on the slopes of the hills, as at *b* and *c* (fig. 14). The general surface of the upland region is covered continuously, for miles in every direction, by loam or brick-earth (No. 4), about five feet thick, devoid of fossils. To the wide extent of this loam the soil of Picardy chiefly owes its great fertility. Here and there we also observe, on the chalk, outlying patches of tertiary sand and clay (No. 5, fig. 14), with Eocene fossils, the remnants of a formation once more extensive, and which probably once spread in a continuous mass over the chalk, before the present system of valleys had begun to be shaped out. It is necessary to allude to these relics of tertiary strata, of which the larger part is missing, because their denudation has contributed largely to furnish the materials of gravels in which the flint implements and bones of extinct mammalia are entombed. From this source have been derived not only the egg-shaped pebbles, so common in the old fluviatile alluvium at all levels, but those huge masses of hard sandstone, several feet in diameter, to which I shall allude in the sequel. The upland loam also (No. 4) has often, in no slight degree, been formed at the expense of the same tertiary sands and clays, as is attested by its becoming more or less sandy or argillaceous, according to the nature of the nearest Eocene outlier in the neighbourhood.

The average width of the Valley of the Somme between Amiens and Abbeville is one mile. The height, therefore, of the hills, in relation to the river-plain, could not be correctly represented in the annexed diagram (fig. 14), as they would have to be reduced in altitude; or if not, it would be necessary to make the space between *c* and *b* four times as great. The dimensions also of the masses of drift or alluvium, 2 and 3, have been exaggerated, in order to render them sufficiently

Fig. 14.

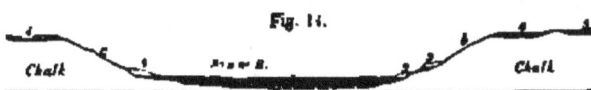

Section across the Valley of the Somme in Picardy.

1 Peat, twenty to thirty feet thick, resting on gravel, *a*.
2 Lower-level gravel with elephants' bones and flint tools, covered with fluviatile loam, twenty to forty feet thick.
3 Higher-level gravel with similar fossils, and with overlying loam, in all thirty feet thick.
4 Upland loam without shells (*Limon des plateaux*), five or six feet thick.
5 Eocene tertiary strata, resting on the chalk in patches.

conspicuous; for, all important as we shall find them to be as geological monuments of the Post-Pliocene period, they form a truly insignificant feature in the general structure of the country, so much so, that they might easily be overlooked in a cursory survey of the district, and are usually unnoticed in geological maps not specially devoted to the superficial formations.

It will be seen by the description given of the section, fig. 14, that No. 2 indicates the lower level gravels, and No. 3 the higher ones, or those rising to elevations of eighty or a hundred feet above the river. Newer than these is the peat No. 1, which is from ten to thirty feet in thickness, and which is not only of later date than the alluvium, 2 and 3, but is also posterior to the denudation of those gravels, or to the time when the valley was excavated through them. Underneath the peat is a bed of gravel, *a*, from three to fourteen feet thick, which rests on undisturbed chalk. This gravel was probably formed, in part at least, when the valley was scooped out to its present depth, since which time no geological change has taken place, except the growth of the peat, and certain oscillations in the general level of the country, to which we shall allude by and by. A thin layer of impervious clay separates the gravel *a* from the peat No. 1, and seems to have been a necessary preliminary to the growth of the peat.

Peat of the Valley of the Somme.

As hitherto, in our retrospective survey, we have been obliged, for the sake of proceeding from the known to the less known, to reverse the natural order of history, and to treat of the newer before the older formations, I shall begin my account of the geological monuments of the Valley of the Somme by saying something of the most modern of all of them, the peat. This substance occupies the lower parts of the valley far above Amiens, and below Abbeville as far as the sea. It has already been stated to be in some places thirty feet thick, and is even occasionally more than thirty feet, corresponding in that respect to the Danish mosses before described (Chap. II.). Like them, it belongs to the recent period; all the embedded mammalia, as well as the shells, being of the same species as those now inhabiting Europe. The bones of quadrupeds are very numerous, as I can bear witness, having seen them brought up from a considerable depth near Abbeville, almost as often as the dredging instrument was used. Besides remains of the beaver, I was shown, in the collection of M. Boucher de Perthes, two perfect lower jaws with teeth of the bear, *Ursus Arctos*; and in the Paris Museum there is another specimen, also from the Abbeville peat.

The list of mammalia already comprises a large proportion of those proper to the Swiss lake-dwellings, and to the shell-mounds and peat of Denmark; but unfortunately as yet no special study has been made of the French fauna, like that by which the Danish and Swiss zoologists and botanists have enabled us to compare the wild and tame animals and the vegetation of the age of stone with that of the age of iron.

Notwithstanding the abundance of mammalian bones in the peat, and the frequency of stone implements of the Neolithic and Gallo-Roman periods, M. Boucher de Perthes has only met with three or four fragments of human skeletons.

At some depth in certain places in the valley near Abbeville, the trunks of alders have been found standing erect as they grew, with their roots fixed in an ancient soil, afterwards covered with peat. Stems of the hazel, and nuts of the same, abound; trunks, also, of the oak and walnut. The peat extends to the coast, and is there seen passing under the sand-dunes and below the sea-level. At the mouth of the river Canche, which joins the sea near the embouchure of the Somme, yew trees, firs, oaks, and hazels have been dug out of peat, which is there worked for fuel, and is about three feet thick.[*] During great storms, large masses of compact peat, enclosing trunks of flattened trees, have been thrown up on the coast at the mouth of the Somme; seeming to indicate that there has been a subsidence of the land and a consequent submergence of what was once a westward continuation of the Valley of the Somme into what is now a part of the British Channel, or La Manche.

Whether the vegetation of the lowest layers of peat differed as to the geographical distribution of some of the trees from the middle, and this from the uppermost peat, as in Denmark, has not yet been ascertained; nor have careful observations been made with a view of calculating the minimum of time which the accumulation of so dense a mass of vegetable matter must have taken. A foot in thickness of highly compressed peat, such as is sometimes reached in the bottom of the bogs, is obviously the equivalent of a much

---

* D'Archiac, Hist. des Progrès, vol. ii. p. 151.

greater thickness of peat of spongy and loose texture, such as
is found near the surface. The workmen who cut peat, or
dredge it up from the bottom of swamps and ponds, declare
that in the course of their lives none of the hollows which
they have found, or caused by extracting peat, have ever
been refilled, even to a small extent. They deny, therefore,
that the peat grows. This, as M. Boucher de Perthes observes,
is a mistake; but it implies that the increase in one generation is not very appreciable by the unscientific.

The antiquary finds near the surface Gallo-Roman remains,
and still deeper, Neolithic weapons of the period. But the
depth at which Roman works of art occur varies in different
places, and is no sure test of age; because in some parts of
the swamps, especially near the river, the peat is often so fluid
that heavy substances may sink through it, carried down by
their own gravity. In one case, however, M. Boucher de
Perthes observed several large flat dishes of Roman pottery,
lying in a horizontal position in the peat, the shape of which
must have prevented them from sinking or penetrating
through the underlying peat. Allowing about fourteen centuries for the growth of the superincumbent vegetable matter,
he calculated that the thickness gained in a hundred years
would be no more than three French centimetres, or about
1·2 English inches.* This rate of increase would demand so
many thousands of years for the formation of the entire
thickness of thirty feet, that we must hesitate before adopting it as a chronometric scale. Yet, by multiplying observations of this kind, and bringing one to bear upon and check
another, we may eventually succeed in obtaining data for
estimating the age of the peaty deposit.

The rate of increase in Denmark may not be applicable
to France; because differences in the humidity of the climate,
or in the intensity and duration of summer's heat and winter's

* Antiquités Celtiques, vol. ii. p. 134.

cold, as well as diversity in the species of plants which most
abound, would cause the peat to grow more or less rapidly,
not only when we compare two distinct countries in Europe,
but the same country at two successive periods.

I have already alluded to some facts which favour the idea
that there has been a change of level on the coast since the
peat began to grow. This conclusion seems confirmed by the
mere thickness of peat at Abbeville, and the occurrence of
alder and hazel-wood near the bottom of it. If thirty feet
of peat were now removed, the sea would flow up and fill the
valley for miles above Abbeville. Yet this vegetable matter
is all of supra-marine origin, for where shells occur in it they
are all of terrestrial or fluviatile kinds, so that it must have
grown above the sea-level when the land was more elevated
than now.

Small as is the progress hitherto made in interpreting
the pages of the peaty record, their importance in the Valley
of the Somme is enhanced by the reflection that, whatever
be the number of centuries to which they relate, they belong
to times posterior to the ancient implement-bearing beds,
which we are next to consider, and are even separated from
them, as we shall see, by an interval far greater than that
which divides the earliest strata of the peat from the latest.

*Flint Implements of the Pleistocene Period in the Valley
of the Somme.*

The alluvium of the Valley of the Somme exhibits nothing
extraordinary or exceptional in its position or its external
appearance, nor in the arrangement or composition of its
materials, nor in its organic remains: in all these characters
it might be matched by the alluvial deposits of a hundred
other valleys in France or England. Its claim to our peculiar
attention is derived from the wonderful number of flint tools,
of a very antique type, which, as above stated, occur in

Fig. 15.

Flint implement from St. Acheul, near Amiens, of the spear-head shape.
Half the size of the original, which is seven and a half inches long.

a Side view.   b Same seen edgewise.

These spear-headed implements have been found in greater number, proportionally to the oval ones, in the upper level gravel at St. Acheul, than in any of the lower gravels in the valley of the Somme. In these last the oval form predominates, especially at Abbeville.

Fig. 16.   Fig. 17.

Flint implements from the Post-Pliocene Drift of Abbeville and Amiens.

Fig 16 a Oval-shaped flint hatchet from Mautort, near Abbeville, half size of original, which is five and a half inches long, from a bed of gravel underlying the fluvio-marine stratum.

> b Same seen edgewise.
>
> c Shows a recent fracture of the edge of the same at the point a, or near the top. The portion of the tool, c, is drawn of the natural size, the black central part being the unaltered flint, the white margin representing the outer coat which has been altered or bleached since the tool was first made.
>
> The entire surface of No. 9 must have been black when first shaped, and the bleaching to such a depth must have been the work of time, whether produced by exposure to the sun and air before it was imbedded, or afterwards when it lay deep in the soil.

Fig. 17 Flint tool from St. Acheul, seen edgewise; original, six and a half inches long, and three inches wide.

> e, f Portion not artificially shaped.
>
> c, d Part chipped into shape, and having a cut edge at a.

undisturbed strata, associated with the bones of extinct
quadrupeds.

As much doubt has been cast on the question, whether
the so-called flint hatchets have really been shaped by the
hands of Man, it will be desirable to begin by satisfying the
reader's mind on that point, before inviting him to study the
details of sections of successive beds of mud, sand, and gravel,
which vary considerably even in contiguous localities.

Since the spring of 1859, I have paid four visits to the
Valley of the Somme, and examined all the principal localities of these flint tools. In my excursions around Abbeville,
I was accompanied by M. Boucher de Perthes, and during
one of my explorations in the Amiens district, by Mr.
Prestwich. The first time I entered the pits at St. Acheul,
I obtained seventy flint instruments, all of them collected
from the drift in the course of the preceding five or six
weeks. The two prevailing forms of these tools are represented in the annexed figures 15 and 16, each of which is
half the size of the originals; the first is the spear-headed
form, varying in length from six to eight inches; the second,
the oval form, which is not unlike some stone implements
used to this day as hatchets and tomahawks by natives of
Australia, but with this difference, that the edge in the
Australian weapons (as in the case of those called 'celts' in
Europe) has been produced by friction, whereas the cutting
edge in the old tools of the Valley of the Somme was always
gained by the simple fracture of the flint, and by the repetition of many dexterous blows.

The oval-shaped Australian weapons, however, differ in
being sharpened at one end only. The other, though reduced
by fracture to the same general form, is left rough, in which
state it is fixed into a cleft stick, which serves as a handle.
To this it is firmly bound by thin straps of kangaroo hide,
cemented by the gum of the eucalyptus. One of these tools,

now in my possession, was given me by Mr. Farquharson, of
Haughton, who saw a native using it in 1854, on the Auburn
river, in Burnet district, North Australia.

Out of more than a hundred flint implements which I
obtained at St. Acheul, not a few had their edges more or
less fractured or worn, either by use as instruments before they
were buried in gravel, or by being rolled in the river's bed.

Some of these tools were probably used as weapons, both
of war and of the chase, others to grub up roots, cut down
trees, and scoop out canoes. Some of them may have served,
as Mr. Prestwich has suggested, for cutting holes in the ice
both for fishing and for obtaining water, as will be explained
in the 9th chapter when we consider the arguments in favour
of the higher level drift having belonged to a period when
the rivers were frozen over for several months every winter.

When the natural form of a chalk flint presented a
suitable handle at one end, as in the specimen fig. 17, that
part was left as found. The portion, for example, between
b and *c* has probably not been altered ; the protuberances
which are fractured having been broken off by river action
before the flint was chipped artificially. The other extremity, *a*, has been worked till it acquired a proper shape
and cutting edge.

Many of the hatchets are stained of an ochreous-yellow
colour, when they have been buried in yellow gravel, others
have acquired white or brown tints, according to the matrix
in which they have been enclosed.

This accordance in the colouring of the flint tools with
the character of the bed from which they have come, indicates, says Mr. Prestwich, not only a real derivation from
such strata, but also a sojourn therein of equal duration to
that of the naturally broken flints forming part of the same
beds.[*]

* Philosophical Transactions, 1861, p. 297.

Mr. Evans considers that the white coating so commonly found on flints which are dark in the interior, can be explained by a chemical change first pointed out to him by M. Meillet, of Poitiers.* Certain flints, he says, are composed of two kinds of silica, the one white and insoluble in water, the other transparent, horny, and soluble, though both in their other properties are chemically the same. It appears then, that in these whitened flints the soluble portion has been removed by the passage of infiltrating water through the body of the flint, while the insoluble portion has been left in a finely divided state, and is consequently white and consists of particles susceptible of disaggregation by moderate force.

The surface of many of the tools is encrusted with a film of carbonate of lime, while others are adorned by those ramifying crystallisations called dendrites (see figs. 18–20)

Fig. 18.   Fig. 19.   Fig. 20.

Dendrites on surfaces of flint hatchets in the drift of St. Acheul, near Amiens.
Fig. 18, a natural size.   Fig. 19, b natural size; c magnified.
Fig. 20, d natural size; e magnified.

usually consisting of the mixed oxides of iron and manganese, forming extremely delicate blackish brown sprigs, resembling the smaller kinds of sea-weed. They are a useful test of antiquity when suspicions are entertained of the workmen having forged the hatchets which they offer for sale. The most general test, however, of the genuineness of the implements obtained by purchase is their superficial varnish-like or vitreous gloss, as contrasted with the dull aspect of freshly

* Recherches Chimiques sur la Patine des Silex taillés. Montauban, 1866.

fractured flints. I also remarked, during each of my three
visits to Amiens, that there were some extensive gravel-pits,
such as those of Moutiers and St. Roch, agreeing in their
geological character with those of St. Acheul, and only a mile
or two distant, where the workmen, although familiar with
the forms, and knowing the marketable value of the articles
above described, assured me that they had never been able
to find a single implement.

Respecting the authenticity of the tools as works of art,
Professor Ramsay, than whom no one could be a more com-
petent judge, observes : ' For more than twenty years, like
others of my craft, I have daily handled stones, whether
fashioned by nature or art ; and the flint hatchets of Amiens
and Abbeville seem to me as clearly works of art as any
Sheffield whittle.'*

Mr. Evans, in 1859, classified the implements under three
heads, two of which, the spear-heads and the oval or almond-
shaped kinds, have already been described. The third form,
fig. 21, consists of flakes, apparently intended for knives, or

Fig. 21.

Flint knife or flake from below the sand containing *Cyrena fluminalis*,
Menchecourt, Abbeville.
d Transverse section along the line of fracture, *b, c.*
(Size, two-thirds of the original.)

some of the smaller ones for arrow-heads. In his work
just published,† he states, that although some additional
types have been brought to light in the last thirteen years,
he sees no cause to alter this classification.

* Athenæum, July 16, 1859.   † Ancient Stone Implements, 1872.

M 2

In regard to their origin, he observes that there is a
uniformity of shape, a correctness of outline, and a sharpness
about the cutting edges and points, which cannot be due to
anything but design.*

Of these knives and flakes, I obtained several specimens
from a pit which I caused to be dug at Abbeville, in sand in
contact with the chalk, and below certain fluvio-marine beds,
which will be alluded to in the next chapter.

There seems to be little doubt that on account of the
sharp-cutting edge presented by flint flakes they were much
used both as cutting and boring instruments. 'Each flint,'
says Mr. Evans, 'when dexterously made, has on either side
a cutting edge, so sharp that it almost might, like the obsidian flakes of Mexico, be used to shave with.' Their
efficiency as borers was proved by the late M. E. Lartet, who
drilled the eye of a bone needle with a flint borer found
in one of the French caves; and Mr. Evans has himself
bored perfectly round and smooth holes through stag's horn
and wood with flint flakes.†

Between the spear-head and oval shapes, there are various
intermediate gradations, and there are also a vast variety of
very rude implements, many of which may have been rejected
as failures, and others struck off as chips in the course of
manufacturing the more perfect ones. Some of these chips
can only be recognised by an experienced eye as bearing
marks of human workmanship.

It has often been asked, how, without the use of metallic
hammers, so many of these oval and spear-headed tools could
have been wrought into so uniform a shape. Mr. Evans, in
order experimentally to illustrate the process, constructed a
stone hammer, by mounting a pebble in a wooden handle,
and with this tool struck off flakes from the edge on both

* Archæologia, vol. xxxviii.
† Evans's Ancient Stone Implements, 1872, pp. 260–288.

sides of a chalk flint, till it acquired precisely the same shape as the oval tool, fig. 16, p. 159.

If I were invited to estimate the probable number of the more perfect tools found in the Valley of the Somme since 1842, rejecting all the knives, and all that might be suspected of being spurious or forged, I should conjecture that they far exceed a thousand. Yet it would be a great mistake to imagine that an antiquary or geologist, who should devote a few weeks to the exploration of such a valley as that of the Somme, would himself be able to detect a single specimen. But few tools were lying on the surface. The rest have been exposed to view by the removal of such a volume of sand, clay, and gravel, that the price of the discovery of one of them could only be estimated by knowing how many hundred labourers have toiled at the fortifications of Abbeville, or in the sand and gravel-pits near that city, and around Amiens, for road materials and other chemical purposes, during the last twenty years.

In the gravel-pits of St. Acheul, and in some others near Amiens, small round bodies, having a tubular cavity in the centre, occur. They are well known as fossils of the white chalk. Dr. Rigollot suggested that they might have been

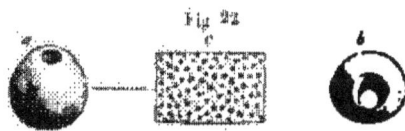

Fig 22.

a, b Coscinopora globularis D'Orb. = Orbitolina concava Parker and Jones.
c Part of the same magnified.

strung together as beads, and he supposed the hole in the middle to have been artificial. Some of these round bodies are found entire in the chalk and in the gravel, others have a hole passing through them, and sometimes one or

two holes penetrating some way in from the surface, but
not extending to the other side.   Others, like *b*, fig. 22,
have a large cavity, which has a very artificial aspect.
It is impossible to decide whether they have or have not
served as personal ornaments, recommended by their globular
form, lightness, and by being less destructible than ordinary
chalk. Granting that there were natural cavities in the axis of
some of them, it does not follow that these may not have been
taken advantage of for stringing them as beads, while others
may have been artificially bored through.   Dr. Rigollot's
argument in favour of their having been used as necklaces or
bracelets, appears to me a sound one.   He says he often found
small heaps or groups of them in one place, all perforated,
just as if, when swept into the river's bed by a flood, the
bond or string which had united them together remained
unbroken.*

* Rigollot, Mémoire sur des Instruments en Silex, etc., p. 16.   Amiens, 1851.

CHAPTER IX.

PLEISTOCENE ALLUVIUM WITH FLINT IMPLEMENTS OF THE VALLEY OF THE SOMME,

Continued.

FLUVIO-MARINE STRATA, WITH FLINT IMPLEMENTS, NEAR ABBEVILLE — MARINE SHELLS IN SAME — CYRENA FLUMINALIS — MAMMALIA — ENTIRE SKELETON OF RHINOCEROS — FLINT IMPLEMENTS, WHY FOUND LOW DOWN IN FLUVIATILE DEPOSITS — RIVERS SHIFTING THEIR CHANNELS—RELATIVE AGES OF HIGHER AND LOWER-LEVEL GRAVELS— SECTION OF ALLUVIUM OF ST. ACHEUL — TWO SPECIES OF ELEPHANT AND HIPPOPOTAMUS COEXISTING WITH MAN IN FRANCE — VOLUME OF DRIFT, PROVING ANTIQUITY OF FLINT IMPLEMENTS — PORTIONS OF A HUMAN SKELETON FOUND IN OLDER ALLUVIUM AT CLICHY, NEAR PARIS —GENERAL ABSENCE OF HUMAN BONES IN TOOL-BEARING ALLUVIUM, HOW EXPLAINED—HUMAN BONES NOT FOUND IN DRAINED LAKE OF HAARLEM.

IN the section of the Valley of the Somme, given at p. 153 (fig. 14), the successive formations newer than the chalk are numbered in chronological order, beginning with the most modern, or the peat, which is marked No. 1, and which has been treated of in the last chapter. Next in the order of antiquity are the lower-level gravels, No. 2, which we have now to describe; after which the alluvium, No. 3, found at higher levels, or about eighty and one hundred feet above the river-plain, will remain to be considered.

I have selected, as illustrating the old alluvium of the Somme occurring at levels slightly elevated above the present river, the sand and gravel-pits of Menchecourt, in the north-west suburbs of Abbeville, to which, as before stated, p. 146, attention was first drawn by M. Boucher de Perthes, in his work on Celtic antiquities. Here, although in every adjoining pit some minor variations in the nature and thickness of the superimposed deposits may be seen, there is yet a general approach to uniformity in the series. The relative age of the gravel marked *a* (fig. 23), underlying the peat, and resting on the chalk, is somewhat doubtful. It is only known by borings, and some of it may be of the same age as No. 3; but I believe it to be for the most part of more modern origin, consisting of the wreck of all the older gravel, including No. 3, and forming during the last hollowing out

Fig. 23.

Section of fluvio-marine strata, containing flint implements and bones of extinct mammalia, at Menchecourt, Abbeville.*

1 Brown clay with angular flints, and occasionally chalk rubble, unstratified, following the slope of the hill, probably of subaërial origin, of very varying thickness, from two to five feet and upwards.

2 Calcareous loam, buff-coloured, resembling loess, for the most part unstratified, in some places with slight traces of stratification, containing fresh-water and land shells, with bones of elephant, &c.; thickness about fifteen feet.

3 Alternations of beds of gravel, marl, and sand, with fresh-water and land shells, and in some of the lower sands, a mixture of marine shells; also bones of elephant, rhinoceros, &c., and flint implements; thickness about twelve feet.

a Gravel underlying peat, age undetermined.

b Layer of impervious clay, separating the gravel from the peat.

and deepening of the valley immediately before the commencement of the growth of peat.

The greater number of flint implements have been dug out of No. 3, often near the bottom, and twenty-five, thirty, or even more than thirty feet below the surface of No. 1.

* For detailed sections and maps of this district, see Prestwich, Philosophical Transactions, 1860, p. 277.

A geologist will perceive by a glance at the section
(fig. 23, p. 168) that the valley of the Somme must have
been excavated nearly to its present depth and width when
the strata of No. 3 were thrown down, and that after the deposits Nos. 3, 2, and 1 had been formed in succession, the
present valley was scooped out, patches only of Nos. 3 and
2 being left For these deposits cannot originally have ended
abruptly as they now do, but must have once been continuous farther towards the centre of the valley.

To begin with the oldest, No. 3, it is made up of a succession of beds, chiefly of fresh-water origin, but occasionally
a mixture of marine and fluviatile shells is observed in it,
proving that the sea sometimes gained upon the river, whether
at high tides or when the fresh water was less in quantity
during the dry season, and sometimes perhaps when the land
was slightly depressed. All these accidents might occur
again and again at the mouth of any river, and give rise
to alternations of fluviatile and marine strata, such as are
seen at Menchecourt.

In the lowest beds of gravel and sand in contact with the
chalk, flint hatchets, some perfect, others much rolled, have
been found; and in a sandy bed in this position some workmen, whom I employed to sink a pit, found four flint flakes.
Above this sand and gravel occur beds of white and siliceous
sand, containing shells of the genera Planorbis, Limnea,
Paludina, Valvata, Cyclas, Cyrena, Helix, and others, all now
natives of the same part of France, except Cyrena fluminalis
(fig. 24, p. 170), which no longer lives in Europe, but inhabits
the Nile, and many parts of Asia, including Cashmere, where
it abounds. No species of Cyrena is now met with in a living
state in Europe. Mr. Prestwich first observed it fossil at
Menchecourt, and it has since been found in two or three
contiguous sandpits, always in the fluvio-marine bed.

The following marine shells occur mixed with the fresh-water species above enumerated :—*Buccinum undatum, Littorina littorea, Nassa reticulata, Purpura lapillus, Tellina solidula, Cardium edule,* and fragments of some others. Several of these I have myself collected entire though much decomposed, lying in the white sand called ' sable nigre ' by the workmen. They are all littoral species now proper to the contiguous coast of France. Their occurrence in a fossil state associated with fresh-water shells at Menchecourt, had been noticed as long ago as 1836 by

Fig. 21.

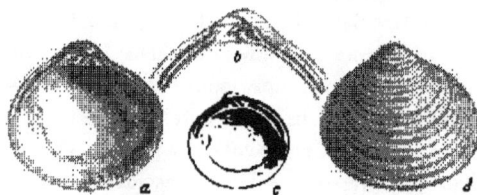

Cyrena (Corbicula) fluminalis, O. F. Müller, sp.*

a Interior of left valve, from Gray's Thurrock, Essex.
b Hinge of same magnified.
c Interior of right valve of a small specimen, from Shacklewell, London.
d Outer surface of right valve, from Erith, Kent.

MM. Ravin and Baillon, before M. Boucher de Perthes commenced the researches which have since made the locality so celebrated.† The numbers since collected and their manner of occurrence preclude all idea of their having been brought inland as eatable shells by the fabricators of the flint hatchets found at the bottom of the fluvio-marine sands. From the same beds, and in marls alternating with the sands, remains of the elephant, rhinoceros and other mammalia, have been exhumed.

* For synonyms, of which there are many, including five different generic names, see S. Woodward, Tibet Shells,

Proc. Zool. Soc., July 8, 1856.
† D'Archiac, Histoire des Progrès, &c., vol ii. p. 154.

Above the fluvio-marine strata are those designated No. 2 in the section (fig. 23, p. 168), which are almost devoid of stratification, and probably formed of mud or sediment thrown down by the waters of the river when they overflowed the ancient alluvial plain of that day. Some land shells, a few river shells, and bones of mammalia, some of them extinct, occur in No. 2. Its upper surface had been deeply furrowed and cut into by the action of water, at the time when the earthy matter of No. 1 was superimposed. The materials of this uppermost deposit are arranged as if they had been the result of land-floods, taking place after the formations 2 and 3 had been raised, or had become exposed to denudation.

The fluvio-marine strata and overlying loam of Menchecourt recur on the opposite or left bank of the alluvial plain of the Somme, at a distance of two or three miles. They are found at Muntort, among other places, and I obtained there the oval flint hatchet figured at p. 159 (fig. 16), of an oval form. It was extracted from gravel, above which were strata containing a mixture of marine and fresh-water shells, precisely like those of Menchecourt. In the alluvium of all parts of the valley, both at high and low levels, rolled bones are sometimes met with in the gravel. Some of the flint tools in the gravel of Abbeville have their angles very perfect, others have been much triturated, as if in the bed of the main river or some of its tributaries.

The mammalia most frequently cited as having been found in the deposits Nos. 2 and 3 at Menchecourt, are the following:—

Elephas primigenius.	*Cervus somonensis.*
Rhinoceros tichorhinus.	*C. Tarandus priscus.*
Equus fossilis.	*Felis spelæa.*
Bos primigenius.	*Hyæna spelæa.*
Bison priscus.	

The *Ursus spelæus* has also been mentioned by some

writers; but M. Lartet says he has sought in vain for it
among the osteological treasures sent from Abbeville to Cuvier
at Paris, and in other collections. The same palæontologist,
after a close scrutiny of the bones sent formerly to the Paris
Museum from the Valley of the Somme, observed that some
of them bore the evident marks of an instrument, agreeing
well with incisions such as a rude flint saw would produce.
Among other bones mentioned as having been thus artificially
cut, are those of a *Rhinoceros tichorhinus*, and the antlers
of *Cervus eomonensis.**

The evidence obtained by naturalists that some of the
extinct mammalia of Menchecourt really lived and died in
this part of France, at the time of the embedding of the flint
tools in fluviatile strata, is most satisfactory; and not the less
so for having been put on record long before any suspicion
was entertained that works of art would ever be detected
in the same beds. Thus M. Baillon, writing in 1834 to
M. Ravin, says, 'They begin to meet with fossil bones at
the depth of ten or twelve feet in the Menchecourt sandpits,
but they find a much greater quantity at the depth of eighteen and twenty feet. Some of them were evidently broken
before they were embedded, others are rounded, having, without doubt, been rolled by running water. It is at the bottom
of the sandpits that the most entire bones occur. Here they
lie without having undergone fracture or friction, and seem
to have been articulated together at the time when they
were covered up. I found in one place a whole hind limb
of a rhinoceros, the bones of which were still in their true
relative position. They must have been joined together by
ligaments, and even surrounded by muscles at the time of
their interment. The entire skeleton of the same species
was lying at a short distance from the spot.'†

* Quarterly Journal of the Geologi-
cal Society, London, vol. xvi. p. 471.   † Société Roy. d'Emulation d'Abbeville, 1834. p. 197.

If we suppose that the greater number of the flint implements occurring in the neighbourhood of Abbeville and Amiens were brought by river action into their present position, we can at once explain why so large a proportion of them are found at considerable depths from the surface, for they would naturally be buried in gravel and not in fine sediment, or what may be termed 'inundation mud,' such as No. 2 (fig. 23, p. 168), a deposit from tranquil water, or where the stream had not sufficient force or velocity to sweep along chalk flints, whether wrought or unwrought. Hence we have almost always to pass down through a mass of incumbent loam with land shells, or through fine sand with fresh-water molluscs, before we get into the beds of gravel containing hatchets. Occasionally a weapon used as a projectile may have fallen into quiet water, or may have dropped from a canoe to the bottom of the river, or may have been floated by ice, as are some stones occasionally by the Thames in severe winters, and carried over the meadows bordering its banks; but such cases are exceptional, though helping to explain how isolated flint tools or pebbles and angular stones are now and then to be seen in the midst of the finest loams.

The endless variety in the sections of the alluvium of the Valley of the Somme may be ascribed to the frequent silting up of the main stream and its tributaries during different stages of the excavation of the valley, probably also during changes in the level of the land. As a rule, when a river attacks and undermines one bank, it throws down gravel and sand on the opposite side of its channel, which is growing somewhere shallower, and is soon destined to be raised so high as to form an addition to the alluvial plain, and to be only occasionally inundated. In this way, after much encroachment on cliff or meadow at certain points, we find at the end of centuries that the width of the channel has not been enlarged, for the new-made ground is raised after a time to the average height of

the older alluvial tract. Sometimes an island is formed in
midstream, the current flowing for a while on both sides of
it, and at length scooping out a deeper channel on one side so
as to leave the other to be gradually filled up during freshets
and afterwards elevated by inundation mud or 'brick-earth.'
During the levelling up of these old channels, a flood sometimes cuts into and partially removes portions of the previously
stratified matter, causing those repeated signs of furrowing
and filling up of cavities, those memorials of doing and
undoing, of which the tool-bearing sands and gravels of
Abbeville and Amiens afford such reiterated illustrations, and
of which a parallel is furnished by the ancient alluvium of
the Thames valley, where similar bones of extinct mammalia,
and shells including *Cyrena fluminalis*, are found.

Professor Noeggerath, of Bonn, informs me that, about
the year 1845, when the bed of the Rhine was deepened artificially by the blasting and removal of rock in the narrows
at Bingerloch, not far from Bingen, several flint hatchets and
an extraordinary number of iron weapons of the Roman
period were brought up by the dredge from the bed of the
great river. The decomposition of the iron had caused much
of the gravel to be cemented together into a conglomerate.
In such a case we have only to suppose the Rhine to deviate
slightly from its course, changing its position, as it has often
done in various parts of its plain in historical times, and then,
when the old channel had been silted up, tools of the stone
and iron periods would be found in gravel at the bottom,
with a great thickness of sand and overlying loam deposited
in them.

Changes in a river plain, such as those above alluded to,
give rise frequently to ponds, swamps, and marshes, marking
the course of old beds or branches of the river not yet filled
up, and in these depressions shells proper both to running
and stagnant water may be preserved, and quadrupeds may

be mired.   The latest and uppermost deposit of the series
will be loam or brick-earth, with land and amphibious shells
(*Helix* and *Succinea*), while below will follow strata containing fresh-water shells, implying continuous submergence;
and lowest of all in most sections will be the coarse gravel
accumulated by a current of considerable strength and
velocity.

When the St. Katharine docks were excavated at London,
and similar works executed on the banks of the Mersey, old
ships were dug out, as I have elsewhere noticed,* showing
how the Thames and Mersey have in modern times been
shifting their channels.   Recently, an old silted-up bed of
the Thames has been discovered by boring at Shoeburyness
at the mouth of the river opposite Sheerness, as I learn from
Mr. Mylne.   The old deserted branch is separated from the
new or present channel of the Thames, by a mass of London
clay which has escaped denudation.   The depth of the old
branch, or the thickness of fluviatile strata with which it has
been filled up, is seventy-five feet.   The actual channel in
the neighbourhood is now sixty feet deep, but there is
probably ten or fifteen feet of stratified sand and gravel at
the bottom; so that, should the river deviate again from its
course, its present bed might be the receptacle of a fluvio-
marine formation seventy-five feet thick, equal to the former
one at Shoeburyness, and more considerable than that of
Abbeville.   It would consist both of fresh-water and marine
strata, as the salt water is carried by the tide far up above
Sheerness; but in order that such deposits should resemble,
in geological position, the Menchecourt beds, they must be
raised ten or fifteen feet above their present level, and be
partially eroded.   Such erosion they would not fail to suffer
during the process of upheaval, because the Thames would

* Principles of Geology.

scour out its bed, and not alter its position relatively to the
sea, while the land was gradually rising.

Before the canal was made at Abbeville, the tide was perceptible in the Somme for some distance above that city. It
would only require, therefore, a slight subsidence to allow
the salt water to reach Menchecourt, as it did in the Pleistocene period. As a stratum containing exclusively land and
fresh-water shells usually underlies the fluvio-marine sands at
Menchecourt, it seems that the river first prevailed there, after
which the land subsided; and then there was an upheaval
which raised the country to a greater height than that at
which it now stands, after which there was a second sinking,
indicated by the position of the peat, as already explained
(p. 157). All these changes happened since Man first inhabited this region.

At several places in the environs of Abbeville, there are
fluviatile deposits at a higher level by fifty feet than the
uppermost beds at Menchecourt, resting in like manner on
the chalk. One of these occurs in the suburbs of the city
at Moulin Quignon, one hundred feet above the Somme and
on the same side of the valley as Menchecourt, and containing flint implements of the same antique type and the bones
of elephants; but no marine shells have been found there,
nor in any gravel or sand at higher elevations than the
Menchecourt marine shells.

It has been a matter of discussion among geologists
whether the higher or the lower sands and gravels of the
Somme valley are the most ancient. As a general rule, when
there are alluvial formations of different ages in the same
valley, those which occupy a more elevated position above the
river plain are the oldest. In Auvergne and Velay, in Central
France, where the bones of fossil quadrupeds occur at all
heights above the present rivers from ten to one thousand
feet, we observe the terrestrial fauna to depart in character

from that now living in proportion as we ascend to higher
terraces and platforms. We pass from the lower alluvium,
containing the mammoth, tichorhine rhinoceros, and reindeer, to various older groups of fossils, till, on a table-land a
thousand feet high (near Le Puy, for example), the abrupt
termination of which overlooks the present valley, we discover
an old extinct river-bed covered by a current of ancient lava,
showing where the lowest level was once situated. In that
elevated alluvium the remains of a tertiary mastodon and
other quadrupeds of like antiquity are embedded.

If the Menchecourt beds had been first formed, and the
valley, after being nearly as deep and wide as it is now, had
subsided, the sea must have advanced inland, causing small
delta-like accumulations at successive heights, wherever the
main river and its tributaries met the sea. Such a movement,
especially if it were intermittent, and interrupted occasionally
by long pauses, would very well account for the accumulation
of stratified débris which we encounter at certain points in
the valley, especially around Abbeville and Amiens. But we
are precluded from adopting this theory by the entire absence
of marine shells, and the presence of fresh-water and land
species, and mammalian bones in considerable abundance in
the drift both of higher and lower levels above Abbeville.
Had there been a total absence of all organic remains, we
might have imagined the former presence of the sea, and the
destruction of such remains might have been ascribed to
carbonic acid or other decomposing causes ; but the Pleistocene and implement-bearing strata can be shown by their
fossils to be of fluviatile origin.

Flint Implements in Gravel near Amiens.
Gravel of St. Acheul.

When we ascend the valley of the Somme, from Abbeville,
to Amiens, a distance of about twenty-five miles, we observe

N

a repetition of all the same alluvial phenomena which we
have seen exhibited at Menchecourt and its neighbourhood,
with the single exception of the absence of marine shells and
of *Cyrena fluminalis.* We find lower-level gravel, such as
No. 12, fig. 14, p. 153, and higher-level alluvium, such as No.
3, the latter rising to one hundred feet above the plain, which
at Amiens is about fifty feet above the level of the river at
Abbeville. In both the upper and lower gravels, as Dr.
Rigollot stated in 1854, flint tools and the bones of extinct
animals, together with river shells and land shells of living
species, abound.

Immediately below Amiens, a great mass of stratified
gravel slightly elevated above the alluvial plain of the Somme,
is seen at St. Roch, and half a mile further down the valley
at Moutiers. Between these two places, a small tributary
stream, called the Celle, joins the Somme. In the gravel at
Moutiers, Mr. Prestwich and I found some flint knives, one
of them flat on one side, but the other carefully worked, and
exhibiting many fractures, clearly produced by blows skilfully applied. Some of these knives were taken from so low
a level as to satisfy us that this great bed of gravel at Moutiers, as well as that of the contiguous quarries of St. Roch,
which I regard as a continuation of the same deposit, may
be referred to the Human period. Dr. Rigollot had already
mentioned flint hatchets as obtained by him from St. Roch,
but as none have been found there of late years, his statement was thought to require confirmation. The discovery,
therefore, of these flint knives in gravel of the same age was
interesting, especially as many tusks of a hippopotamus have
been obtained from the gravel of St. Roch—some of these
recently by Mr. Prestwich; while M. Garnier, of Amiens, has
procured a fine elephant's molar from the same pits, which
Dr. Falconer refers to *Elephas antiquus,* see fig. 26, p. 179.
Hence I infer that both these animals coexisted with Man.

The alluvial formations of Moutiers are very instructive
in another point of view. If, leaving the lower gravel of that
place, which is topped with loam or brick-earth (of which
the upper portion is about thirty feet above the level of the
Somme), we ascend the chalky slope to the height of about

Fig. 25.*

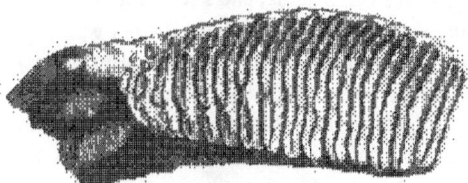

Elephas primigenius.
Penultimate molar, lower jaw, right side, one third of natural size. Pleistocene.
Coexisted with Man.

Fig. 26.

Elephas antiquus Falconer,
Penultimate molar, lower jaw, right side, size one-third of nature, Pleistocene
and Newer Pliocene. Coexisted with Man.

eighty feet, another deposit of gravel and sand, with fluviatile
shells in a perfect condition, occurs, indicating most clearly
an ancient river-bed, the waters of which ran habitually at
that higher level before the valley had been scooped out to
its present depth. This superior deposit is on the same side

* Fig. 25 will be found in M. Lartet's paper in Bulletin de la Société
Géologique de France, Mars 1859. Fig.
26 is from Fauna Sivalensis, Falconer
and Cautley. For a molar of *Elephas
Meridionalis*, see chap. xii.

B 3

of the Somme, and about as high, as the lowest part of the
celebrated formation of St. Acheul, two or three miles distant,
to which I shall now allude.

The terrace of St. Acheul may be described as a gently
sloping ledge of chalk, covered with gravel, topped as usual
with loam or fine sediment, the surface of the loam being
100 feet above the Somme, and about 150 above the sea.

Many stone coffins of the Gallo-Roman period have been
dug out of the upper portion of this alluvial mass. The
trenches made for burying them sometimes penetrate to the
depth of eight or nine feet from the surface, entering the
upper part of No. 3 of the sections figs. 27 and 28. They
prove that when the Romans were in Gaul they found this
terrace in the same condition as it is now, or rather as it
was before the removal of so much gravel, sand, clay, and
loam, for repairing roads, and for making bricks and pottery.

In the annexed section, which I observed during my last
visit in 1860, it will be seen that a fragment of an elephant's
tooth is noticed as having been dug out of unstratified sandy
loam at the point a, eleven feet from the surface. This was
found at the time of my visit; and at a lower point, at b,
eighteen feet from the surface, a large nearly entire and unrolled molar of the same species was obtained, which is now
in my possession. It has been pronounced by Dr. Falconer
to belong to *Elephas primigenius.*

A stone hatchet of an oval form, like that represented at
fig. 16, p. 159, was discovered at the same time, about one
foot lower down, at c, in densely compressed gravel. In the
chalky sand, sometimes occurring in interstices between the
separate fragments of flint constituting the coarse gravel No.
4, entire as well as broken fresh-water shells are often met
with. To some it may appear enigmatical how such fragile
objects could have escaped annihilation in a river-bed, when
flint tools and much gravel were shoved along the bottom;

but I have seen the dredging instrument employed in the
Thames, above and below London Bridge, to deepen the river,
and worked by steam power, scoop up gravel and sand from

Fig. 27.

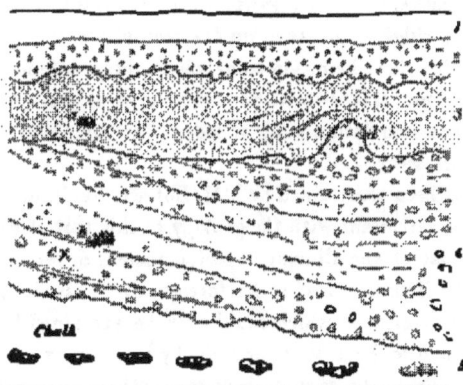

Section of a gravel-pit containing flint implements at St. Acheul, near
Amiens, observed in July 1860.

1 Vegetable soil and made ground—two to three feet thick.
2 Brown loam with some angular flints, in parts passing into ochreous
  gravel, filling up indentations on the surface of No. 3.—three
  feet thick.
3 White silicoous sand with layers of chalky marl, and included
  fragments of chalk, for the most part unstratified—nine feet.
4 Flint-gravel, and whitish chalky sand, flints sub-angular, average
  size of fragments three inches diameter, but with some large
  unbroken chalk flints intermixed, cross stratification in parts.
  Bones of mammalia, grinder of elephant at *b*, and flint implement
  at *c*—ten to fourteen feet.
5 Chalk with flints.

  a Part of elephant's molar, eleven feet from the surface.
  b Entire molar of *E. primigenius*, seventeen feet from surface.
  c Position of flint hatchet, eighteen feet from surface.
  d Projecting ridge of the gravel, No. 4.

the bottom, and then pour the contents pell-mell into the
boat, and still many specimens of Limnea, Planorbis, Paludina, Cyclas, and other shells might be taken out uninjured
from the gravel.

It will be observed that the gravel No. 4 is obliquely stratified, and that its surface had undergone denudation before the white sandy loam, No. 3, was superimposed. The materials of the gravel at *d* must have been cemented or frozen together into a somewhat coherent mass to allow the projecting ridge, *d*, to stand up five feet above the general surface, the sides being in some places perpendicular. In No. 3 we probably behold an example of a passage from river-silt to inundation mud, or loess. In some parts of it' land shells occur.

It has been ascertained by MM. Buteux, Ravin, and other observers conversant with the geology of this part of France, that in none of the alluvial deposits, ancient or modern, are there any fragments of rocks foreign to the basin of the Somme—no erratics which could only be explained by supposing them to have been brought by ice, during a general submergence of the country, from some other hydrographical basin.

But in some of the pits at St. Acheul there are seen in the beds No. 4, fig. 27, not only well-rounded tertiary pebbles, but great blocks of hard sandstone, of the kind called in the south of England 'greywethers,' some of which are three or four feet and upwards in diameter. They are usually angular, and when spherical owe their shape generally to an original concretionary structure, and not to trituration in a river's bed. These large fragments of stone abound both in the higher and lower level gravels round Amiens and at the higher level at Abbeville. They have also been traced far up the valley above Amiens, wherever patches of the old alluvium occur. They have all been derived from the tertiary strata which once covered the chalk. Their dimensions are such that it is impossible to imagine a river like the present Somme, flowing through a flat country, with a gentle fall towards the sea, to have carried

them for miles down its channel, unless ice co-operated as
a transporting power.   Their angularity also favours the
supposition of their having been floated by ice, or rendered
so buoyant by it as to have escaped much of the wear and
tear which blocks propelled along the bottom of a river
channel would otherwise suffer.   We must remember that the
present mildness of the winters in Picardy and the north-
west of Europe generally does not extend over the whole of

Fig. 26.

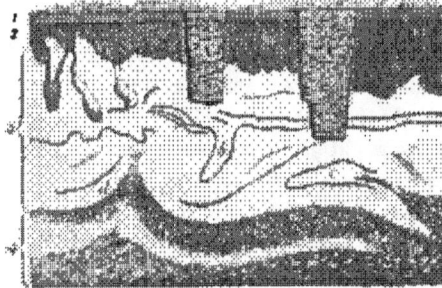

Contorted fluviatile strata at St. Acheul (Prestwich, Phil. Trans. 1861, p. 299).

1 Surface soil.
2 Brown loam, as in fig. 27, p. 181—thickness, six feet.
3 White sand with bent and folded layers of marl—thickness, six feet.
4 Gravel, as in fig. 27, p. 181, with bones of mammalia and flint im-
    plements.
a Gravel filled with made ground and human bones.
b and c Seams of laminated marl often bent round upon themselves.
d Beds of gravel with sharp curves.

the northern hemisphere, and that large fragments of granite,
sandstone and limestone are now carried annually by ice
down the Canadian rivers in latitudes further south than
Paris.*

Another sign of ice agency, of which Mr. Prestwich has
given a good illustration in one of his published sections, and
which I myself observed in several pits at St. Acheul, deserves

* Principles of Geology, 11th ed. p. 360.

notice. It consists in flexures and contortions of the strata
of sand, marl, and gravel (as seen at *b, c,* and *d,* in beds 3
and 4, fig. 28), which they have evidently undergone since
their original deposition, and from which both the underlying
chalk and part of the overlying beds of sand are usually
exempt.

In my former writings I have attributed this kind of
derangement to two causes : first, the pressure of ice running
aground on yielding banks of mud and sand ; and, secondly,
the melting of masses of ice and snow of unequal thickness,
on which horizontal layers of mud, sand, and other fine and
coarse materials had accumulated. The late Mr. Trimmer
first pointed out in what manner the unequal failure of support caused by the liquefaction of underlying or intercalated
snow and ice might give rise to such complicated foldings.*

When 'ice-jams' occur on the St. Lawrence and other
Canadian rivers (lat. 46° N.), the sheets of ice, which become
packed or forced under or over one another, assume in most
cases a highly inclined and sometimes even a vertical position.
They are often observed to be coated on one side with mud,
sand, or gravel frozen on to them, derived from shallows in
the river on which they rested when congelation first reached
the bottom.

As often as portions of these packs melt near the margin
of the river, the layers of mud, sand, and gravel, which result
from their liquefaction, cannot fail to assume a very abnormal
arrangement,—very perplexing to a geologist who should
undertake to interpret them without having the ice-clue in
his mind.

Mr. Prestwich has suggested that ground-ice may have
had its influence in modifying the ancient alluvium of the
Somme.† It is certain that ice in this form plays an active

* See Chapter XII.
† Prestwich, Memoir read to Royal Society, April 1862.

part every winter in giving motion to stones and gravel in
the beds of rivers in European Russia and Siberia. It appears
that when in those countries the streams are reduced nearly
to the freezing point, congelation begins frequently at the
bottom ; the reason being, according to Arago, that the current
is slowest there, and the gravel and large stones, having parted
with much of their heat by radiation, acquire a temperature
below the average of the main body of the river. It is,
therefore, when the water is clear, and the sky free from
clouds, that ground-ice forms most readily, and oftener on
pebbly than on muddy bottoms. Fragments of such ice,
rising occasionally to the surface, bring up with them gravel,
and even large stones.

Without dwelling longer on the various ways in which
ice may affect the forms of stratification in drift, so as to
cause bendings and foldings in which the underlying or overlying strata do not participate, a subject to which I shall have
occasion again to allude in the sequel, I will state in this
place that such contortions, whether explicable or not, are
very characteristic of glacial formations. They have also no
necessary connection with the transportation of large blocks
of stone, and they therefore afford, as Mr. Prestwich remarks,
independent proof of ice-action in the Pleistocene gravel of
the Somme.

Let us, then, suppose that, at the time when flint hatchets
were embedded in great numbers in the ancient gravel which
now forms the terrace of St. Acheul, the main river and its
tributaries were annually frozen over for several months in
winter. In that case, the primitive people may, as Mr.
Prestwich hints, have resembled in their mode of life those
American Indians who now inhabit the country between
Hudson's Bay and the Polar Sea. The habits of those Indians
have been well described by Hearne, who spent some years
among them. As often as deer and other game become

scarce on the land, they betake themselves to fishing in the
rivers; and for this purpose, and also to obtain water for
drinking, they are in the constant practice of cutting round
holes in the ice, a foot or more in diameter, through which
they throw baited hooks or nets. Often they pitch their tent
on the ice, and then cut such holes through it, using ice-
chisels of metal when they can get copper or iron, but when
not, employing tools of flint or hornstone.

The great accumulation of gravel at St. Acheul has taken
place in part of the valley where the tributary streams,
the Noye and the Arve, now join the Somme. These tributaries, as well as the main river, must have been running at
the height first of a hundred feet, and afterwards at various
lower levels above the present valley-plain, in those earlier
times when the flint tools of the antique type were buried
in successive river-beds. I have said at various levels, because there are, here and there, patches of drift at heights
intermediate between the higher and lower gravel, and also
some deposits, showing that the river once flowed at elevations
above as well as below the level of the platform of St. Acheul.
As yet, however, no patch of gravel skirting the valley at
heights exceeding one hundred feet above the Somme have
yielded flint tools or other signs of the former sojourn of
Man in this region.

Possibly, in the earlier geographical condition of this
country, the confluence of tributaries with the Somme afforded
inducements to a hunting and fishing tribe to settle there,
and some of the same natural advantages may have caused
the first inhabitants of Amiens and Abbeville to fix on the
same sites for their dwellings. If the early hunting and
fishing tribes frequented the same spots for hundreds or
thousands of years in succession, the number of the stone
implements lost in the bed of the river need not surprise us.
Ice-chisels, flint hatchets, and spear-heads may have slipped

accidentally through holes kept constantly open, and the recovery of a lost treasure once sunk in the bed of the ice-bound stream, inevitably swept away with gravel on the breaking up of the ice in the spring, would be hopeless. During a long winter, in a country affording abundance of flint, the manufacture of tools would be continually in progress; and, if so, thousands of chips and flakes would be purposely thrown into the ice-hole, besides a great number of implements having flaws, or rejected as too unskilfully made to be worth preserving.

As to the fossil fauna of the drift, considered in relation to the climate, when, in 1859, I took a collection which I had made of all the more common species of land and fresh-water shells from the Amiens and Abbeville drift, to my friend M. Deshayes at Paris, he declared them to be, without exception, the same as those now living in the basin of the Seine. This fact may seem at first sight to imply that the climate had not altered since the flint tools were fabricated; but it appears that all these species of molluscs now range as far north as Norway and Finland, and may therefore have flourished in the valley of the Somme when the river was frozen over annually in winter.[*]

In regard to the accompanying mammalia, some of them, like the mammoth and tichorhine rhinoceros, may have been able to endure the rigours of a northern winter as well as the reindeer, which we find fossil in the same gravel. It is a more difficult point to determine whether the climate of the lower gravels (those of Menchecourt, for example) was more genial than that of the higher ones. Mr. Prestwich inclines to this opinion. None of those contortions of the strata above described (p. 183) have as yet been observed in the lower drift. It contains large blocks of tertiary sandstone and grit, which may have required the aid of ice to convey

* See Prestwich, Paper read to Royal Society in 1862.

them to their present sites; but as such blocks already
abounded in the older and higher alluvium, they may simply
be monuments of its destruction, having been let down successively to lower and lower levels without making much
seaward progress.

The *Cyrena fluminalis* of Menchecourt and the hippopotamus of St. Roch seem to be in favour of a less severe
temperature in winter; but so many of the species of
mammalia, as well as of the land and fresh-water shells, are
common to both formations, and our information respecting
the entire fauna is still so imperfect, that it would be prema-
ture to pretend to settle this question in the present state of
our knowledge. We must be content with the conclusion
(and it is one of no small interest), that when Man first
inhabited this part of Europe, at the time that the St. Acheul
drift was formed, the climate as well as the physical geography
of the country differed considerably from the state of things
now established there.

Among the elephant remains from St. Acheul, in M.
Garnier's collection, Dr. Falconer recognised a molar of the
Elephas antiquus, fig. 26, the same species which has been
already mentioned as having been found in the lower-level
gravels of St. Roch. This species, therefore, endured while
important changes took place in the geographical condition
of the Valley of the Somme. Assuming the lower-level
gravel to be the newer, it follows that the *Elephas antiquus*
and the hippopotamus of St. Roch continued to flourish long
after the introduction of the mammoth, a well characterised
tooth of which, as I have before stated, was found at St.
Acheul at the time of my visit in 1860.

As flint hatchets and knives have been discovered in the
alluvial deposits both at high and low levels, we may safely
affirm that Man inhabited this region together with all
the fossil quadrupeds above enumerated, a conclusion which

is independent of any difference of opinion as to the relative age of the higher and lower gravels.

The disappearance of many large pachyderms and beasts of prey from Europe has often been attributed to the intervention of Man, and no doubt he played his part in hastening the era of their extinction; but there is good reason for suspecting that other causes co-operated to the same end. No naturalist would for a moment suppose that the extermination of the *Cyrena fluminalis* throughout the whole of Europe— a species which coexisted with our race in the valley of the Somme, and which was very abundant in the waters of the Thames at the time when the elephant, rhinoceros, and hippopotamus flourished on its banks—was accelerated by human agency. The great modification in climate and in other conditions of existence which affected this aquatic mollusc, may have mainly contributed to the gradual dying out of many of the large mammalia.

We have already seen that the peat of the Valley of the Somme is a formation which, in all likelihood, took a very long time for its growth. Yet no change of a marked character has occurred in the mammalian fauna since it began to accumulate. The fauna of the ancient alluvium differs almost as much from the fauna of the oldest peat as from the existing fauna, the memorials of Man being common to the whole series, hence we may infer that the interval of time which separated the era of the large extinct mammalia from that of the earliest peat, was of far longer duration than that of the entire growth of the peat. Yet we by no means need the evidence of the ancient fossil fauna to establish the antiquity of Man in this part of France. The various heights at which the drift occurs would alone suffice to demonstrate a vast lapse of time during which heaps of shingle, derived both from the Eocene and the cretaceous rocks, were thrown down in a succession of river channels. We observe thousands of rounded and half-rounded flints, and a vast number of angular ones, with rounded pieces of white chalk of various sizes, testifying to a prodigious amount of mechanical action, accompanying the repeated widening and deepening of the valley, before it became the receptacle of peat ; and the position of many of the flint tools leaves no doubt on the mind of the geologist that their fabrication preceded at least the greater part of this reiterated denudation.

On the absence of Human Bones in the Alluvium of the Somme.

It is naturally a matter of no small surprise that, after we have collected many hundred flint implements (including knives, many thousands), not a single human bone has yet been met with in the old alluvial sand and gravel of the Somme, for after much discussion and investigation we must, I think, give up all claims to authenticity for the celebrated jaw of Moulin Quignon. Yet in these same formations there is no want of bones of mammalia belonging to extinct and living species. In the course of the last quarter of a century, thousands of them have been submitted to the examination of skilful osteologists, and they have been unable to detect among them one fragment of a human skeleton, not even a tooth. Yet Cuvier pointed out long ago, that the bones of Man found buried in ancient battle-fields were not more decayed than those of horses interred in the same graves. We have seen that in the Liége caverns, the skulls jaws, and teeth, with other bones of the human race, were preserved in the same condition as those of the cave-bear, tiger, and mammoth.

In a former edition of this work (1863) I expressed my expectation that ere long, now that curiosity had been so much excited on the subject, human remains would be de-

tected in the older alluvium of European valleys, and this
prediction has been to a certain extent justified by the discovery, in 1868, of portions of the human skeleton by M.
Bertrand and M. Reboux in the valley of the Seine at Clichy,
in the suburbs of Paris, in the same beds in which implements of true Palæolithic types were present.* The remains
were found at a depth of about seventeen feet, in a bed of
red sand underlying grey drift (*diluvium gris*) which was
about three and a half feet thick, and covered in its turn by
yellow sands, red alluvium and vegetable earth, in all about
fourteen feet thick. The skull, supposed to be that of a
female, exhibited very striking characters of inferiority,
such as the enormous thickness of the frontal bone, and the
low shape of the skull, which was narrow and slanting from
the front to the back.† The absence, with this single exception, of all vestige of the bones which belonged to that
population by which so many weapons were designed and
executed, affords a most striking and instructive lesson in
regard to the value of negative evidence, when adduced in
proof of the non-existence of certain classes of terrestrial
animals at given periods of the past. It is a new and
emphatic illustration of the extreme imperfection of the
geological record, of which even they who are constantly
working in the field cannot easily bring themselves to form
a just conception.

We must not forget that Dr. Schmerling, after finding
extinct mammalia and *flint tools* in forty-two Belgian
caverns, was only rewarded by the discovery of human
bones in three or four of those rich repositories of osseous
remains. In like manner, it was not till the year 1855
that the first skull of the musk buffalo (*Bubalus moschatus*

---

* Evans's Stone Implements, p. 617.
† Hamy, Paléontologie Humaine, 1870, p. 210.

Owen)* was detected in the fossiliferous gravel of the
Thames; and not till 1860, as will be seen in the next
chapter, that the same quadruped was proved to have coexisted
in France with the mammoth. The same theory which will
explain the comparative rarity of such species would, no
doubt, account for the still greater scarcity of human bones,
as well as for our general ignorance of the Pleistocene terrestrial fauna, with the exception of that part of it which
is revealed to us by cavern researches.

In valley drift we meet commonly with the bones of
quadrupeds which graze on plains bordering rivers. Carnivorous beasts, attracted to the same ground in search of
their prey, sometimes leave their remains in the same
deposits, but more rarely. The whole assemblage of fossil
quadrupeds at present obtained from the alluvium of
Picardy is obviously a mere fraction of the entire fauna
which flourished contemporaneously with the primitive
people by whom the flint hatchets were made.

Instead of its being part of the plan of nature to store
up enduring records of a large number of the individual
plants and animals which have lived on the surface, it seems
to be her chief care to provide the means of disencumbering
the habitable areas lying above and below the waters of
those myriads of solid skeletons of animals, and those
massive trunks of trees, which would otherwise soon choke
up every river, and fill every valley. To prevent this
inconvenience, she employs the heat and moisture of the
sun and atmosphere, the dissolving power of carbonic and
other acids, the grinding teeth and gastric juices of quadrupeds, birds, reptiles, and fish, and the agency of many of
the invertebrata. We are all familiar with the efficacy of
these and other causes on the land; and as to the bottoms of

* Musk sheep (*Ovibos moschatus*) of M. de Blainville.

seas, we have only to read the published reports of Mr.
MacAndrew, the late Edward Forbes, and other experienced
dredgers, who, while they failed utterly in drawing up from
the deep a single human bone, declared that they scarcely
ever met with a work of art, even after counting tens of
thousands of shells and zoophytes, collected on a coast line
of several hundred miles in extent, where they often
approached within less than half a mile of a land peopled
by millions of human beings.

Lake of Haarlem.

It is not many years since the Government of Holland
resolved to lay dry that great sheet of water formerly called
the Lake of Haarlem, extending over 45,000 acres. They
succeeded, in 1853, in turning it into dry land, by means
of powerful pumps constantly worked by steam, which raised
the water and discharged it into a canal running for twenty
or thirty miles round the newly-gained land. This land was
depressed thirteen feet beneath the mean level of the ocean.
I travelled, in 1859, over part of the bed of this old lake,
and found it already converted into arable land, and peopled
by an agricultural population of 5,000 souls. Mr. Staring,
who had been for some years employed by the Dutch Government in constructing a geological map of Holland, was my
companion and guide. He informed me that he and his
associates had searched in vain for human bones in the deposits which had constituted for three centuries the bed of
the great lake.

There had been many a shipwreck, and many a naval fight,
in those waters, and hundreds of Dutch and Spanish soldiers
and sailors had met there with a watery grave. The population which lived on the borders of this ancient sheet of water
numbered between thirty and forty thousand souls.   In

o

digging the great canal, a fine section had been laid open, about thirty miles long, of the deposits which formed the ancient bottom of the lake. Trenches, also, innumerable, several feet deep, had been freshly dug on all the farms, and their united length must have amounted to thousands of miles. In some of the sandy soil recently thrown out of the trenches, I observed specimens of fresh and brackish water shells, such as Unio and Dreissena, of living species; and in clay brought up from below the sand, shells of Tellina, Lutraria, and Cardium, all of species now inhabiting the adjoining sea.

As the Dreissena is believed by modern conchologists to have been introduced into Western Europe in very modern times, brought with foreign timber in the holds of vessels from the Wolga and other rivers flowing into the Black Sea, the layer of sand, containing it in the Haarlem lake is probably not more than a hundred years old.

One or two wrecked Spanish vessels, and arms of the same period, have rewarded the antiquaries who had been watching the draining operations in the hope of a richer harvest, and who were not a little disappointed at the result. In a peaty tract on the margin of one part of the lake a few coins were dug up; but if history had been silent, and if there had been a controversy whether Man was already a denizen of this planet at the time when the area of the Haarlem lake was under water, the archæologist, in order to answer this question, must have appealed, as in the case of the valley of the Somme, not to fossil bones, but to works of art embedded in the superficial strata.

Mr. Staring, in his valuable memoir on the ' Geological Map of Holland,' has attributed the general scarcity of human bones in Dutch peat, notwithstanding the many works of art preserved in it, to the power of the humic and sulphuric acids to dissolve bones, the peat in question being plentifully impregnated with such acids. His theory may be correct, but it is not applicable to the gravel of the Valley of the Somme, in which the bones of fossil mammalia are frequent, nor to the uppermost fresh-water strata forming the bottom of a large part of the Haarlem lake, in which it is not pretended that such acids occur.

The primitive inhabitants of the Valley of the Somme may have been too wary and sagacious to be often surprised and drowned by floods, which swept away many an incautious elephant and rhinoceros, horse and ox. But even if those rude hunters had cherished a superstitious veneration for the Somme, and had regarded it as a sacred river (as the modern Hindoos revere the Ganges), and had been in the habit of committing the bodies of their dead or dying to its waters—even had such funeral rites prevailed, it by no means follows that the bones of many individuals would have been preserved to our time.

A corpse cast into the stream first sinks, and unless it be almost immediately overspread with sediment of a certain weight, it will rise again when distended with gases, and float perhaps to the sea before it finally sinks. It may then be attacked by fish of marine species, some of which are capable of digesting bones. If, before being carried into the sea and devoured, it is enveloped with fluviatile mud and sand, the next flood, if it lie in mid channel, may tear it out again, scatter all the bones, roll some of them into pebbles, and leave others exposed to destroying agencies; and this may be repeated annually, till all restiges of the skeleton disappear. On the other hand, a bone washed through a rent into a subterranean cavity, even through a rarer contingency, may have a greater chance of escaping destruction, especially if there be stalactite dropping from the roof of the cave or walls of a rent, and if the cave be not constantly traversed by too strong a current of engulfed water.

CHAPTER X.

WORKS OF ART IN PLEISTOCENE ALLUVIUM OF FRANCE AND ENGLAND.

FLINT IMPLEMENTS IN ANCIENT ALLUVIUM OF THE BASIN OF THE SEINE—BONES OF MAN AND OF EXTINCT MAMMALIA IN THE CAVE OF ARCY—EXTINCT MAMMALIA IN THE VALLEY OF THE OISE—FLINT IMPLEMENT IN GRAVEL OF SAME VALLEY—WORKS OF ART IN PLEISTOCENE DRIFT IN VALLEY OF THE THAMES—MUSK-BUFFALO—MEETING OF NORTHERN AND SOUTHERN FAUNA—MIGRATIONS OF QUADRUPEDS —MAMMALS OF AMOORLAND—MIGRATIONS OF THE HIPPOPOTAMUS— CHRONOLOGICAL RELATION OF THE OLDER ALLUVIUM OF THE THAMES TO THE GLACIAL DRIFT—FLINT IMPLEMENTS OF PLEISTOCENE PERIOD IN SURREY, MIDDLESEX, KENT, BEDFORDSHIRE, AND SUFFOLK—FLINT IMPLEMENTS IN THE DRIFT OF THE SOUTH OF HAMPSHIRE.

Flint Implements in Pleistocene Alluvium in the Basin of the Seine.

IN the ancient alluvium of the valleys of the Seine and its principal tributaries, the same assemblage of fossil animals which has been alluded to in the last chapter as characterising the gravel of Picardy, has long been known: but it was not till the year 1860, and when diligent search had been expressly made for them, that flint implements of the Amiens type were discovered in this part of France.

In the neighbourhood of Paris, deposits of drift occur answering both to those of the higher and lower levels of the basin of the Somme before described.* In both are found, mingled with the wreck of the tertiary and cretaceous rocks of the vicinity, a large quantity of granitic sand, and pebbles, and occasionally large blocks of granite, from a few inches

* Prestwich, Proceedings of Roy. Soc. 1862.

to a foot or more in diameter. These blocks are peculiarly
abundant in the lower drift commonly called the 'diluvium
gris.' The granitic materials are traceable to a chain of hills
called the Morvan, where the head waters of the Yonne take
their rise, 150 miles SSE. of Paris.

It was in the lowest gravel that M. H. T. Gosse, of
Geneva, found, in April, 1860, in the suburbs of Paris, at La
Motte Piquet, on the left bank of the Seine, one or two well-
formed flint implements of the Amiens type, accompanied
by a great number of ruder tools or attempts at tools. I
visited the spot in 1861 with M. Hébert, and saw the stratum
from which the worked flints had been extracted, twenty feet
below the surface, and near the bottom of the 'grey diluvium,' a bed of gravel from which I have myself, in and
near Paris, frequently collected the bones of the elephant,
horse, and other mammalia.

In 1862, M. Lartet discovered at Clichy, in the environs
of Paris, in the same lower gravel, a well-shaped flint implement of the Amiens type, together with remains both of
Elephas primigenius and *E. antiquus*; and I have already
mentioned (p. 191) that human remains have since been
found in these beds. No tools have yet been met with in
any of the gravel occurring at the higher levels of the valley
of the Seine; but no importance can be attached to this
negative fact, as so little search has yet been made for them.

Mr. Prestwich has observed contortions indicative of ice-
action, of the same kind as those near Amiens (see p. 183),
in the higher level drift at Charonne, near Paris; but as yet
no similar derangement has been seen in the lower gravels—
a fact, so far as it goes, in unison with the phenomena observed in Picardy.

In the cavern of Arcy-sur-Yonne a series of deposits have
lately been investigated by the Marquis de Vibraye, who
discovered human bones in the lowest of them, mixed with

remains of quadrupeds of extinct and recent species. This cavern occurs in Jurassic limestone, at a slight elevation above the Cure, a small tributary of the Yonne, which last joins the Seine near Fontainebleau, about forty miles south of Paris. The lowest formation in the cavern resembles the ' diluvium gris ' of Paris, being composed of granitic materials, and like it derived chiefly from the waste of the crystalline rocks of the Morvan. In it have been found the two branches of a human lower jaw with teeth well-preserved, and the bones of the *Elephas primigenius, Rhinoceros tichorhinus, Ursus spelæus, Hyæna spelæa,* and *Cervus Tarandus,* all specifically determined by M. Lartet. I have been shown this collection of fossils by M. de Vibraye, and remarked that the human and other remains were in the same condition and of the same colour.

Above the grey gravel is a bed of red alluvium, made up of fragments of Jura limestone, in a red argillaceous matrix, in which were embedded several flint knives, with bones of the reindeer and horse, but no extinct mammalia. Over this, in a higher bed of alluvium, were several polished hatchets of the more modern type called 'celts,' and above all, loam or cave-mud, in which were Gallo-Roman antiquities.*

The French geologists have made as yet too little progress in identifying the age of the successive deposits of ancient alluvium of various parts of the basin of the Seine, to enable us to speculate with confidence as to the coincidence in date of the granitic gravel with human bones of the Grotte d'Arcy and the stone hatchets buried in 'grey diluvium' of La Motte Piquet, before mentioned ; but as the associated extinct mammalia are of the same species in both localities, I feel strongly inclined to believe that the stone hatchets found by M. Gosse at Paris, and the human bones discovered by M. de Vibraye, may be referable to the same period.

* Bulletin de la Société Géologique de France, 1860.

Valley of the Oise.

A flint hatchet, of the old Abbeville and Amiens type, was found lately by M. Peigné Delacourt at Précy, near Creil, on the Oise, in gravel resembling, in its geological position, the lower level gravels of Moutiers, near Amiens, already described (p. 178). I visited these extensive gravel-pits in 1861, in company with Mr. Prestwich; but we remained there too short a time to entitle us to expect to find a flint implement, even if they had been as abundant as at St. Acheul.

In 1859, I examined, in a higher part of the same valley of the Oise, near Chauny and Noyon, some fine railway cuttings, which passed continuously through alluvium of the Pleistocene period for half a mile. All this alluvium was evidently of fluviatile origin, for, in the interstices between the pebbles, the *Ancylus fluviatilis* and other fresh-water shells were abundant. My companion, the Abbé E. Lambert, had collected from the gravel a great many fossil bones, among which M. Lartet has recognised both *Elephas primigenius* and *E. antiquus*, besides a species of hippopotamus (*H. major?*), also the reindeer, horse, and the musk-sheep (*Ovibos moschatus*). The latter seems never to have been seen before in the old alluvium of France.[*] Over the gravel above mentioned, near Chauny, are seen dense masses of loam, like the loess of the Rhine, containing shells of the genera Helix and Succinea. We may suppose that the gravel containing the flint hatchet at Précy is of the same age as that of Chauny, with which it is continuous, and that both of them are coeval with the tool-bearing beds of Amiens, for the basins of the Oise and the Somme are only separated by a narrow watershed, and the same fossil quadrupeds occur in both.

[*] Lartet, Annales des Sciences Naturelles, tom. xv. p. 224.

The alluvium of the Seine and its tributaries, like that of the Somme, contains no fragments of rocks brought from any other hydrographical basin ; yet the shape of the land, or fall of the river, or the climate, or all these conditions, must have been very different when the grey alluvium in which the flint tools occur at Paris was formed. The great size of some of the blocks of granite, and the distance which they have travelled, imply a power in the river which it no longer possesses. We can hardly doubt that river-ice once played a much more active part than now in the transportation of such blocks, one of which may be seen in the Museum of the École des Mines at Paris, three or four feet in diameter.

Vallée de l'Infernet, Pyrenees.

M. Noullet found in the gravel of the Vallée de l'Infernet, near Clermont, roughly-shaped instruments of quartz and quartzite, resembling in form and size those from St. Acheul. Associated with them were the bones of *E. primigenius*, *Cervus megaceros*, *C. Tarandus*, *Felis spelæa*, &c., which he showed to myself and Mr. Mc K. Hughes when we visited the Pyrenees in 1872.

Thus we have in the basin of the Garonne, close under the Pyrenees, terraces of river gravel of considerable antiquity, as shown by their height above the river, containing the same group of animals as are found in the older gravels of the Somme, and similar stone weapons, but made of the rocks of the Pyrenees instead of the flint of Picardy.

Pleistocene Alluvium of England, containing Works of Art.

In the ancient alluvium of the basin of the Thames, at moderate heights above the main river, and its tributaries, we find fossil bones of the same species of extinct and living

mammalia, accompanied by recent species of land and
fresh-water shells, as we have shown to be characteristic of
the basins of the Somme and the Seine. We can scarcely
therefore doubt that these quadrupeds, during some part of
the Pleistocene period, ranged freely from the continent
of Europe to England, at a time when there was an uninterrupted communication by land between the two
countries. The reader will not therefore be surprised to
learn that flint implements of the same antique type as
those of the valley of the Somme have been detected in
British alluvium.

The most marked feature of this alluvium in the Thames
valley is that great bed of ochreous gravel, composed chiefly
of broken and slightly worn chalk flints, on which a great
part of London is built. It extends from above Maidenhead through the metropolis to the sea, a distance from west
to east of fifty miles, having a width varying from two to
nine miles. Its thickness ranges commonly from five to
fifteen feet.* Interstratified with this gravel, in many
places, are beds of sand, loam, and clay, the whole containing occasionally remains of the mammoth and other extinct
quadrupeds. Fine sections have been exposed to view, at
different periods, at Brentford and Kew bridge, others in
London itself, and below it at Erith in Kent, on the right
bank of the Thames, and at Ilford and Gray's Thurrock in
Essex, on the left bank. The united thickness of the beds
of sand, gravel, and loam amounts sometimes to forty or
even sixty feet. They are for the most part elevated above,
but in some cases they descend below, the present level of
the overflowed plain of the Thames. At the bottom of a
bed of gravel at Turnham Green (called by him a mid-
level gravel) Colonel Lane Fox found a large quantity of
animal remains, among which Mr. Busk recognised the

* Prestwich, Geological Quarterly Journal, vol. xii. p. 131.

bones of *Hippopotamus major*, one of them the left frontal
of a very young animal ; but no implements have yet been
found in this particular bed.*

If the reader will refer to the section of the Pleistocene
sands and gravels of Menchecourt, near Abbeville, given
at p. 168, he will perfectly understand the relations of the
ancient Thames alluvium to the modern channel and plain
of the river, and their relation, on the other hand to the
boundary formations of older date, whether tertiary or
cretaceous.

So far as they are known, the fossil mollusca and
mammalia of the two districts also agree very closely, the
Cyrena fluminalis being common to both, and being the
only extra-European shell, this and all the species of testacea being recent. Of this agreement with the living
fauna there is a fine illustration in Essex, for the determination of which we are indebted to the late Mr. John
Brown, F.G.S., who collected at Copford, in Essex, from a
deposit containing bones of the mammoth, a large bear
(probably *Ursus spelæus*), a beaver, stag, and aurochs, no
less than sixty-nine species of land and fresh-water shells.
Forty-eight of these were terrestrial, and two of them,
Helix incarnata and *H. ruderata*, no longer inhabit the
British Isles, but are still living on the continent, *H.
ruderata* in high northern latitudes.† The *Cyrena fluminalis* and the *Unio littoralis*, to which last I shall presently
allude, were not among the number.

I long ago suggested the hypothesis, that in the basin of
the Thames there are indications of a meeting in the
Pleistocene period of a northern and southern fauna. To
the northern group may have belonged the mammoth

* Lane Fox, Quart. Geol. Journal,
vol. xxviii. p. 467, 1872.
† Quarterly Geological Journal,
vol. viii. p. 190, 1852.

Mr. Brown calls them extinct species,
which may mislead some readers, but
he merely meant extinct in England.
See also Jeffreys, Brit. Conch. p. 174.

(*Elephas primigenius*) and the *Rhinoceros tichorhinus*,
both of which Pallas found in Siberia, preserved with their
flesh in the ice. With these are associated the reindeer,
and occasionally, though much less frequently, the glutton or
wolverine (*Gulo luscus*, Sabine).* In 1855, the skull of the
musk-sheep (*Ovibos moschatus*) was also found in the
ochreous gravel of Maidenhead, by the Rev. C. Kingsley
and Sir J. Lubbock ; the identification of this fossil with
the living species being made by Professor Owen. A
second fossil skull of the same arctic animal was afterwards
found by Sir J. Lubbock near Bromley, in the valley of a
small tributary of the Thames ; and two other skulls, those
of a bull and a cow, were dug up near Bathenston from the
gravel of the valley of the Avon, by Mr. Charles Moore.
Professor Owen has truly said, that, 'as this quadruped has
a constitution fitting it at present to inhabit the high
northern regions of America, we can hardly doubt that its
former companions, the warmly-clad mammoth and the
two-horned woolly rhinoceros (*R. tichorhinus*), were in
like manner capable of supporting life in a cold climate.'‡

I have alluded at p. 199 to the recent discovery of this
same musk-sheep near Chauny, in the valley of the Oise, in
France : and in 1856 I found a skull of it preserved in the
museum at Berlin, which Professor Quenstedt, the curator,
had correctly named so long ago as 1836, when the fossil
was dug out of drift, in the hill called the Kreuzberg, in
the southern suburbs of that city. By an account published
at the time, we find that the mammalia which accompanied
the musk-sheep were the mammoth and the tichorhine
rhinoceros, with the horse and ox ;‡ but I can find no
record of the occurrence of a Hippopotamus, nor of *Elephas*

* Dawkins and Sanford, Pleistocene
Mammalia, part i. p. xli., Pal. Soc. ;
also 5th Report on Kent's Cavern,
Brit. Assoc., 1869, p. 207.

† Geological Quarterly Journal,
vol. xii. p. 124.
‡ Leonhard and Bronn's Jahrbuch,
1836, p. 213.

antiquus or *Rhinoceros leptorhinus*, in the drift of the north of Germany, bordering the Baltic.

On the other hand, in another locality in the same drift of North Germany, Dr. Hensel, of Berlin, detected, near Quedlinburg, the Norwegian Lemming (*Myodes Lemmus*), and another species of the same family, called by Pallas, *Myodes torquatus* (by Hensel, *Misothermus torquatus*)—a still more arctic quadruped, found by Parry in latitude 82°, and which never strays further south than the northern borders of the woody region. Professor Beyrich also informs me that the remains of the *Rhinoceros tichorhinus* were obtained at the same place.[*]

As an example of what may possibly have constituted a more southern fauna in the valley of the Thames, I may allude to the fossil remains found in the fluviatile alluvium of Gray's Thurrock, in Essex, situated on the left bank of the river, twenty-one miles below London. The strata of brick-earth, loam, and gravel exposed to view in artificial excavations in that spot, are precisely such as would be formed by the silting up of an old river channel. Among the mammalia are *Elephas antiquus*, *Rhinoceros leptorhinus* (*R. megarhinus*, Christol), *Hippopotamus major*, species of horse, bear, ox, stag, &c., and, among the accompanying shells, *Cyrena fluminalis*, which is extremely abundant, instead of being scarce, as at Abbeville. It is associated with *Unio littoralis*, fig. 29, also in great numbers, and with both valves united. This conspicuous fresh-water mussel is no longer an inhabitant of the British Isles, but still lives in the Seine, and is still more abundant in the Loire. The fresh-water univalve (*Paludina marginata*, Michaud), which is not now found in Britain, but is common in the south of France, likewise occurs, together with a peculiar

[*] Zeitschrift der Deutschen Geologischen Gesellschaft, vol. vii. 1855, p. 407, &c.

variety of *Cyclas amnica*, which by some naturalists has been regarded as a distinct species. With these, moreover, is found a peculiar variety of *Valvata piscinalis*.

If we consult Dr. Von Schrenck's account of the living mammalia of Amoorland, lying between lat. 45° and 55° North, we learn that, in that part of North-Eastern Asia

Fig. 29.

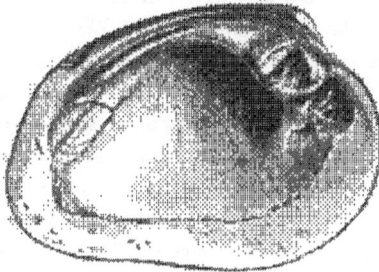

Unio littoralis, Gray's Thurrock, Essex: extinct in British Isles, living in France.

recently annexed to the Russian empire, no less than thirty-four out of fifty-eight living quadrupeds are identical with European species, while some of those which do not extend their range to Europe are arctic, others tropical forms. The Bengal tiger ranges northwards occasionally to lat. 52° North, where he chiefly subsists on the flesh of the reindeer, and the same tiger abounds in lat. 48°, to which the small tailless hare, or pika, a polar resident, sometimes wanders southwards.* We may readily conceive that the countries now drained by the Thames, the Somme, and the Seine, were, in the Pleistocene period, on the borders of two distinct zoological provinces, one lying to the north, the other to the south, in which case many species belonging to each fauna endowed with migratory habits, like the living

* Mammalia of Amoorland, Natural History Review, vol. i. p. 12, 1861.

musk-sheep or the Bengal tiger, may have been ready to
take advantage of any, even the slightest, change in their
favour to invade the neighbouring province, whether
in the summer or winter months, or permanently for a
series of years, or centuries. The *Elephas antiquus* and
its associated *Rhinoceros leptorhinus* may have preceded
the mammoth and tichorhine rhinoceros in the valley of
the Thames, or both may have alternately prevailed in the
same area in the Pleistocene period.

In attempting to settle the chronology of fluviatile
deposits, it is almost equally difficult to avail ourselves of
the evidence of organic remains, or of the superposition of
the strata, for we may find two old river-beds on the same
level in juxtaposition, one of them perhaps many thousands
of years posterior in date to the other. I have seen an
example of this at Ilford, where the Thames, or a tributary
stream, has at some former period cut through sands containing *Cyrena fluminalis*, and again filled up the channel
with argillaceous matter, evidently derived from the waste
of the tertiary London clay. Such shiftings of the site of
the main channel of the river, the frequent removal of
gravel and sand previously deposited, and the throwing
down of new alluvium, the flooding of tributaries, the
rising and sinking of the land, fluctuations in the cold and
heat of the climate—all these changes seem to have given
rise to that complexity in the fluviatile deposits of the
Thames, which accounts for the small progress we have
hitherto made in determining their order of succession, and
that of the embedded groups of quadrupeds. It may
happen, as at Brentford and Ilford, that sandpits in two
adjoining fields may each contain distinct species of elephant and rhinoceros ; and the fossil remains in both cases
may occur at the same depth from the surface, yet may be
specially referable to different parts of the Pleistocene epoch,
separated by thousands of years.

Climate and Habits of the Hippopotamus.

As I have alluded already, and shall have to do so several times in future, to the occurrence of the remains of the hippopotamus in places where there are now no rivers, not even a rill of water, and as other bones of the same genus have been met with in the lower level gravels of the Somme (p. 178), where large blocks of sandstone seem to imply that ice once played a part in their transportation, it may be well to consider, before proceeding further, what geographical and climatal conditions are indicated by the presence of these fossil pachyderms.

It is now very generally conceded that the mammoth and tichorhine rhinoceros were fitted to inhabit northern regions, and it is therefore natural to begin by asking whether the extinct hippopotamus may not in like manner have flourished in a cold climate. In answer to this inquiry, it has been remarked, that the living hippopotami, anatomically speaking, so closely allied to the extinct species, are so aquatic and fluviatile in their habits, as to make it difficult to conceive that their congeners could have thriven all the year round in regions where, during winter, the rivers were frozen over for months. Moreover, I have been unable to learn that, in any instance, bones of the hippopotamus have been found in the drift of northern Germany associated with the remains of the mammoth, tichorhine rhinoceros, musk-buffalo, reindeer, lemming, and other arctic quadrupeds, before alluded to (p. 204); yet, though not proved to have ever made a part of such a fauna, the presence of the fossil hippopotamus north of the fiftieth parallel of latitude naturally tempts us to speculate on the migratory powers and instincts of some of the extinct species of the genus. They may have resembled, in this respect, the living musk-buffalo, herds of which pass for hundreds of miles over the ice to the rich pastures of

Melville Island, and then return again to southern latitudes
before the ice breaks up.

I am indebted to Dr. Falconer for having called my
attention to the account given by an experienced zoologist,
Sir Andrew Smith,* of the migratory habits of the living
hippopotamus of Southern Africa (*H. amphibius*, Linn).

He states that, when the Dutch first colonized the Cape
of Good Hope, this animal abounded in all the great rivers,
as far south as the land extends; whereas, in 1849, they had
all disappeared, scarcely one remaining, even within a moderate distance of the colony. He also tells us that this species
evinces great sagacity in changing its quarters whenever
danger threatens, quitting every district invaded by settlers
bearing fire-arms. Bulky as they are, they can travel
speedily for miles over land from one pool of a dried-up
river to another; but it is by water that their powers of
locomotion are surpassingly great, not only in rivers, but in
the sea, for they are far from confining themselves to fresh
water. Indeed, Sir A. Smith finds it 'difficult to decide
whether, during the daytime and when not feeding, they
prefer the pools of rivers or the waters of the ocean for their
abode.' In districts where they have been disturbed by man,
they feed almost entirely in the night, chiefly on certain
kinds of grass, but also on brushwood. Sir A. Smith relates
that, in an expedition which he made north of Port Natal,
he found them swarming in all the rivers about the tropic
of Capricorn. Here they were often seen to have left their
foot-prints on the sands, entering or coming out of the salt
water; and on one occasion Smith's party tried in vain to
intercept a female with her young as she was making her
way to the sea. Another female, which they had wounded
on her precipitate retreat to the sea, was afterwards shot in
that element.

* Illustrations of the Zoology of South Africa: art. 'Hippopotamus.'

The geologist, therefore, may freely speculate on the time when herds of hippopotami issued from North African rivers, such as the Nile, and swam northwards in summer along the coasts of the Mediterranean, or even occasionally visited islands near the shore.   Here and there they may have tarried awhile where they landed to graze or browse, and afterwards have continued their course northwards.   Others may have swum in a few summer days from rivers in the south of Spain or France to the Somme, Thames, or Severn. making timely retreat to the south before the snow and ice set in.

The relation of the glacial period to alluvial deposits, such as that of Gray's Thurrock, where the *Cyrena fluminalis*, *Unio littoralis*, and the hippopotamus seem rather to imply a warmer climate, has been a matter of long and animated discussion ; and some geologists are inclined to ascribe the presence of these more southern forms to the prevalence of an interglacial warm period, supposed to have preceded the era of the most prolonged and intense cold.*   But we can, I think, hardly overestimate the difficulties which we encounter in these speculations on the changes of the fauna in valleys such as those of the Thames, Somme, and Seine. We must bear in mind how much of the geography of some of the lower and more fertile regions may be altered by changes of level of less than one hundred fathoms, or even feet (see Chap. XIV.), and, what is still more difficult to allow for, that a very general change of climate may be caused by geographical revolutions in the southern hemisphere, while the configurations of sea and land in this hemisphere may be unchanged.   Patches of the northern drift, at elevations of about two hundred feet above the Thames, occur in the neighbourhood of London, as at Muswell Hill, near High-

* James Geikie, Geol. Magazine, vol. ix. p. 259, 1872.

P

gate. In this drift, blocks of granite, syenite, greenstone, coal-measure sandstone with its fossils, and other palæozoic rocks, and the wreck of chalk and oolite, occur confusedly mixed together. The same glacial formation is also found capping some of the Essex hills farther to the east, and extending some way down their southern slopes towards the valley of the Thames. Although no fragments washed out of these older and upland drifts have been found in the gravel of the Thames containing elephants' bones, yet it seems fair to presume, as Mr. Prestwich has contended,[*] that the glacial formation is the older of the two, and that it originated, as we shall see in a future chapter, when the greater part of England was submerged beneath the sea. In short, I believe we must suppose that the basin of the Thames and all its fluviatile deposits are post-glacial, in the modified sense of that term; i.e. that they were subsequent to the marine drift of the central and northern counties, and to the period of its emergence above the level of the sea.

Having offered these general remarks on the alluvium of the Thames, I may now say something of the implements hitherto discovered in it. In the British Museum there is a flint weapon of the spear-headed form, such as is represented in fig. 15, p. 158, which we are told was found with an elephant's tooth at Black Mary's, near Gray's Inn Lane, London. In a letter dated 1715, printed in Herne's edition of 'Leland's Collectanea,' vol. i. p. 73, it is stated to have been found in the presence of Mr. Conyers, with the skeleton of an elephant.[†] So many bones of the elephant, rhinoceros, and hippopotamus have been found in the gravel on which London stands, that there is no reason to doubt the statement as handed down to us. Fossil remains of all these three genera have been dug up, within the memory of

* Prestwich, Quart. Geol. Journal,   ibid, 1861, p. 446.
1855, p. 110; ibid, 1856, p. 132; and   † Evans, Archæologia, 1860.

persons now living, on the site of Waterloo Place, St. James's
Square, Charing Cross, the London Docks, Limehouse, Beth-
nal Green, and other places. In the gravel and sand of
Shacklewell, in the north-east district of London, I have
myself collected specimens of the *Cyrena fluminalis* in
great numbers, (fig. 24 c, p. 170,) with the bones of deer
and other mammalia.

In the alluvium also of the Wey, near Guildford, in a
place called Pease Marsh, a wedge-shaped flint implement,
resembling one brought from St. Acheul by Mr. Prestwich,
and compared by some antiquaries to a sling-stone, was
obtained in 1836 by Mr. Whitburn, four feet deep in sand
and gravel, in which the teeth and tusks of elephants had
been found.[*] The Wey flows through the gorge of the
North Downs at Guildford to join the Thames. Mr. Godwin
Austen infers the great antiquity of the drift in this gorge
from the relative position of the newer and older gravel-
beds.[†]

Among other places where flint tools of the Paleolithic
type have been met with in the course of the last three
years, I may mention one of an oval form found by Mr.
Whitaker in the valley of the Darent, in Kent, and another
which Mr. Evans found lying on the shore at Swalecliff, near
Whitstable, in the same county, where Mr. Prestwich had
previously described a fresh-water deposit, resting on the
London clay, and consisting chiefly of gravel, in which an
elephant's tooth and the bones of a bear were embedded.
The flint implement was deeply discoloured and of a peculiar
bright light brown colour, similar to that of the old fluviatile
gravel in the cliff.

Another flint implement was found in 1860, by Mr. T.
Leech, at the foot of the cliff between Herne Bay and the

* Prestwich. Quart. Geol. Journal,   † Quarterly Geological Journal,
vol. xvii., 1861, p. 367.   1851, vol. vii. p. 27

Reculvers, and on further search five other specimens of the
spear-head pattern so common at Amiens. 'One of these,'
says Mr. Evans, 'is of considerable interest, as having been
formed from a Lower Tertiary pebble, and not from a flint
derived directly from the chalk.' Messrs. Prestwich and
Evans have since found three other similar tools on the
beach, at the base of the same wasting cliff, which consists
of sandy Eocene strata, covered by a deposit of gravel,
about fifty feet above the sea-level, from which the flint
weapons must have been derived. These gravelly beds now
capping the cliffs of Kent may be the old alluvial deposits
of tributaries of the Thames before the sea encroached to
its present position and widened its estuary. On following
up one of them westward of the Reculvers, Mr. Prestwich
found in it, at Chislet, near Grove Ferry, the *Cyrena
fluminalis* among other shells.

The changes which have taken place in the physical
geography of this part of England during, or since, the
Pleistocene period, have consisted partly of such encroach-
ments of the sea on the coast as are now going on, and
partly of a general subsidence of the land. Among the
signs of the latter movement may be mentioned a fresh-
water formation at Faversham, below the level of the sea.
The gravel there contains exclusively land and fluviatile
shells, of the same species as those of other localities of the
Pleistocene alluvium before mentioned, and was probably
formed when the river was at a higher level and when it
extended farther east. At that era the Thames was probably
a tributary of the Rhine, as represented by the late Mr.
Trimmer in his ideal restoration of the geography of the
olden time.* For England was then united to the continent,
and what is now the German Ocean was land. It is well

* Quarterly Geological Journal, vol. ix. pl. 13, No. 4.

known that in many places, especially near the coast of
Holland, elephants' tusks and other bones are often dredged
up from the bed of that shallow sea, and the reader will see
in the map given in Chap. XIV. how vast would be the
area of sea converted into land by an upheaval of 600 feet.
Vertical movements of much less than half that amount
would account for the annexation of England to the
continent, and the extension of the Thames and its valley
far to the north-east, and the flowing of rivers from the
easternmost parts of Kent and Essex into the Thames,
instead of emptying themselves into its estuary.

Since the discovery of the Reculvers implement many
other flint weapons of the Amiens type have been found in
the basin of the Thames.   Mr. T. Mc K. Hughes, writing in
1868, mentions some found by him in the river-gravels near
Chatham, which are now in the Museum at Jermyn Street.
One of these he took with his own hands from the bottom
of a brick-earth which is seen in a continuous section to pass
at Otterham into the bed in which the remains of the cave-
lion, reindeer, and elephant occur ; [*] and Colonel Lane Fox
in 1869 also brought to light several implements of well-
marked Palæolithic types in the gravels of Ealing Dean
and Acton.   One of these implements from Acton, of an
oval form, was found under seven feet of stratified sand and
gravel, and resting on the clay beneath ; another, of pointed
form, was found in the middle of the gravel, about ten feet
from the surface, and beneath beds of sand eight feet in
thickness.   A few mammalian remains have been found in
the Acton gravels, including a tooth of *Elephas primigenius*,
but as yet no land or fresh-water shells have been discovered.[†]

* Geol. and Nat. History Repertory,
1868, p. 131.
† Col. Lane Fox, Brit. Assoc. Report, 1869, p. 130; and Evans, Ancient
Stone Implements, p. 526.

Flint Implements of the Valley of the Ouse, near Bedford.

The ancient fluviatile gravel of the valley of the Ouse, around Bedford, has been noted for the last thirty years for yielding to collectors a rich harvest of the bones of extinct mammalia. By observations made in 1854 and 1858, Mr. Prestwich had ascertained that the valley was bounded on both sides by oolitic strata, capped by boulder clay, and that the gravel No. 3, fig. 30, p. 215, contained bones of the elephant, rhinoceros, hippopotamus, bos, equus, and cervus, which animals he therefore inferred must have been posterior in date to the boulder clay, through which, as well as the subjacent oolite, the valley had been excavated. Mr. Evans had found in the same gravel many land and fresh-water shells, and these discoveries induced Mr. James Wyatt, of Bedford, to pay two visits to St. Acheul, in order to compare the implement-bearing gravels of the Somme with the drift of the valley of the Ouse. After his return he resolved to watch carefully the excavation of the gravel-pits at Biddenham, two miles WNW. of Bedford, in the hope of finding there similar works of art. With this view he paid almost daily visits for months in succession to those pits, and was at last rewarded by the discovery of two well-formed implements, one of the spear-head and the other of the oval shape, perfect counterparts of the two prevailing French types figured at pp. 158, 159. Both specimens were thrown out by the workmen on the same day from the lowest bed of stratified gravel and sand, thirteen feet thick, containing bones of the elephant, deer, and ox, and many fresh-water shells. The two implements occurred at the depth of thirteen feet from the surface of the soil, and rested immediately on solid beds of oolitic limestone, as represented in the accompanying section.

Having been invited by Mr. Wyatt to verify these facts,

I went to Biddenham within a fortnight of the date of his
discovery (April 1861), and, for the first time, saw evidence
which satisfied me of the chronological relations of those
three phenomena, the antique tools, the extinct mammalia,
and the glacial formation. On that occasion I examined
the pits in company with Messrs. Prestwich, Evans, and
Wyatt, and we collected ten species of shells from the
stratified drift No. 3, or the beds overlying the lowest gravel
from which the flint implements had been exhumed. They
were all of common fluviatile and land species now living in

Fig. 30.

Section across the Valley of the Ouse, two miles WNW. of Bedford.[*]

1 Oolitic strata,
2 Boulder clay, or marine northern drift, rising to about ninety feet
   above the Ouse.
3 Ancient gravel, with elephant bones, fresh-water shells, and flint
   implements.
4 Modern alluvium of the Ouse.
a Biddenham gravel pits, at the bottom of which flint tools were
   found.

the same part of England. Since our visit Mr. Wyatt has
added to them *Paludina marginata*, Michaud (*Hydrobia*
of some authors, see fig. 38, p. 269), a species of the South of
France no longer inhabiting the British Isles. The same
geologist has also found, since we were at Biddenham, several
other flint tools of corresponding type, both there and at
other localities in the valley of the Ouse, near Bedford.

The boulder clay, No. 2, extends for miles in all directions, and was evidently once continuous from *b* to *c*, before

* Prestwich, Quarterly Geological   Wyatt, 'Geologist,' 1861, p. 212.
Society, vol. xvii. p. 364. 1861 and

the valley was scooped out. It is a portion of the great
marine glacial drift of the midland counties of England, and
contains blocks, some of large size, not only of the oolite of
the neighbourhood, but of chalk and other rocks transported
from still greater distances, such as syenite, basalt, quartz,
and new red sandstone. These erratic blocks of foreign
origin are often polished and striated, having undergone
what is called glaciation, of which more will be said by and
by. Blocks of the same mineral character, embedded at
Biddenham in the gravel No. 3, have lost all signs of this
striation by the friction to which they were subjected in the
old river-bed.

The great width of the valley of the Ouse, which is
sometimes two miles, has not been expressed in the diagram.
It may have been shaped out by the joint action of the
river and the tides when this part of England was emerging
from the waters of the glacial sea, the boulder clay being
first cut through, and then an equal thickness of underlying
oolite. After this denudation, which may have accompanied
the emergence of the land, the country was inhabited by the
primitive people who fashioned the flint tools. The old
river, aided perhaps by the continued upheaval of the whole
country, or by oscillations in its level, went on widening and
deepening the valley, often shifting its channel, until at
length a broad area was covered by a succession of deposits, which may have corresponded in age to the higher
and lower gravels of the Valley of the Somme, already
described, p. 153.

At Biddenham, and elsewhere in the same gravel,
remains of *Elephas antiquus* have been discovered, and
Mr. Wyatt obtained, Jan. 1863, a flint implement associated
with bones and teeth of hippopotamus from gravel at
Summerhouse Hill, which lies east of Bedford, lower down
the valley of the Ouse, and four miles from Biddenham.

One step at least we gain by the Bedford sections, which those of Amiens and Abbeville had not enabled us to make. They teach us that the fabricators of the antique tools, and the extinct mammalia coeval with them, were post-glacial, or, in other words, posterior to the grand submergence of central England beneath the waters of the glacial sea.

Flint Implements in a Fresh-water Deposit at Hoxne in Suffolk.

So early as the first year of the present century, a remarkable paper was communicated to the Society of Antiquaries by Mr. John Frere,* in which he gave a clear description of the discovery at Hoxne, in Suffolk, of flint tools of the type since found at Amiens, adding at the same time good geological reasons for presuming that their antiquity was very great, or, as he expressed it, beyond that of the present world, meaning the actual state of the physical geography of that region. 'The flints,' he said, 'were evidently weapons of war, fabricated and used by a people who had not the use of metals. They lay in great numbers at the depth of about twelve feet in a stratified soil which was dug into for the purpose of raising clay for bricks. Under a foot and a half of vegetable earth was clay seven and a half feet thick, and beneath this one foot of sand with shells, and under this two feet of gravel, in which the shaped flints were found generally at the rate of five or six in a square yard. In the sandy beds with shells were found the jaw-bone and teeth of an enormous unknown animal. The manner in which the flint weapons lay would lead to the persuasion that it was a place of their manufacture, and not of their accidental deposit. Their numbers were so great that the man who carried on the brick-work

* Frere, Archæologia for 1800, vol. xiii. p. 206.

told me that before he was aware of their being objects of
curiosity, he had emptied baskets full of them into the ruts
of the adjoining road.'

Mr. Frere then goes on to explain that the strata in
which the flints occur are disposed horizontally, and do not
lie at the foot of any higher ground, so that portions of
them must have been removed when the adjoining valley
was hollowed out. If the author had not mistaken the
fresh-water shells associated with the tools for marine species,
there would have been nothing to correct in his account of
the geology of the district, for he distinctly perceived that
the strata in which the implements were embedded had,
since that time, undergone very extensive denudation.
Specimens of the flint spear-heads, sent to London by Mr.
Frere, are still preserved in the British Museum, and others
are in the collection of the Society of Antiquaries.

Mr. Prestwich's attention was called by Mr. Evans to
those weapons, as well as to Mr. Frere's memoir after his
return from Amiens in 1859, and he lost no time in visiting
Hoxne, a village five miles eastward of Diss, which stands
on the other, or Norfolk, side of the river Waveney. It is
not a little remarkable that he should have found, after a
lapse of sixty years, that the extraction of clay was still
going on in the same brick-pit. Only a few months before
his arrival, two flint instruments had been dug out of the
clay, one from a depth of seven and the other of ten feet
from the surface. Others have since been disinterred from
undisturbed beds of gravel in the same pit. Mr. Amyot, of
Diss, has also obtained from the underlying fresh-water strata
the astragalus of an elephant, and bones of the deer and
horse; but although many of the old implements have
recently been discovered *in situ* in regular strata and preserved by Sir Edward Kerrison, no bones of extinct mammalia

seem as yet to have been found, as they were by Mr. Frere, in strata overlying the flint implements.

By reference to the annexed section, the geologist will see that the basin-shaped hollow *a b c*, after having been excavated out of the more ancient boulder clay, No. 6, has been filled up gradually with the fresh-water strata 3, 4, 5.

Fig. 31.

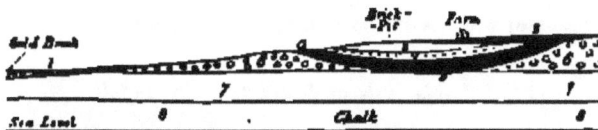

Section showing the position of the flint weapons at Hoxne, in Suffolk. (See Prestwich, Philosophical Transactions, Pl. 11, 1860.)

1 Gravel of Gold Brook, a tributary to the Waveney.
2 Higher-level gravel overlying the implement-bearing deposit.
3 and 4. Sand and gravel, with fresh-water shells, and flint implements, and bones of mammalia.
5 Peaty and clayey beds, with same fossils.
6 Boulder clay or glacial drift.
7 Sand and gravel below boulder clay.
8 Chalk with flints.

The relative position of these formations will be better understood when I have described in the Twelfth Chapter the structure of Norfolk and Suffolk as laid open in the sea-cliffs at Mundesley, about thirty miles distant from Hoxne, in a north-north-east direction.

I examined the deposits at Hoxne in 1860, when I had the advantage of being accompanied by the Rev. J. Gunn, and the Rev. S. W. King. In the loamy beds 3 and 4, fig. 31, we observed the common river shell *Valvata pisci-nalis* in great numbers. With it, but much more rare, were *Limnea palustris, Planorbis albus, P. spirorbis, Succinea putris, Bithinia tentaculata, Cyclas cornea*; and Mr. Prestwich mentions *Cyclas amnica* and fragments of a

Unio, besides several land shells. In the black peaty mass
No. 5, fragments of wood of the oak, yew, and fir have been
recognised. The flint weapons which I have seen from
Hoxne, are so much more perfect, and have their cutting
edge so much sharper than those from the Valley of the
Somme, that they seem neither to have been used by man,
nor to have been rolled in the bed of a river. The opinion
of Mr. Frere, therefore, that there may have been a manufactory of weapons on the spot, appears probable.

Flint Implements at Icklingham in Suffolk.

In another part of Suffolk, at Icklingham, in the Valley of
the Lark, below Bury St. Edmund's, there is a bed of gravel,
in which two flints of a lance-head form have been found
at the depth of four feet from the surface. I have visited
the spot, which has been described by Mr. Prestwich.[*]

The section of the Bedford tool-bearing alluvium, given at
p. 215, may serve to illustrate that of Icklingham, if we sub-
stitute chalk for oolite, and the river Lark for the Ouse. In
both cases, the present bed of the river is about thirty feet
below the level of the old gravel, and the chalk hill, which
bounds the Valley of the Lark on the right side, is capped
like the oolite of Biddenham by boulder clay, which rises to
the height of one hundred feet above the Lark. About
twelve years ago, a large erratic block, above four feet in
diameter, consisting of a hard siliceous schist, was dug out
of the boulder clay at Icklingham. It was apparently a
Silurian rock, which must have come from a remote region.
The tool-bearing gravel here, as in the case to which it has
been compared near Bedford, is proved to be newer than the
glacial drift, by containing pebbles of basalt and other rocks
derived from that formation.

[*] Quarterly Geological Journal, 1861, vol. xvii. p. 364.

Flint Implements in the Drift of the South of Hampshire.

Flint implements of the normal type of the Palæolithic period have been lately found in the south of Hampshire, not in caves nor in old river-gravels within the limits of existing valleys, but in a tabular mass of drift which caps the tertiary strata, and which is intersected both by the Solent and by the valleys of all the rivers which flow into that channel of the sea. The position of these implements, to which the archæologists of Salisbury have called our attention, attests perhaps in a more striking manner the antiquity of prehistoric man in Europe than any other monument of the earlier stone age yet discovered. The great bed of gravel resting on Eocene tertiary strata in which these implements have been found, consists in most places of half-rolled or of semi-angular chalk-flints, mixed with rounded pebbles washed out of the Tertiary strata. But this drift, although often continuous over wide areas, is not everywhere present, nor does it always present the same characters. The first flint implements found in it were discovered mid-way between Gosport and Southampton, by Mr. James Brown, of Salisbury, in May 1864, imbedded in gravel from 8 to 12 feet thick, capping a cliff which at its greatest height is 35 feet above high-water mark. I have visited this spot, which had previously been seen by Messrs. Prestwich and Evans. The flint-tools exactly resemble those found at Abbeville and Amiens In France, being some of them of the oval, and others of the lanceolate form. Many of them exhibit the same colours and ochreous stain as do the flints in the gravel in which they lay. A fine series of these implements, from the Hampshire cliffs, may now be seen in the Blackmore Museum at Salisbury.

In the gravel capping the cliffs alluded to are blocks of sandstone of various sizes, some of enormous dimensions, more than 20 feet in circumference and from 1 to 2½ feet thick. They have probably not travelled far, being a portion of the wreck of the Eocene strata which have suffered much denudation. Nevertheless, to explain how they and the stone implements became enveloped in the débris of chalk-flints, we must have recourse to ice, which may have been frozen on to them in winter, so as to give them buoyancy and enable rivers or the sea to transport them to slight distances from their original site. An extreme climate, causing a vast accumulation of snow during a cold winter, and great annual floods when this snow was suddenly melted in the beginning of the warm season, may best account for the destruction of large masses of chalk in the upland country, and the spreading over the ancient surface of the flinty material originally dispersed in layers through the soft chalk. The occasional occurrence of unrolled chalk-flints in the gravel, in places where they must have travelled twelve miles from their nearest source, also implies the aid of ice-action. The transverse valleys now intersecting the region near the coast where the flint tools are found, near Gosport, must have been cut through the tertiary strata after the overlying gravel had been superimposed, for this last forms a flat table-land between the valleys.

On the whole we may infer that not only the valleys of the smaller streams near Gosport, but those of the Test (or South-ampton river) and of the stream which enters at Lymington, and those of the rivers Avon and Stour, which reach the Solent at Christchurch, as well as the Bournemouth valley, have all been excavated since Palæolithic Man inhabited this region; for not only at various points east of the Southampton estuary, but west of it also, on both sides of the opening at Bournemouth, flint tools of the ancient type have been met with in the gravel capping the cliffs. The gravel from which the flint tool was taken at Bournemouth is about 100 feet

above the level of the sea, as I ascertained after examining
the spot in 1867.*

The gravel consists in great part of pebbles derived from
tertiary strata; and if it was originally spread out by rivers,
the course of the drainage must have since been altered to
such an extent that it is not easy to trace any connection
between the old watercourses and those of the existing
valleys. In the same manner, at Milford Hill, in the neighbourhood of Salisbury, which forms a spur at the confluence
of the Avon and the Bourne, the capping of gravel in
which Palæolithic implements occur is at a height of 100
feet above the valleys on either side, and it is evident that
the whole period of the excavation of these valleys must have
elapsed since the deposition of the gravel, since no river,
however great, flowing through the valley could have deposited the heavy gravel on the top of a hill.

I learn from Mr. Evans that Mr. Thomas Codrington discovered in 1868 an oval flint implement in gravel at the top
of the Foreland cliff on the most eastern point of the Isle of
Wight five miles south-east of Ryde. It is of the true
Palæolithic type, and the gravel in which it is imbedded at
the height of about 80 feet above the level of the sea, may,
as Mr. Evans suggests, have once extended to the cliffs near
Gosport; in which case we should have to infer that the
channel called the Solent had not yet been scooped out when
this region was inhabited by Palæolithic man. The gravel
found at Freshwater, at the west end of the Isle of Wight, in
which the remains of the mammoth have been detected, is
probably of the same date. Mr. Evans has given a graphic
picture of the evidence afforded by the Palæolithic tools at
the summit of the cliffs at Bournemouth, Gosport, and other

* Mr. Alfred Stevens first dug out a
hatchet (April 1868) from this gravel
at the top of the sea-cliff east of the
Bournemouth opening. Dr. Blackmore
sora afterwards obtained two other
similar implements from gravel west
of the Bournemouth valley.

banks of the Solent, in favour of the former existence of a large river once flowing from the west before Dorsetshire was divided from the Isle of Wight, as suggested in 1862 by the Rev. W. Fox. Admitting that the early part of the denudation by the principal river and its tributaries was accelerated by a greater rainfall and by the sudden melting of snow during parts of the year, Mr. Evans insists with reason on the vast lapse of time required for the extensive geographical changes, as well as the successive modifications of the terrestrial fauna which have occurred since Palæolithic man inhabited the southern counties of England.*

If we ascend the Avon from Christchurch to Salisbury about 30 miles to the north, we find in gravels at various heights above the river, and in old fluviatile alluvium, flint tools of the same Palæolithic type. The first of these was found by Dr. Blackmore in 1863 at Bemerton, and since then a large number have been found in that locality. But the most important, on account of the animal remains associated with it, was one taken out in 1864 by Dr. Blackmore from beneath the remains of a mammoth, at Fisherton, near Salisbury. The remains of no less than 21 species of mammalia have also been detected at the same place, the greatest number, perhaps, obtained in any one spot in Great Britain. The associated land and fresh-water shells belong to 31 species, and are all still living in England, although the quadrupeds imply a colder climate. Among these are the mammoth and woolly-haired rhinoceros, the reindeer, and Norwegian lemming, the Greenland lemming, and another species of the same family, the Spermophilus, allied to the marmot. Of this last 13 individuals have been found, some of the skeletons being perfect, and lying, as remarked by Dr. Blackmore, in the curved attitude of hibernation, as may now be seen in the Blackmore Museum.

* Evans, 'Ancient Stone Implements,' 1872, p. 605.

Besides the bones of quadrupeds, the femur and caracoid
bones of the wild goose (*Anser palustris*) have been met with,
and some egg-shells corresponding in size with the eggs of
the wild goose and wild duck. These shells are in part
covered with superficial incrustations. As the wild goose
now resorts to arctic regions in the breeding season, the
occurrence of its eggs at Fisherton seems to imply a cold
climate such as would have suited the lemming and
marmot.* Below the brick-earth containing these remains,
and separated from it by a bed of large unrolled chalk flints,
was a bed of white chalk, decomposed and rubbly at the top.
The flints and rubble seem to indicate the waste of the
chalk by the dissolving action of water charged with carbonic acid, a waste which may have gone on after the old
alluvium was deposited. Mr. J. Evans, F.G.S., has suggested to me that the deepening of many valleys in the
chalk by gently flowing streams which have no mechanical
power of erosion is constantly carried on by the dissolving
power of the water, which conveys annually a considerable
volume of carbonate of lime to the sea, representing in the
course of ages enormous masses of solid matter gently and
insensibly subtracted, and chiefly from the lower part of each
valley. He has calculated from experiments made in a
chalk district that a rainfall of twenty-six inches annually
will carry away 140 tons of carbonate of lime from each
square mile, showing 'how great are the solvent powers of
water charged with carbonic acid, and the extent to which,
in the course of centuries, it might remove the calcareous
rocks with which it came in contact.' †

To conclude, there are three independent classes of evidence, which in this part of Hampshire point distinctly to
the vast antiquity of Palæolithic man.   First, the great

* Evans, Geol. Quart. Journal,   † Evans, 'Ancient Stone Imple-
p. 193, Aug. 1864.   ments,' 1872. p. 430.

denudation of the chalk and tertiary strata, and the important changes in the shape and depth of the valleys and the contour of the sea-coast which have since occurred in Hampshire; secondly, a marked change in the fauna, by the dying out of so many conspicuous species of quadrupeds; and thirdly, the change of climate from a colder to a warmer temperature, implied by the former presence of northern animals, and by the ice-borne erratics of the drift.[*]

* This account of the Hampshire Palæolithic Implements was first published in my Principles of Geology, 10th ed., 1868.

CHAPTER XI.

AGE OF HUMAN FOSSILS OF LE PUY IN CENTRAL FRANCE AND OF NATCHEZ ON THE MISSISSIPPI.

QUESTION AS TO THE AUTHENTICITY OF THE FOSSIL MAN OF DENISE, NEAR LE PUY-EN-VELAY, CONSIDERED — ANTIQUITY OF THE HUMAN RACE IMPLIED BY THAT FOSSIL — SUCCESSIVE PERIODS OF VOLCANIC ACTION IN CENTRAL FRANCE — WITH WHAT CHANGES IN THE MAMMALIAN FAUNA THEY CORRESPOND — THE ELEPHAS MERIDIONALIS ANTERIOR IN TIME TO THE IMPLEMENT-BEARING GRAVEL OF ST. ACHEUL —THE MARKED AND NOTCHED BONES FOUND AT SAINT PREST ASSOCIATED WITH ELEPHAS MERIDIONALIS—HUMAN FOSSIL OF NATCHEZ ON THE MISSISSIPPI—HUMAN REMAINS IN THE LOESS NEAR MAESTRICHT —HUMAN REMAINS IN THE LOESS NEAR STRASBURG.

AMONG the fossil remains of the human species supposed to have claims to high antiquity, and which have for many years attracted attention, two of the most prominent examples are—

First,—'The fossil man of Denise,' comprising the remains of more than one skeleton, found in a volcanic breccia near the town of Le Puy-en-Velay, in Central France.

Secondly,—The fossil human bone of Natchez, on the Mississippi, supposed to have been derived from a deposit, containing remains of mastodon and megalonyx.

Having carefully examined the sites of both of these celebrated fossils, I shall proceed to consider the nature of the evidence from which the remote date of their entombment is inferred.

Fossil Man of Denise.

An account of the fossil remains of Denise was first published in 1844 by M. Aymard, of Le Puy, a writer of deservedly high authority both as a palæontologist and archæologist.* The late M. Pictet, after visiting Le Puy and investigating the site of the alleged discovery, was satisfied that the fossil bones belonged to the period of the last volcanic eruptions of Velay; but expressly stated in his important treatise on palæontology that this conclusion, though it might imply that Man had coexisted with the extinct elephant, did not draw with it the admission that the human race was anterior in date to the filling of the caverns of France and Belgium with the bones of extinct mammalia.†

At a meeting of the 'Scientific Congress' of France, held at Le Puy in 1856, the question of the age of the Denise fossil bones was fully gone into, and in the report of their proceedings published in that year, the opinions of some of the most skilful osteologists respecting the point in controversy are recorded. The late Abbé Croizet, a most experienced collector of fossil bones in the volcanic regions of Central France, and an able naturalist, and the late M. Laurillard, of Paris, who assisted Cuvier in modelling many fossil bones, and in the arrangement of the museum of the Jardin, declared their opinion that the specimen preserved in the museum of Le Puy is no counterfeit. They believed the human bones to have been enveloped by natural causes in the tufaceous matrix in which we now see them.

In the year 1859, Professor Hébert and M. Lartet visited Le Puy, expressly to investigate the same specimen, and to inquire into the authenticity of the bones and their geological age. Later in the same year, I went myself to Le Puy,

* Bulletin de la Société Géologique   † Traité de Paléontologie, tom. i.
de France, 1844, 1845, 1847.   p. 162. 1853.

having the same object in view, and had the good fortune to meet there my friend Mr. Poulett Scrope, with whom I examined the Montagne de Denise, where a peasant related to us how he had dug out the specimen with his own hands and in his own vineyard, not far from the summit of the volcano. I employed a labourer to make under his directions some fresh excavations, following up those which had been made a month earlier by MM. Hébert and Lartet, in the hope of verifying the true position of the fossils, but all of us without success. We failed even to find *in situ* any exact counterpart of the stone of the Le Puy Museum.

The osseous remains of the specimen consist of a frontal and some other parts of the skull, including the upper jaw with teeth, both of an adult and young individual ; also a radius, some lumbar vertebræ, and some metatarsal bones. They are all embedded in a light porous tuff, resembling in colour and mineral composition the ejectamenta of several of the latest eruptions of Denise. But none of the bones penetrate into another part of the same specimen, which consists of a more compact rock thickly laminated. Nevertheless, I agree with the Abbé Croizet and M. Aymard, that it is not conceivable even that the less coherent part of the museum specimen which envelopes the human bones should have been artificially put together, whatever may have been the origin of certain other slabs of tuff which were afterwards sold as coming from the same place, and which also contained human remains. Whether some of these were spurious or not is a question more difficult to decide. One of them, now in the possession of M. Pichot-Dumazel, an advocate of Le Puy, was suspected of having had some plaster of Paris introduced into it to bind the bones more firmly together in the loose volcanic tuff. But an expert anatomist remarked to me that it would far exceed the skill, whether of the peasant who owned the vineyard or of the

dealer above mentioned, to put together in their true
position all the thirty-eight bones of the hand and fingers,
or the sixteen of the wrist, without making any mistake,
and especially without mixing those of the right with the
homologous bones of the left hand, assuming that they had
brought bones from some other spot, and then artificially
introduced them into a mixture of volcanic tuff and plaster
of Paris.

Granting, however, that the high prices given for 'human
fossils' at Le Puy may have led to the perpetration of
some frauds, it is still an interesting question to consider
whether the admission of the genuineness of a single fossil,
such as that now in the museum at Le Puy, would lead us
to assign a higher antiquity to the existence of Man in
France than is deducible from many other facts explained in
the last seven chapters. In reference to this point, I may
observe that, although I was not able to fix with precision
the exact bed in the volcanic mountain from which the rock
containing the human bones was taken, M. Félix Robert has,
nevertheless, after studying 'the volcanic alluviums' of
Denise, ascertained that, on the side of Cheyrac and the
village of Malouteyre, blocks of tuff frequently occur
exactly like the one in the museum. That tuff he considers
a product of the latest eruption of the volcano. In it have
been found the remains of *Hyæna spelæa* and *Hippopotamus
major*. The eruptions of steam and gaseous matter which
burst forth from the crater of Denise broke through laminated tertiary clays, small pieces of which, some of them
scarcely altered, others half converted into scoriæ, were cast
out in abundance, while other portions must have been in a
state of mud. Floods of such materials would be styled
by the Neapolitans 'aqueous lava' or 'lava d'aqua,' and we
may well suppose that some human individuals, if any existed, would, together with wild animals, be occasionally over-

whelmed in these tuffs. From near the place on the mountain whence the block with human bones now in the museum is said to have come, a stream of lava, well marked by its tabular structure, flowed down the flanks of the hill, within a few feet of the alluvial plain of the Borne, a small tributary of the Loire, on the opposite bank of which stands the town of Le Puy. Its continuous extension to so low a level clearly shows that the valley had already been excavated to within a few feet of its present depth at the time of the flowing of the lava.

We know that the alluvium of the same district, having a similar relation to the present geographical outline of the valleys, is of Pleistoeene (post-Pliocene) date, for it contains around Le Puy the bones of *Elephas primigenius* and *Rhinoceros tichorhinus*; and this affords us a palæontological test of the age of the human skeleton of Denise, if the latter be assumed to be coeval with the lava stream above referred to.

It is important to dwell on this point, because some geologists have felt disinclined to believe in the genuineness of the 'fossil man of Denise,' on the ground that, if conceded, it would imply that the human race was contemporary with an older fauna, or that of the *Elephas meridionalis*. Such a fauna is found fossil in another layer of tuff covering the slope of Denise, opposite to that where the museum specimen was exhumed.   The quadrupeds obtained from that more ancient tuff comprise *Elephas meridionalis*, *Hippopotamus major*, *Rhinoceros megarhinus*, *Antilope torticornis*, *Hyæna brevirostris*, and twelve others of the genera horse, ox, stag, goat, tiger, &c., all supposed to be of extinct species. This tuff, found between Malouteyre and Polignac, M. Robert regards as the product of a much older eruption, and referable to the neighbouring Montagne de St. Anne, a volcano in a much more wasted and denuded state than Denise, and classed by M. Bertrand de Doue as of intermediate age between the ancient and modern cones of Velay.

The fauna to which *Elephas meridionalis* and its associates belong, can be shown to be of anterior date, in the north of France, to the flint implements of St. Acheul, by the following train of reasoning. The Valley of the Seine is not only geographically contiguous to the Valley of the Somme, but its ancient alluvium contains the same mammoth and other fossil species. The Eure, one of the tributaries of the Seine, in its way to join that river, flows in a valley which follows a line of fault in the chalk; and this valley is seen to be comparatively modern, because it intersects at Saint-Prest, four miles below Chartres, an older valley belonging to an anterior system of drainage, and which has been filled by a more ancient fluviatile alluvium, consisting of sand and gravel, ninety feet thick. I have examined the site of this older drift, and the fossils were determined by the late Dr. Falconer. They comprise *Elephas meridionalis,* a species of rhinoceros (not *R. tichorhinus*), and other mammalia differing from those of the implement-bearing gravels of the Seine and Somme; which gravels, as they belong to the period of the mammoth, might very well have been contemporary with the modern volcanic eruptions of Central France, and we may presume, even without the aid of the Denise fossil, that Man may have witnessed these. But the tuffs and gravels in which the *Elephas meridionalis* are embedded were synchronous with an older epoch of volcanic action, to which the cone of St. Anne, near Le Puy, and many other mountains of M. Bertrand de Doue's middle period belong, having cones and craters which have undergone much waste by aqueous erosion. We have as yet no proof that Man witnessed the origin of these hills of lava and scoriæ of the middle phase of volcanic action.

Some surprise was expressed in 1856, by several of the assembled naturalists of Le Puy, that the skull of the ' fossil man of Denise,' although contemporary with the mammoth, and coeval with the last eruptions of the Le Puy volcanoes, should be of the ordinary Caucasian or European type; but the observations of Professor Huxley on the Engis skull, cited in the fifth chapter (p. 92), showing the near approach of that ancient cranium to the European standard, will help to remove this source of perplexity.

Supposed Proofs of Man's Coexistence with Elephas Meridionalis.

In 1863 M. Desnoyers found in the stratified sand and gravel of Saint-Prest certain fossil bones on which furrows, cuts, notches, and other markings were observable. These, if the work of man, as he then believed them to be, would have been evidence of man's coexistence with the *Elephas meridionalis* which occurs in these beds. I was, however, afterwards convinced, as indeed was M. Desnoyers to a great extent himself, that these marks might possibly be attributed to the teeth of animals; and the experiments by means of which we arrived at this conclusion are so instructive that I will give them at full length.

The remains of a large extinct rodent of the beaver family (*Trogontherium*) had been found associated with the *Elephas meridionalis* at Saint-Prest, and I therefore was desirous of ascertaining whether the teeth of such an animal could have given rise to some of the furrows and excisions alluded to; I accordingly proposed to Mr. Bartlett, of the London Zoological Gardens, to make experiments for me with a view of testing the point. Two bones, namely, the radius (or rather the united radius and ulna) of a horse and the humerus of an ox, both entire and perfectly free from any superficial scratches, were placed in the cage in

which four porcupines, two of *Histrix cristata* and two
of *Histrix javanica*, were kept. They were supplied with
their ordinary allowance of vegetable food, more than they
usually consume, and at the end of ten days the two bones were
taken out. The large ball-headed extremity of the humerus
of the ox was all eaten away, and more than half the marrow
extracted from the bone. Numerous grooves were also cut,
some an inch long, some oblique, but most of them exactly
transverse to the length of the bone, and some bending
slightly round its convex surface. On the radius of the horse
I counted nearly a hundred transverse grooves, scratches, and
tooth-marks from a quarter of an inch to an inch in length,
some continuous round a small part of the curvature of the
bone. Many of the furrows exhibited several straight and
very fine parallel striæ exactly like those to which the
uneven edge of a flint tool might give rise. The porcupines
had also gnawed off a portion of the prominent ridge of the
radius of the horse where the bone was very hard, and had
left an elliptical scar an inch and a half long, three-fourths of
an inch broad, and the fifth of an inch deep in the middle,
singularly resembling in shape and general appearance
several of the cuts which are conspicuous on a few of the
fossil bones of Saint-Prest, but with this difference, that the
separate tooth-marks transverse to the length of the bone
were distinctly visible on the recent bone. The *Histrix
cristata* was seen in the act of gnawing one of the bones,
and so few splinters were left behind on the floor that he
must have swallowed most of the bony matter. The cage is
surrounded by iron bars, intended to exclude rats, and I
do not believe that any even of the smaller erosions are
ascribable to their intrusion; but assuming this to be the
case, it would not affect the bearing of the experiments now
under consideration, as it is immaterial to what species of
rodent the marks in question may be ascribable. I observed

some of the parallel grooves cut by the porcupines crossing at an angle of 40 degrees other older and parallel ones.

No one has acquired more skill in deciphering the true meaning and origin of the various marks and incisions so often seen on bones found in tumuli, and on others derived from the drift, than the late M. Lartet, whose authority is often cited in the Memoir of M. Desnoyers.*

When I showed him, at the time of his visit to London in 1863, two of the bones partially eaten by the porcupines, he said that the cuts, although not exactly identical with any of those at Saint-Prest, resembled some of those observed on cave-bones in the south of France, which he had been disposed to attribute to human agency. He begged me therefore to follow up the experiment by placing in the porcupines' cage some shed horns of deer, which are extremely hard. This was done by Mr. Bartlett, an antler of the Javanese *Cervus rusa*, Müller, and another of the East Indian *Cervus Barrasinga*, being placed in the rodents' cage together with some fresh bones of the horse and ox. The latter were treated as before, except that on this occasion nearly the whole of the marrow was extracted from the humerus of the ox. At the same time, the dry, hard, and marrowless stag's horns were equally gnawed and eaten away during the four days they were left in the cage. The antler of the Rusa deer had three branches, one of which was shortened by being sliced off obliquely; the other two were so cut that their extremities, originally blunt, were brought to a sharp point, so as to prick almost like pins. Near the base of the same horn, where the beam was about four inches in circumference, three flat sides were produced by gnawing, two of them meeting at a sharp angle. Had we found it in Pompeii, and supposed it to be the work of a cutler, we might naturally have imagined that he had intended to give

* Comptes Rendus, etc. Inst. Imp. de France, 1863.

the lower part of the horn a pentagonal instead of a cylindrical shape. Many parts of the Barrasinga's horn were also gnawed off, and one half of the brow antler cut away in such a manner as to present a flat surface on its upper side, while the other half retained its original rotundity. But in all these excisions the separate tooth-marks could be perceived.

Several other examples of marked bones similar to those of Saint-Prest have been found in the Val d'Arno and elsewhere. Some of these are of interest as being on bones too large for a rodent to grasp so as to produce the long and straight striæ; but until we have further evidence of Man in these deposits the experiments just cited must cause us to hesitate before assuming that these marks have been the result of human agency.

Human Fossil of Natchez on the Mississippi.

I have already alluded to Dr. Dowler's attempt to calculate, in years, the antiquity of the human skeleton said to have been buried under four cypress forests in the delta of the Mississippi, near New Orleans (see page 46). In that case no remains of extinct animals were found associated with those of Man ; but in another part of the basin of the Mississippi, at Natchez, a human bone, accompanied by bones of the mastodon and megalonyx, is supposed to have been washed out of a more ancient alluvial deposit.

After visiting the spot in 1846, I described the geological position of the bones, and discussed their probable age, with a stronger bias, I must confess, as to the antecedent improbability of the contemporaneous entombment of Man and the mastodon than any geologist would now be justified in entertaining. Natchez is about eighty miles in a straight line south of Vicksburg, on the same left bank of the Mississippi. Here there is a bluff, the upper sixty feet of which consists of a continuous portion of the same calcareous loam

as at Vicksburg, equally resembling the Rhenish loess in
mineral character and in being sometimes barren of fossils,
sometimes so full of them that bleached land-shells stand
out conspicuously in relief in the vertical and weathered
face of cliffs which form the banks of streams, everywhere
intersecting the loam.

So numerous are the shells that I was able to collect at
Natchez, in a few hours, in 1846, no less than twenty species
of the genera *Helix, Helicina, Pupa, Cyclostoma, Achatina*,
and *Succinea*, all identical with shells now living in the same
country; and in one place I observed (as happens also occasionally in the valley of the Rhine) a passage of the loam
with land-shells into an underlying marly deposit of subaqueous origin, in which were embedded shells of the genera
Limnea, Planorbis, Paludina, Physa, and *Cyclas*, all consisting of recent American species.   Such deposits, more
distinctly stratified than the loam containing land-shells,
are produced, as before stated, p. 174, in all great alluvial plains, where the river shifts its position, and where
marshes, ponds, and lakes are formed in its old deserted
channels.   In this part of America, however, it may have
happened that some of these lakes were caused by partial
subsidences, such as were witnessed, during the earthquakes
of 1811-12, around New Madrid, in the valley of the
Mississippi.

Owing to the destructible nature of the yellow loam,
every streamlet flowing over the platform has cut for
itself, in its way to the Mississippi, a deep gully or ravine;
and this erosion has of late years, especially since 1812, proceeded with accelerated speed, ascribable in some degree to
the partial clearing of the native forest, but partly also to
the effects of the earthquake of 1811-12.   By that convulsion the region around Natchez was rudely shaken and
much fissured.   One of the narrow valleys near Natchez, due

to this fissuring, is now called the 'Mammoth Ravine' on
account of the remains of the mammoth found in it. Though
no less than seven miles long, and in some parts sixty feet
deep, I was assured by a resident proprietor, Colonel Wiley,
that it had no existence before 1812. With its numerous
ramifications, it is said to have been entirely formed since
the earthquake at New Madrid. Before that event, Colonel
Wiley had ploughed some of the land exactly over a spot
now traversed by part of this water-course.

I satisfied myself that the ravine had been considerably
enlarged and lengthened a short time before my visit, and it
was then freshly undermined and undergoing constant waste.
From a clayey deposit immediately below the yellow loam,
bones of the *Mastodon ohioticus*, a species of megalonyx,
bones of the genera *Equus*, *Bos*, and others, some of extinct
and others presumed to be of living species, had been
detached, and had fallen to the base of the cliffs. Mingled
with the rest, the pelvic bone of a man, *os innominatum*,
was obtained by Dr. Dickeson of Natchez, in whose collection
I saw it. It appeared to be quite in the same state of preservation, and was of the same black colour as the other
fossils, and was believed to have come like them from a depth
of about thirty feet from the surface. In my 'Second Visit
to America,' in 1846,[*] I suggested, as a possible explanation
of this association of a human bone with remains of a mastodon
and megalonyx, that the former may possibly have been
derived from the vegetable soil at the top of the cliff, whereas
the remains of extinct mammalia were dislodged from a lower
position, and both may have fallen into the same heap or talus
at the bottom of the ravine. The pelvic bone might, I conceived, have acquired its black colour by having lain for
years or centuries in a dark superficial peaty soil, common
in that region. I was informed that there were many human

bones, in old Indian graves in the same district, stained of as
black a dye. On suggesting this hypothesis to Colonel Wiley,
of Natchez, I found that the same idea had already occurred
to his mind. No doubt, had the pelvic bone belonged to any
recent mammifer other than Man, such a theory would never
have been resorted to; but so long as we have only one
isolated case, and are without the testimony of a geologist who
was present to behold the bone when still engaged in the
matrix, and to extract it with his own hands, it is necessary to
suspend our judgment as to the high antiquity of the fossil.

In regard to other parts of America, which I myself have
not visited, I have not as yet been able to obtain authentic
proof of the coexistence of Man with the mastodon, though
it is highly probable that such proofs will eventually be
brought to light. Professor Whitney, indeed, points out that
'amid the foot-hills of the Sierra, works of Man have been
frequently found among the recent deposits of auriferous
gravel, in close connection with the bones of the mastodon
and elephant,' * but I have not yet had an opportunity of
examining fully into the evidence.

Human Remains in Loess near Maestricht.

The banks of the Meuse at Maestricht, like those of the
Rhine at Bonn and Cologne, are slightly elevated above the
level of the alluvial plain. On the right bank of the Meuse,
opposite Maestricht, the difference of level is so marked, that
a bridge, with many arches, has been constructed to keep up,
during the flood season, a communication between the river
banks and the hills or bluffs which bound the alluvial
plain. This plain is composed of loess, the origin of
which, probably from the melting of the Alpine glaciers,
when they retreated at the close of the glacial period, will
be treated of in the sixteenth chapter. It consists of a brown

* Geology of California, vol. i. p. 252.

friable loam, almost totally devoid of any appearance of
stratification, and containing, here and there, land and amphibious shells. It extends over large portions of Germany
and Austria, wrapping over the inequalities of the ground
and attaining an enormous thickness in the river valleys.
It is extensively worked for brick-earth to the depth of about
eight feet. The bluffs before alluded to often consist of a
terrace of gravel, from thirty to forty feet in thickness,
covered by an older loess, which is continuous as we ascend
the valley to Liége. In the suburbs of that city, patches
of loess are seen at the height of two hundred feet above
the level of the Meuse. The table-land in that region, composed of Carboniferous and Devonian rocks, is about four
hundred and fifty feet high, and is not overspread with
loess.

A terrace of gravel covered with loess has been mentioned
as existing on the right bank of the Meuse at Maestricht. Answering to it another is also seen on the left bank
below that city, and a promontory of it projecting into the
alluvial plain of the Meuse, and approaching to within a
hundred yards of the river, was cut through during the excavation of a canal running from Maestricht to Hocht,
between the years 1815 and 1823. This section occurs at
the village of Smeermas, and is about sixty feet deep, the
lower forty feet consisting of stratified gravel, and the upper
of twenty feet of loess. The number of molars, tusks, and
bones (probably parts of entire skeletons) of elephants obtained during these diggings, was extraordinary. Not a few
of them are still preserved in the museums of Maestricht and
Leyden, together with some horns of deer, bones of the ox-
tribe, and other mammalia, and a human lower jaw, with
teeth. According to Professor Crahay, who published an
account of it at the time, this jaw, which is now preserved
at Leyden, was found at the depth of nineteen feet from the

surface, where the loess joins the under-lying gravel, in a stratum of sandy loam resting on gravel, and overlaid by some pebbly and sandy beds. The stratum is said to have been intact and undisturbed, but the human jaw was isolated, the nearest tusk of an elephant being six yards removed from it in horizontal distance.

Most of the other mammalian bones were found, like these human remains, in or near the gravel, but some of the tusks and teeth of elephants were met with much nearer the surface. I visited the site of these fossils in 1860, in company with M. van Binkhorst, and we found the description of the ground published by the late Professor Crahay of Louvain to be very correct.* The projecting portion of the terrace, which was cut through in making the canal, is called the hill of Caberg, which is flat-topped, sixty feet high, and has a steep slope on both sides towards the alluvial plain. M. van Binkhorst (who is the author of some valuable works on the paleontology of the Maestricht chalk) has recently visited Leyden, and ascertained that the human fossil above mentioned is still entire in the museum of the university. Although we had no opportunity of verifying the authenticity of Professor Crahay's statements, we could see no reason for suspecting the human jaw to belong to a different geological period from that of the extinct elephant. If this were granted, it might have no claims to a higher antiquity than the human remains which Dr. Schmerling disentombed from the Belgian caverns; but the fact of their occurring in a pleistocene alluvial deposit in the open plains, would be one of the first examples on record of such a phenomenon. The top of the hill of Caberg is not so high above the Meuse as is the terrace of St. Acheul, with its flint implements above

* M. van Binkhorst has shown me the original MS. read to the Maestricht Athenæum in 1823. The me- / moir was published in 1836 in the Bulletin de l'Académie Royale de Belgique, tom. III. p. 43.

R

the Somme, but at St. Acheul no human bones have yet
been detected.

In the museum at Maestricht are preserved a human
frontal and a pelvic bone, stained of a dark peaty colour :
the frontal very remarkable for its lowness, and the prominence of the superciliary ridges, which resemble those of the
Borreby skull, figured at p. 91. These remains may be the
same as those alluded to by Professor Crabay in his memoir,
where he says, that in a deposit in the suburbs of Hocht, of a
black colour, were found leaves, nuts, and freshwater shells
in a very perfect state, and a human skull of a dark colour.
They were of an age long posterior to that of the loess containing the bones of elephants, and in which the human jaw
now at Leyden is said to have been embedded.

Human Remains in the Loess near Strasburg.

In the year 1823 the late M. Ami Boué disinterred with
his own hands, many bones of a human skeleton from ancient
undisturbed loess at Lahr, nearly opposite Strasburg, on the
right side of the great valley of the Rhine, four miles distant
from the river and about one hundred feet above the water
level. It is situated near the point where the tributary
valley, drained by the small stream called the Schutter,
flowing from the Black Forest, joins the great alluvial plain
of the Rhine. In this part of that plain the loess is at least
two hundred feet thick, and small hills and valleys have
been excavated in it. A portion of the formation passes up
from the principal into the tributary valley, the sides of which
it skirts, rising to the height of eighty feet or more above the
Schutter. It has been denuded at Lahr, so as to form a
succession of terraces on the right bank of the small stream.
On examining the lowest of these terraces, M. Boué saw, in
the face of a perpendicular cliff of loess, about five feet high,
a large bone projecting, which proved afterwards to be a

human femur. On digging into the cliff the bones of nearly half a skeleton were obtained, consisting of the femur, tibia, fibula, ribs, vertebræ, metatarsals, and others ; but no skull. They lay in a nearly horizontal position, but not as if they were part of a corpse which had been buried there.

The enveloping loess was solid, not like loess-mud washed down by rain and then reconsolidated. The beds immediately below the bones contained some pebbles, and still lower down was gravel with rounded stones of Bunter sandstone and gneiss from the Black Forest. In the inferior beds of loess, on a level with the bones, shells of the genera *Lymnea*, *Pupa*, *Physa*, *Clausilia*, *Helix*, and more rarely *Cyclostoma*, occurred. But as to the *Lymnea*, mentioned by M. Boué in his paper, he now thinks it may possibly have been the prevailing *Succinea oblonga* of the loess.

M. Boué conceives that, before the loess was denuded in this valley by the Schutter, a thickness of at least eighty feet of it must have been superimposed on the human bones. He considers the loamy deposit at Lahr to be continuous with the loess of the Rhine, and to have come from the same source, and not to belong properly to the alluvium of the Schutter. He ascribes great antiquity to the bones partly because of their position so low down in the loess, and partly, because in loess of the same age in the vicinity, the remains of extinct mammalia had been detected.

When M. Boué, accompanied by M. Cordier, first showed the bones in Paris to Cuvier, that naturalist at once pronounced them to be human ; but to the surprise of the two geologists, declared his belief that they came from a burial-ground. The same notion was afterwards adopted by M. Alexandre Brongniart, who supposed them to have been interred in modern river-mud of the Schutter.* Even after

* Annales des Sciences Naturelles. 1829, vol. xviii. Revue Bibliogr. p. 160.

R 2

M. Boué had revisited the locality in 1829, and confirmed his first observations, the judgment of so experienced a geologist went for nothing against the preconceived ideas then generally entertained as to the geological date of man's origin.[*]

The precious collection of Lahr-bones, filling a box, was left by M. Boué in M. Cuvier's care, and having been neglected, is now lost. As to their age, I see no reason for supposing that they were more ancient than those found by Schmerling in the Liége caverns. But if the views which will be set forth in the sixteenth chapter are sound, some extensive continental movements of elevation and depression, which happened immediately after the retreat of the great Alpine glaciers, were of date posterior to the embedding of these bones in the ancient mud of the Rhine.

* Akademie der Wissenschaften, Sitzungsberichte, Band 8, p. 89, 1852. Dr. A. Boué, Erläuterungen über die von mir im Lorse des Rheinthales im Jahre 1823 aufgefundenen Menschenknochen.

PART II.

THE GLACIAL PERIOD CONSIDERED WITH REFERENCE TO THE ANTIQUITY OF MAN

——

CHAPTER XII.

ANTIQUITY OF MAN RELATIVELY TO THE GLACIAL PERIOD AND TO THE EXISTING FAUNA AND FLORA.

CHRONOLOGICAL RELATION OF THE GLACIAL PERIOD, AND THE EARLIEST KNOWN SIGNS OF MAN'S APPEARANCE IN EUROPE—SERIES OF TERTIARY DEPOSITS IN NORFOLK AND SUFFOLK IMMEDIATELY ANTECEDENT TO THE GLACIAL PERIOD—GRADUAL REFRIGERATION OF CLIMATE PROVED BY THE MARINE SHELLS OF SUCCESSIVE DEPOSITS—FOREST BED OF CROMER AND FLUVIO-MARINE STRATA—FOSSIL PLANTS AND MAMMALIA OF THE SAME—TABLE OF THE SUCCESSION OF BEDS ON THE NORFOLK COAST—ARCTIC PLANTS IN THE LIGNITE LAYERS OVERLYING THE FOREST BED—OVERLYING CONTORTED DRIFT, MID-GLACIAL SANDS, AND BOULDER CLAY—NEWER FRESHWATER FORMATION OF MUNDESLEY COMPARED TO THAT OF HOXNE—GREAT OSCILLATIONS OF LEVEL IMPLIED BY THE SERIES OF STRATA IN THE NORFOLK CLIFFS—EARLIEST KNOWN DATE OF MAN LONG SUBSEQUENT TO THE APPEARANCE OF THE EXISTING FAUNA AND FLORA.

FREQUENT allusions have been made in the preceding pages to a period called the glacial, to which no reference is made in the Chronological Table of Formations given at p. 5. It comprises a long series of ages, chiefly of post-tertiary date, during which the power of cold, whether exerted by glaciers on the land, or by floating ice on the sea, was greater in the northern hemisphere, and extended to more southern latitudes than now.

It often happens that when in any given region we have
pushed back our geological investigations as far as we can, in
search of evidence of the first appearance of Man in Europe,
we are stopped by arriving at what is called the 'boulder-
clay' or 'northern drift.' This formation is usually quite
destitute of organic remains, so that the thread of our inquiry into the history of the animate creation, as well as of
man, is abruptly cut short. The interruption, however, is by
no means encountered at the same point of time in every
district. In the case of the Danish peat, for example, we
get no farther back than the recent period of our Chronological Table (p. 5), and then meet with the boulder-clay,
and it is the same in the valley of the Clyde, where the
marine strata contain the ancient canoes before described
(p. 51), and where nothing intervenes between that recent
formation and the glacial drift. But we have seen that, in
the neighbourhood of Bedford (p. 214), the memorials of
Man can be traced much farther back into the past, namely,
into the pleistocene epoch, when the human race was contemporary with the mammoth and many other species of
mammalia now extinct. Nevertheless, in Bedfordshire as in
Denmark, the formation next antecedent in date to that
containing the human implements is still a member of the
glacial drift, with its erratic blocks.

If the reader remembers what was stated in the ninth
chapter, p. 190, as to the absence or extreme scarcity of
human bones and works of art in all strata, whether marine
or freshwater, even in those formed in the immediate
proximity of land inhabited by millions of human beings,
he will be prepared for the general dearth of human memorials in glacial formations, whether recent, pleistocene or
of more ancient date. If there were a few wanderers over
lands covered with glaciers, or over seas infested with icebergs, and if a few of them left their bones or weapons in

moraines or in marine drift, the chances, after the lapse of
thousands of years, of a geologist meeting with one of them
must be infinitesimally small.

It is natural, therefore, to encounter a gap in the regular
sequence of geological monuments bearing on the past history
of Man, wherever we have proofs of glacial action having
prevailed with intensity, as it has done over large parts of
Europe and North America, in the pleistocene period. As
we advance into more southern latitudes approaching the
50th parallel of latitude in Europe, and the 40th in North
America, this disturbing cause ceases to oppose a bar to our
inquiries; but even then, in consequence of the fragmentary
nature of all geological annals, our progress is inevitably
slow in constructing anything like a connected chain of
history, which can only be effected by bringing the links of
the chain found in one area to supply the information which
is wanting in another.

The least interrupted series of consecutive documents to
which we can refer in the British Islands, when we desire to
connect the tertiary with the post-tertiary periods, are found
along the eastern coast in the counties of Essex, Norfolk,
Suffolk, Lincolnshire, and Yorkshire; and I shall treat of
them now, as they have a direct bearing on the relations of
the human and glacial periods, which will be the subject of
several of the following chapters. The fossil shells of the
deposits in question clearly point to a gradual refrigeration
of climate, from a temperature somewhat warmer than that
now prevailing in our latitudes, to one of intense cold; and
the successive steps which have marked the coming on of
the increasing cold are matters of no small geological interest.

It will be seen in the Table at p. 5, that next before the
post-tertiary period stands the pliocene, divided into the
older and newer. The shelly and sandy beds representing

these periods in Norfolk and Suffolk are termed provincially Crag, having under that name been long used in agriculture to fertilise soils deficient in calcareous matter, or to render them less stiff and impervious.   In South Suffolk, the older pliocene strata called Crag are divisible into the Coralline and the Red Crags, the former being the older of the two. In Norfolk and North Suffolk, a fluvio-marine formation, commonly termed the 'Norwich Crag,' occurs, together with some beds of clay and sand called the Chillesford beds, and these formations are all referable to the newer pliocene period.

Climate of the Coralline, Red, and Norwich Crags.

We were first indebted to Mr. Searles Wood, F.G.S., for an admirable monograph on the fossil shells of these British pliocene formations, published between the years 1848 and 1856, to which he is now adding a valuable supplement. The number of Crag mollusca, exclusive of freshwater and land species, amounts to more than 450.   Of these, the number unknown as living, commonly called extinct, is greatest in the oldest or Coralline Crag, less in the Red Crag, and least in the Norwich, or fluvio-marine series ; or, in other words, the fauna shows a decided tendency to assimilate with that now living in proportion as it approaches in age to the present time.   But there is also another change observable in the shells as they are traced from older to newer formations, for they indicate a gradual refrigeration of temperature, the Coralline Crag representing a climate like that now prevailing in the Mediterranean, the Red Crag one more resembling that of the present British seas, and the Norwich, or fluvio-marine series, being almost entitled to be styled arctic.

These deposits have long been objects of great interest among geologists, and have occasioned no little controversy

in arranging the disconnected portions in chronological order
and determining the physical conditions under which each
was accumulated. The reader, when studying the different
memoirs on the fossil shells and other organic remains contained in them, may easily be so perplexed by the conflicting
opinions expressed, and by the want of agreement as to the
limits of particular species and varieties, as to regard the
whole as a mass of confusion. But it will be found in reality
that there is a steady approach to agreement; even after all
those modifications of opinion have been taken into account
which arise from the occasional dredging up in deep seas,
and far from land, of some species and varieties previously
supposed to be extinct.

The two following Tables (p. 250), one published by Mr.
Prestwich and the other kindly made for me by Mr. Searles
Wood, are particularly instructive, because, although founded
upon somewhat different data, they illustrate in a very analogous way successive modifications of climate. I have therefore substituted them for similar Tables given in my former
edition on the authority of the late Mr. S. P. Woodward,
which I could not have repeated without considerable
alteration, owing to the increase of our knowledge of the
fossil species in the Norfolk and Suffolk crags, as well as of
the range of living species brought to light in recent deep-
sea explorations.

In Mr. Prestwich's table such species are excluded as are
now living in British seas, and he has confined his comparison
to the relative distribution of the exclusively Northern (or
Scandinavian and Arctic) and Southern (or Mediterranean)
species throughout the three crags. A glance suffices to
show that there is a steady decrease of southern forms as we
pass upward from the older, or Coralline, to the newer, or
Norwich Crag, and though the corresponding increase of
northern forms is not so clear at first sight, it becomes

Relative distribution in the Crags of species not now living in British seas.
Prestwich.[a]

	Total living species	Species now restricted to	
		Northern Seas	Southern Seas
NORWICH CRAG .	130	19 (14·6 p. c.)	11 (8·4 p. c.)
RED CRAG . .	216	23 (10·7 p. c.)	32 (14·8 p. c.)
CORALLINE CRAG	261	14 (5·3 p. c.)	65 (24·6 p. c.)

Table showing increase of cold conditions which took place between the formation of the oldest and newest portions of the Crag, given by Mr. Searles Wood.

	Number of known species both recent and extinct (exclusive of derivative shells)	Species now found living in the Mediterranean	Mediterranean species which do not range to British Seas	Percentage of Mediterranean species which do not range to British Seas
FLUVIO-MARINE CRAG (inclusive of the Chillesford Beds .	124	63	6	8 per cent.
RED CRAG .	255†	119†	22†	18·5 per cent.
CORALLINE CRAG	275	202	48	23·2 per cent.

apparent when we remember that the 23 northern species in the Red Crag, occurring in a total of 216 living species, are in a proportion of only about 10 per cent., while the 19 of the Norwich Crag occur in a total of only 130, which is about 14 per cent.

Mr. Searles Wood in his Table bases the argument as to climate upon the living Mediterranean species contained in

* Prestwich, Crag Beds of Norfolk and Suffolk, Quart. Geol. Journal, vol. xxvii. p. 474.

† These numbers would be lower if the Walton red crag shells were not included in the 255 species of the red crag, and Mr. Searles Wood is of opinion that the Walton crag has its affinities essentially with the coralline crag both in the number of extinct forms and in the Mediterranean aspect of the living ones.

each division of the Crag and the number of those species which do not now range so far north as the Bristol Channel, thus showing, not only that the more southern species diminish in the newer formations, but also that while nearly a fourth part of the Mediterranean forms of the Coralline Crag are entirely confined to that region, there are only five in the Fluvio-marine Crag which do not range to British seas. Nor does this entirely express the difference in the faunas; for three out of the five Norwich Crag species are so exceptional that they were long considered to be extinct, and two of them, *Chemnitzia internodula* and *Ringicula ventricosa*, have been dredged up by Mr. Gwyn Jeffreys from the deep water of the Mediterranean: the latter shell, *R. ventricosa*, is confined to great depths. When we consider, therefore, the many arctic species lately obtained from the abyssmal depths of the Atlantic in temperate latitudes, this shell does not carry with it the weight it would do if it was a common Mediterranean shell of ordinary depths, for its southern range may in this case be connected with the colder temperature of greater depth rather than with latitude.

Moreover the Norwich Crag contains many exclusively arctic shells which are quite unknown in the Coralline Crag, such as *Astarte borealis*, *Leda lanceolata*, *Cardium Grœnlandicum*, and *Tellina calcarea*, besides several others, such as *Scalaria Grœnlandica*, *Natica clausa*, &c., which being purely arctic are only classed as British shells because an occasional living specimen has been dredged up off the Shetlands or other extreme northern points of the British Isles. The presence of these northern shells cannot be explained away by supposing that they were inhabitants of the deep parts of the sea: for some of them, such as *Tellina calcarea* and *Astarte borealis*, occur plentifully, and sometimes with the valves united by their ligament, in company with other littoral shells, such as *Mya arenaria* and *Littorina rudis*, and were evidently not thrown up from deep water. The predominance, therefore, of these peculiarly arctic forms and their large size, together with the stunted specimens of shells common to southern seas, is conclusive evidence to the mind of a conchologist as to the arctic character of the Norwich Crag. In like manner it is the presence of such genera as *Pyrula, Cancellaria, Terebra, Cassidaria, Pholadomya, Lingula, Chama, Discina, Pyramidella, Fossarus,* and others which give a southern aspect to the Coralline Crag shells.*

But to draw safe conclusions from the molluscan fauna as to the temperature of the sea during the accumulation of the Crag deposits it is necessary to bear in mind the physical changes which have taken place in Western Europe since that remote epoch. Many of the shells, such as *Voluta Lamberti* for example, were in all probability remnants of the warm seas of the preceding miocene epoch. They may long have striven against the increasing cold, and when driven out at length would migrate to lower latitudes if not entirely extinguished. On the return of a warmer temperature they would find their old haunts occupied by strangers who had come in the cold time, and were too firmly established to be expelled. At the same time it must be remembered that the late explorations in the Atlantic have shown that where arctic and tropical marine currents meet and pass one another it is possible to have on one side of the line of junction an assemblage of living forms characteristic of warm seas, and on the other, within a few hundred yards, an essentially arctic fauna.

Yet, notwithstanding these variations of marine temperature, there is sufficient evidence to show that the cold must have gone on increasing from the time of the Coralline to that of the Norwich Crag, and continued to become more and

* See S. V. Wood, Quart. Geol. Journal, vol. xiii. p. 541.

more severe until it reached its maximum in what has been called the glacial period, some of the monuments of which in the East of England we shall presently consider. The refrigeration of climate from the time of the older to that of the newer pliocene strata is not now announced for the first time, as it was inferred from a study of the Crag shells in 1846 by the late Edward Forbes, although the data on which to found such a generalisation were far more scanty then than they are now.*

Mammalia of the Crags.

The only remains which have been found of the land animals of the date of the Crags were obtained from the basement portion of each of the three divisions (the Coralline, Red, and Norwich) into which this series is commonly divided. These mark periods of denudation when older strata were swept away and their less perishable constituents re-embedded in the first-formed portion of the new deposit. Such seems to be the true account of the origin of the copro-lite beds at the base of the Red and Coralline Crags. They consist in great part of clay nodules containing phosphate of lime interspersed with rolled bones and teeth of land and marine animals. One of the most remarkable peculiarities of these beds is the number of large flints preserving their white coating as if freshly extracted from the chalk, and per-fectly unrolled even in the Red Crag which is so largely made up of shingle beaches composed of rounded pebbles. The transportation of these flints from their original source in the chalk was probably effected by ice, to whose agency may also be ascribed the presence here of fragments of granite and porphyry and the debris of several other formations.

Professor Owen determined many of the mammalia from the Red and Coralline Crags to be miocene forms; others,

* Memoirs of Geological Survey, London, 1846, p. 391.

however, are of pliocene age, and may be truly contemporaneous with the marine shells among which they lie. Such
in all probability is *Mastodon arvernensis*, an older pliocene
fossil of the South of France, and also *Elephas meridionalis*,
Tapirus priscus, and others. But these are all extinct
species, and have no generic affinity to the present indigenous
land animals of the British Isles. The mammalia obtained
from a bed of unworn chalk flints at the base of the
Norwich Crag have a somewhat different character. The
miocene forms are not present, and there is a greater
abundance of well-known pliocene forms, so that the evidence
of the land and associated marine animals are more in
harmony with each other than in the older members of the
Crag.

Cromer Forest-Bed.

While the Norwich Crag and overlying Chillesford beds
supply a tolerably complete account of the inhabitants of
the sea, and show that the principal living salt-water
molluscs of Great Britain were already in existence, the
land animals and plants of the same period may be most
advantageously studied on the neighbouring coast of Norfolk,
where the so-called 'Forest-Bed' of Cromer contains much
that is wanting to complete an account, though an imperfect
one, of the natural history of England at that time.

The annexed section will give a general idea of the
succession of beds as exhibited in the cliff at Cromer. It
is there seen that the base of the cliff is a floor of solid
chalk (No. 1), on which was deposited the clay that once
supported a forest of pine trees (No. 2), and may now be
seen to enclose their stumps and spreading roots.

Thirty years ago, when I first examined this bed, I saw
many trees, with their roots in the old soil, laid open at the
base of the cliff near Happisburgh; and long before my
visit, other observers, and among them the late Mr. J. C.

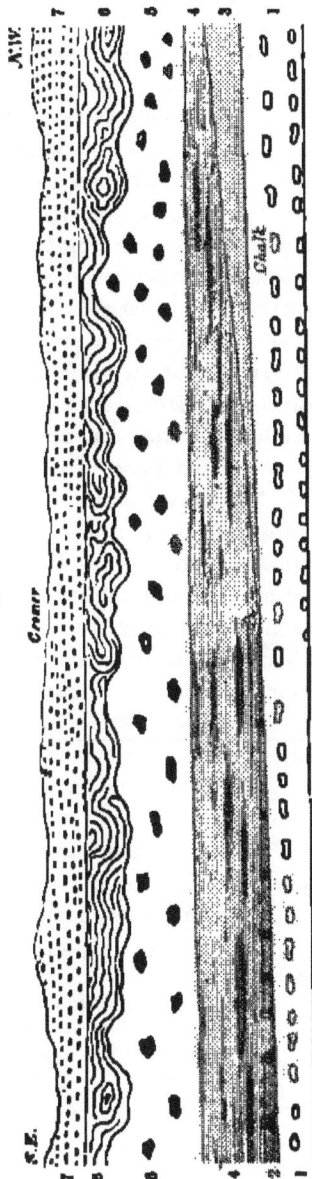

Fig. 32.

Generalised Diagram to illustrate the succession of the Strata in the Norfolk Cliffs, extending several miles N.W. and N.E. of Cromer.

A Site of Cromer Jetty.
1 Upper chalk with flints in regular stratification.
2 'Forest Bed,' with stumps of trees *in situ* and remains of *Elephas meridionalis, E. primigenius, E. antiquus, Rhinoceros etruscus*, &c. This bed increases in depth and thickness eastward.
3 Weybourne sands.
4 Fluvio-marine pebbly sands and clays with abundant lignite beds and mammalian remains, and with cones of the Scotch and spruce firs and other wood. The Bure-valley beds of S.

Wood, and Westleton beds of Prestwich. At the top of these beds, in the lignite layers, arctic plants have been found. East of Cromer this formation becomes more marine, and is underlain at Runton by a bone of peaty clay containing fresh-water shells of the age of the forest-bed.
5 Till of the Norfolk coast boulder-clay of glacial period, with far transported erratics, some of them polished and scratched, twenty to eighty feet in thickness.
6 Contorted Drift.
7 Superficial gravel and mud of undetermined age.

Taylor, had noticed the buried forest. It has been traced for more than forty miles from Cromer to Kessingland, but only portions are visible at once, as in order to expose the stumps to view, a vast body of sand and shingle must be cleared away by the force of the waves. As the sea is always gaining on the land, new sets of trees are brought to light from time to time, so that the breadth as well as the length of the area of ancient forest land seems to have been considerable.

This forest must have existed for a long time, since, besides the erect trunks, some of them two to three feet in diameter, there is a vast accumulation of vegetable matter in the immediately overlying clays. Between the stumps of the buried forest, and in the lignite above them, are many well-preserved cones of the Scotch and spruce firs, *Pinus sylvestris* and *Pinus Abies*. The specific names of these fossils were determined for me in 1840, by a botanist of no less authority than the late Robert Brown; and Professor Heer has lately examined a large collection from the same stratum, and recognised among the cones of the spruce some which had only the central part or axis remaining, the rest having been bitten off, precisely in the same manner as when in our woods the squirrel has been feeding on the seeds. There is also in the forest-bed a great quantity of resin in lumps, resembling that gathered for use, according to Professor Heer, in Switzerland, from beneath spruce firs.

The following is a list of some of the plants and seeds which were collected by the Rev. S. W. King, in 1861, from the forest-bed at Happisburgh, and named by Professor Heer:—

Plants and Seeds of the Forest and Lignite Beds below the Glacial Drift of the Norfolk Cliffs.

Pinus sylvestris, Scotch fir.	*Prunus spinosa*, common sloe.
Pinus Abies, spruce fir.	*Menyanthes trifoliata*, buckbean.
Taxus baccata, yew.	*Nymphœa alba*, white water-lily.
Nuphar luteum, yellow water-lily.	*Alnus*, alder.
Ceratophyllum demersum, hornwort.	*Quercus*, oak.
Potamogeton, pondweed.	*Betula*, birch.

The insects, so far as they are known, including several
species of *Donacia*, are, like the plants and freshwater shells,
of living species. It may be remarked, however, that the
Scotch fir has been confined in historical times to the northern
parts of the British Isles, and the spruce fir is nowhere indigenous in Great Britain. The other plants are such as
might now be found in Norfolk, and many of them indicate
fenny or marshy ground. Mr. King discovered in 1863, in
the forest-bed, several rhizomas of the large British fern
Osmunda Regalis, of such dimensions as they are known to
attain in marshy places. They are distinguishable from those
of other British ferns by the peculiar arrangement of the
vessels, as seen under the microscope in a cross section.

The following is a list of the mammalia which have been
found in these beds. The land species are taken from a paper
by Mr. Boyd Dawkins.[*] The marine are on the authority
of Professor Owen.

l. *living*.   ex. *extinct*.

ex. *Elephas meridionalis*, Nesti
ex. — *antiquus*, Falconer
ex. — *primigenius*, Blum.
ex. *Rhinoceros etruscus*, Falconer
ex. — *megarhinus*, Christol.
l. *Equus caballus*, L. Common horse
ex. *Hippopotamus major*, Nesti
l. *Sus scrofa*, L. Common wild pig
ex. *Ursus spelæus*, Blum. Cave bear
ex. — *arvernensis*
l. *Canis lupus*, L. Wolf
l. — *vulpes*, Briss. Fox
ex. *Machairodus sp.* (*latidens?*)
ex. *Bos primigenius*, Boj.
ex. *Cervus megaceros*, Owen. Gigantic
Irish deer
l. *Cervus capreolus*, L. Roe deer
l. — *elaphus*, L. Red deer

ex. *Cervus Polignacus*, Falconer
ex. — *cornutorum*, Boyd Dawkins
ex. — *verticornis*, Boyd Dawkins
ex. — *Sedgwickii*, Ouen.
l. *Castor fiber* (*Europæus*), Owen.
Beaver
ex. *Trogontherium Cuvieri*, Fischer
l. *Talpa Europæa*, Schm. Common
mole
l. *Sorex moschatus*, L.
l. — *vulgaris*, Owen. Common
shrew
l. *Arvicola amphibia*, Owen. Water-
rat
l. *Trichecus rosmarus*, Walrus
l. *Monodon monoceros*, Narwhal
l. *Balænoptera* (various);

When we consider the familiar aspect of the flora, the
accompanying mammalia are certainly most extraordinary.

* Quart. Geol. Journ., vol. xxviii, p. 417, 1872.

There are no less than three elephants, a rhinoceros and hippopotamus, a large extinct beaver, and several large estuarine and marine mammalia, such as the walrus, the narwhal, and the whale.

Apart from the abundance of so many large animals, the interest attaching to these remains is greatly increased when we reflect how many of them are elsewhere found in association with worked implements of human origin. Two of the three species of elephant are known to have lived with Man, one of them, the *Elephas antiquus*, in the more ancient river gravels of England and France, the other, *E. primigenius*, seen for the first time in the Cromer forest-bed, and continuing as the constant associate of Man down to the

Fig. 33.*

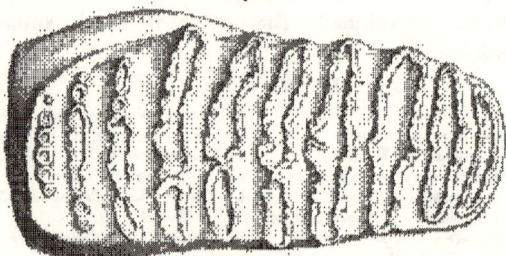

Elephas meridionalis, Nesti.

Penultimate molar, lower jaw, right side, size one-third of original, Norwich Crag, and newer pliocene, Saint Prest, near Chartres. Not yet proved to have coexisted with man.

dawn of historic times. The third species, *Elephas meridionalis*, perhaps the largest of his race, has never yet been found in company with human implements, and this together with the presence of Trogontherium and two species of deer, only known from this deposit in England, render it still

* For this fig. I am indebted to the late M. Lartet.

doubtful whether the forest-bed may not be of older date
than the first introduction of Man into Great Britain. Of
the other animals, the *Rhinoceros etruscus* and *Ursus arvernensis* are peculiar in England to this bed, while *Rhinoceros
megarhinus*, and the cave-bear *Ursus spelæus*, are also
common to the cave-deposits and older river-drifts associated
with flint implements. The freshwater shells accompanying
the fossil quadrupeds above enumerated, are such as now
inhabit rivers and ponds in England, equalling them in point
of size and vigorous development; but among them, as at
Runton, where a small patch of forest-bed occurs buried in
the Weybourne sands, next to be mentioned, a remarkable

Fig. 34.

Cyclas (Pisidium) amnica var.?
The two middle figures are of the natural size.

variety of *Pisidium amnicum* (fig. 34) occurs, identical
with that which accompanies *Elephas antiquus* at Ilford
and Grays, in the valley of the Thames.

The northernmost extension of the Cromer forest-bed is
marked by the Cromer jetty (A, fig. 32), to the north-west of
which it is replaced by a marine deposit called the Weybourne Sands (No. 3), which rest like the forest-bed
immediately upon the chalk forming at the line of junction
a layer of yellow sand and gravel, locally called the 'pan.'
These sands, the relative position of which in the series
will be best understood by consulting the following Table,
are very closely allied to the Norwich and Chillesford Crag,
but are distinguished from it by the presence of *Tellina*

A 2

balthica, a shell which does not occur, in the Norwich Crag, but appears here for the first time, and continues to be abundant throughout the superincumbent glacial deposits.*

Glacial and Post-Glacial Formations succeeding the Cromer Forest-Bed.

	MARINE DEPOSITS.	FRESHWATER AND TERRES-TRIAL DEPOSITS.
PLEISTOCENE	. . . .	Hoxne freshwater deposit, with Palæolithic implements, and Mundesley beds with fish-scales, sects, &c. (pp. 219, 207).
	UPPER GLACIAL: Bridlington Beds Chalky Boulder-Clay	
	MID-GLACIAL: Mid-Glacial Sands	
	LOWER GLACIAL: Contorted Drift Till of the Cromer Cliffs	
NEWER PLIOCENE	Pebbly Beds of Bure Valley and Westleton	Lignite layers, with Arctic plants, *Salix Polaris*, and *Hypnum turgescens* (p. 261).
	WEYBOURNE SANDS . .	Cromer Forest-Bed, with *Elephas meridionalis* and other extinct and living mammalia (see p. 257).
	CHILLESFORD AND NORWICH CRAG	

The Weybourne beds appear on the whole to be coeval with the land forest-bed, or but slightly newer, for in the 'pan' formed at the line of junction, the teeth of elephants are embedded.

* Mr. Searles Wood, jun., and Mr. Harmer, who have worked out the relations of these beds, place them on account of the occurrence of *Tellina balthica*, at the base of the Lower Glacial, but they are so closely allied with the Norwich Crag, that it seems to me safer to keep them as an intermediate formation— as I did in my previous edition.

Glacial formations succeeding the Cromer Forest.

Overlying both the forest-bed and the marine sands, we find a series of sands and clays (No. 4, p. 255), with lignite sometimes thirty feet thick, and containing alternations of fluviatile and marine strata, implying that the old forest-land had sunk down so as to be covered by the waters of an estuary. That the marine shells included in these beds lived and died on the spot, and were not thrown up by the waves during a storm, is proved, as Mr. King has remarked, by the fact that at West Runton, N.W. of Cromer, the *Mya truncata* and *Leda myalis* are found with both valves united and erect in the mud (No. 4), all with their posterior or siphuncular extremities uppermost. This attitude affords as good evidence to the conchologist that those mollusca lived and died on the spot, as the upright position of the trees proves to the botanist that there was a forest over the chalk, east of Cromer.

These fluvio-marine beds have been identified by Mr. Searles Wood, jun., with beds containing *Tellina balthica* in the valley of the Bure near Norwich, and by Mr. Prestwich with certain pebbly beds at Westleton, near Dunwich. At Happisburgh and elsewhere along the Cromer coast they show a gradual passage into the overlying boulder-clay or glacial till (No. 5, p. 255), in which smoothed and striated stones of all sizes are included in a stiff unstratified clayey matrix, affording unmistakeable evidence of glaciers and the prevalence of great cold.

Reasoning from these data it occurred to Mr. Nathorst, a skilful Swedish geologist who visited Cromer section in the autumn of 1872, that the lignite beds of the laminated sands and clays (No. 4, p. 255), ought to exhibit in their vegetable remains a transition from the comparatively mild climate of the

forest-bed to the severe cold indicated by the till, and he was
fortunate enough to find the remains of plants becoming more
stunted as they occurred higher in the beds, until within
half a foot of the boulder-clay he found *Salix Polaris*, now
only known within the arctic circle, together with a moss
which has been referred by the eminent bryologist Berggren
to *Hypnum turgescens*, an arctic moss only found living in
temperate latitudes on the extreme heights of the Alps.
Among the fragments of rock included in the till(No. 5, p. 255)
are blocks of granite and other rocks foreign to this part of
England.   Some of the erratics have been supposed to have
travelled from Scandinavia, and I came to the conclusion in
1834, that they had really come from Norway and Sweden,
having myself traced, in that year, the course of a continuous
stream of such blocks from those countries to Denmark, and
across the Elbe, through Westphalia, to the borders of
Holland.   It is not surprising that they should then reappear
on our eastern coast between the Tweed and the Thames,
regions not half so remote from parts of Norway as are many
Russian erratics from the sources whence they came.

Fig. 35.

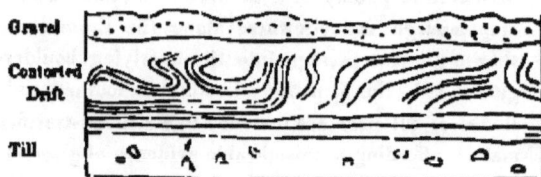

Gravel

Contorted
Drift

Till

Cliff 50 feet high between Bacton Gap and Mundesley.

The overlying strata of silt and mud (No. 6, p. 255), are
commonly called 'Contorted Drift,' from the singular derangement often displayed by their stratification, which in
many places seems to have a very intimate relation to the

irregularities of outline in the subjacent till. There are some cases, however, where the upper strata are much bent, while the lower beds of the same series have continued horizontal. Thus the annexed section (fig. 35) represents a cliff about fifty feet high, at the bottom of which is till, or unstratified clay, containing boulders, having an even horizontal surface, on which repose comformably beds of laminated clay and sand about five feet thick, which, in their turn, are succeeded by vertical, bent, and contorted layers of sand and loam twenty feet thick, the whole being covered by flint gravel. The curves of the variously coloured beds of loose sand, loam, and pebbles, are so complicated that not only may we sometimes find portions of them which maintain their verticality to a height of ten or fifteen feet, but they

Fig. 36.

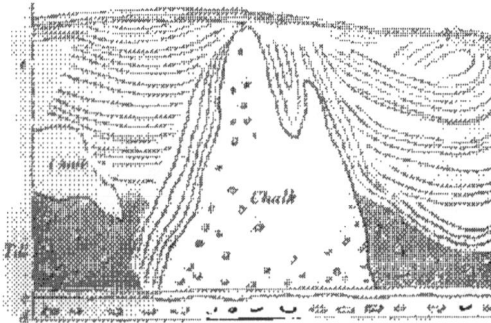

Included pinnacle of chalk at Old Hythe point, west of Sherringham.

d Chalk with regular layers of chalk flints.
c Layer called 'the pan,' of chalk, flints, and marine shells of recent species, cemented by oxide of iron.
c Contorted drift.

have also been folded upon themselves in such a manner that continuous layers might be thrice pierced in one vertical boring.

To the north of Cromer fragments of chalk and crushed masses of the same rock many yards in diameter often occur. Several striking examples may be seen both in the sea-cliffs and inland in pits worked in the drift. I saw one conspicuous example in 1829, in the sea-cliff west of Sherringham where an enormous pinnacle of chalk, between 70 and 80 feet high, was to be seen flanked on both sides by vertical layers of loam clay and gravel (fig. 36). The flints of this huge erratic formed a continuous layer, each flint being unbroken and retaining its original form. The mass, though much altered, was still visible in 1839, but when I visited the spot in 1869 it had disappeared.

This chalky fragment is only one of many detached masses which have been included in the drift, and forced along with it into their present position. Below the drift the level surface of the chalk *in situ* (*d*) may be traced for miles along the coast, where it has escaped the violent movements to which the superincumbent strata have been exposed.[*]

We are called upon, then, to explain how any force can have been exerted against the upper masses, so as to produce movements in which the subjacent strata have not participated. It may be answered that, if we conceive the till and its boulders to have been drifted to their present place by ice, the lateral pressure may have been supplied by the stranding of ice-islands. We learn, from the observations of Messrs. Dease and Simpson in the polar regions, that such islands, when they run aground, push before them large mounds of shingle and sand. It is therefore probable that they often cause great alterations in the arrangement of pliant and incoherent strata forming the upper part of shoals or submerged banks, the inferior portions of the same remaining unmoved. Or many of the complicated curvatures

[*] For a full account of the drift of East Norfolk, see a paper by the author, Philosophical Magazine, No. 101, May, 1840.

of these layers of loose sand and gravel may have been due
to another cause, the melting on the spot of icebergs and
coast ice in which successive deposits of pebbles, sand, ice,
snow, and mud, together with huge masses of rock fallen
from cliffs, may have become interstratified. Ice-islands
so constituted often capsize when afloat, and gravel, once
horizontal, may have assumed, before the associated ice was
melted, an inclined or vertical position. The packing of ice
forced up on a coast may lead to a similar derangement in a
frozen conglomerate of sand or shingle, and, as Mr. Trimmer
has suggested,* alternate layers of earthy matter may have
sunk down slowly during the liquefaction of the intercalated
ice so as to assume the most fantastic and anomalous positions,
while the strata below, and those afterwards thrown down
above, may be perfectly horizontal (see above, p. 262).

In most cases where the principal contortions of the layers
of gravel and sand have a decided correspondence with deep
indentations in the underlying till, the hypothesis of the
melting of large lumps and masses of ice once mixed up with
the till affords the most natural explanation of the phenomena.
The quantity of ice now seen in the cliffs near Behring's
Straits, in which the remains of fossil elephants are common,
and the huge fragments of solid ice which Meyendorf discovered in Siberia, after piercing through a considerable
thickness of incumbent soil, free from ice, is in favour of
such an hypothesis, the partial failure of support necessarily
giving rise to foldings in the overlying and previously horizontal layers, as in the case of creeps in coal mines.

The few fossils that have hitherto been obtained from the
contorted drift are all species found in the Weybourne sands
beneath, but in the cliff between Yarmouth and Lowestoft,
there occur sands capped by boulder-clay full of jurassic and
cretaceous débris, but which as Mr. Searles Wood and Harmer

* Quarterly Journal, Geological Society, vol. vii. pp. 22, 30.

have shown are proved both by their position and the shells
contained in them to be of later date than the contorted drift.
These beds, termed by Mr. Searles Wood 'mid-glacial,'[*]
because they occur between two glacial formations, show in
their fauna a remarkable amelioration of marine temperature,
probably produced by a current from some warmer part of
the sea. Several of the forms of the warm Coralline period,
such as *Turritella incrassata, Cardita corbis, Nassa granu-
lata,* and several others, reappear, while many of those species
which are common to the Norwich and Red Crags are such
as do not now range to northern seas. There is also a total
want of transported blocks, such as are to be found in the
contorted drift below and the chalky boulder-clay above.

This last, which so distinctly reposes on the mid-glacial
sands between Yarmouth and Lowestoft, showing a return of
the ice, is a wide-spread formation covering a great part of
the country between the Wash and the Thames. It is almost
wholly devoid of fossils, containing many fragments of lias
limestone, and other rocks whose place *in situ* is only to be
found considerably to the north-west of their present resting
place. The Bridlington beds, which on the coast of
Yorkshire near Flamborough Head exhibit the next approach
to more modern times, contain the most arctic shells of any
yet found in England. Out of a total of seventy species of
Mollusca, of which only two or three are extinct, and one of
these *Nucula Cobboldiæ* is considered by some as only a
variety of a living species, there are no less than thirteen Arctic
forms which are not present in the glacial drifts of Norfolk or
Suffolk or in any of the preceding crags; and these sands
and shells are buried in a vast accumulation of till extending
along the east coast from Northumberland to Lincolnshire.

With these beds we conclude the proofs of a long series
of changes of climate which can be traced from the warm

* Geology of East Anglia, Intro-   Palæontographical, vol. for 1871,
duction to Crag mollusca supplement,   p. xxii., 1872.

climate of the Coralline Crag to the intense cold of the
Glacial period when the boulder-clay was accumulated, which
limits so far as our knowledge yet extends the appearance of
Man in England. The formation of this boulder-clay and
the connection with it (supposed by Messrs. Searles Wood
and Harmer) of a sheet of land-ice in as low a latitude as
52° N., and without any lofty mountains in the neighbourhood, will be discussed in the thirteenth chapter, when I shall
endeavour to show that such an hypothesis is by no means
inconsistent with the amount of cold experienced in the
present state of the globe.

Mundesley Post-glacial Freshwater Formation.

In the range of cliffs above described, at Mundesley about
two miles south-east of Cromer, a fine example is seen of a
freshwater formation, newer than all those already mentioned,
a deposit which has filled up a depression hollowed out of
all the older beds 4, 5, and 6, of the section, p. 255.

When I examined this line of coast in 1839, the section
alluded to was not so clearly laid open to view as it has
been of late years, and finding at that period not a few of the
fossils in the lignite beds, No. 4, above the forest bed, identical in species with those from the post-glacial deposits, b c,
I supposed the whole to have been of contemporaneous origin,
and so described them in my paper on the Norfolk cliffs.[*]

Mr. Gunn was the first to perceive this mistake, which he
explained to me on the spot when I revisited Mundesley in
the autumn of 1859, in company with Dr. Hooker and
Mr. King. The last-named geologist had the kindness to
draw up for me the annexed diagram of the various beds
which he had studied in detail.[†]

* Philosophical Magazine, vol. xvi.
May 1840, p. 345.
† Mr. Prestwich has given a correct
account of this section in a paper read
to the British Association, Oxford,
1860. See Geologist's Magazine,
vol. iv. 1861.

The formations 4, 5, and 6, already described, p. 255, were evidently once continuous, for they may be followed for miles N.W. and S.E. without a break, and always in the same

Fig. 37.

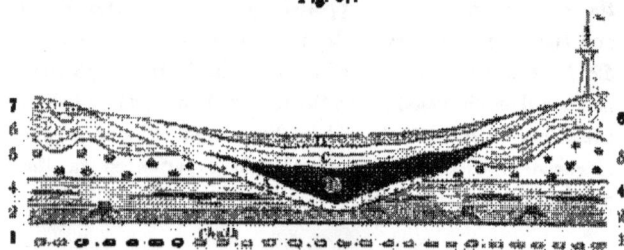

Section of the newer freshwater formation in the cliffs at Mundesley, two miles S.E. of Cromer, drawn up by the Rev. S. W. King.

Height of cliff where lowest, 35 feet above high water.

Older Series.

1 Fundamental chalk, below the beach line.
2 Forest-bed, with elephant, rhinoceros, stag, &c., and with tree roots and stumps, also below the beach line.
4 Finely laminated sands and clays, with thin layers of lignite.
5 Glacial boulder-till.
6 Contorted drift.
7 Gravel overlying contorted drift.

N.B.—No. 3 of the section, fig. 32, at p. 255, is wanting here.

Newer Freshwater Beds.

a Coarse river gravel, in layers inclined against the till and laminated sands.
b Black peaty deposit, with shells of Anodon, Valvata, Cyclas, Succinea, Limnea, Paludina, &c., seeds of Ceratophyllum demersum, Nuphar lutea, scales and bones of pike, perch, salmon, &c., elytra of Donacia, Copris, Harpalus, and other beetles.
c Yellow sands.
d Drift gravel.

order. A valley or river channel was cut through them, probably during the gradual upheaval of the country, and the hollow became afterwards the receptacle of the comparatively modern freshwater beds, a, b, c, and d. They may well represent a silted-up river-channel, which remained for a time in the state of a lake or mere, and in which the black peaty mass B, accumulated by a very slow growth over the gravel of the river-bed A. In B we find remnins of some of the same plants which were enumerated as common in the ancient lignite in 4, such as the yellow water-lily and hornwort (p. 256), together with some freshwater shells which occur in the same fluvio-marine series 4. The only shell which I found not referable to a British species is the minute paludina, fig. 38, already alluded to, p. 215.

Fig. 38.

Paludina marginata, Michaud.   (*P. minuta* Strickland.)
Hydrobia marginata.[*]
The middle figure is of the natural size.

When I showed the scales and teeth of the pike, perch, roach, and salmon, which I obtained from this formation, to Mr. Agassiz, he thought they varied so much from their nearest living representatives that they might rank as distinct species; but Mr. Yarrell doubted the propriety of so distinguishing them. The insects, like the shells and plants, are identical, so far as they are known, with living British species. No progress has yet been made at Mundesley in discovering the contemporary mammalia.

By referring to the description and section of the freshwater deposit at p. 210, the reader will at once perceive the

[*] This shell is said to have a subspiral operculum (not a concentric one, as in Paludina), and therefore to be referable to the Hydrobia, a subgenus of Rissoa. But this species is always associated with freshwater shells, while the Rissoæ frequent marine and brackish waters.

striking analogy of the Mundesley deposits, and those at Hoxne thirty miles to the S.S.W., the latter so productive of flint implements of the Amiens type. Both of them, like the Bedford gravel with flint tools and the bones of extinct mammalia (noticed at p. 215), are post-glacial. It will also be seen that a long series of events, accompanied by changes in physical geography, intervened between the 'forest bed,' No. 2, fig. 32, p. 255, when the *Elephas meridionalis* flourished, and the period of the Mundesley fluviatile beds A, B, C; just as in France I have shown, p. 232, that the same *E. meridionalis* belonged to a system of drainage different from and anterior to that with which the flint implements of the old alluvium of the Somme and the Seine were connected.

Before the growth of the ancient forest, No. 2, fig. 37, the *Mastodon arvernensis*, a large proboscidian, characteristic of the crag, appears to have died out, or to have become scarce, as no remains of it have yet been found in the Norfolk cliffs. There was, no doubt, time for other modifications in the mammalian fauna between the era of the marine beds, No. 3, p. 255 (the shells of which imply permanent submergence beneath the sea), and the accumulation of the uppermost of the fluvio-marine, and lignite beds, No. 4. In the interval we must suppose repeated changes of level, during which land covered with trees, an estuary with its freshwater shells, and the sea with its *Mya truncata* and other mollusca still retaining their erect position, gained by turns the ascendency. These changes were accompanied by some denudation followed by a grand submergence of several hundred feet, probably brought about slowly, and when floating ice aided in transporting erratic blocks from great distances. The glacial till, No. 5, then originated, and the gravel and sands, No. 6, were afterwards superimposed on the boulder-clay, first in horizontal beds, which became sub-

sequently contorted. These were covered in their turn by
other layers of gravel and sand, No. 7, pp. 255 and 268, the
downward movement still continuing.

The entire thickness of the beds above the chalk at
some points near the coast, and the height to which they are
now raised, are such as to lead us to infer that the subsidence
of the country after the growth of the forest-bed exceeded
four hundred feet. The re-elevation must have amounted
to nearly as many feet, as the site of the ancient forest,
originally subaërial, has been brought up again to within a
few feet of high-water mark. Lastly, after all these events,
and probably during the final process of emergence, the
valley was scooped out in which the newer freshwater strata
of Mundesley, fig. 37, p. 268, were gradually deposited.

Throughout the whole of this succession of geographical
changes, the flora and invertebrate fauna of Europe appear
to have undergone no important revolution in their specific
characters. The plants of the forest-bed belonged already
to what has been called the Germanic flora. The mollusca,
the insects, and even some of the mammalia, such as the
European beaver and roebuck, were the same as those now
coexisting with Man. Yet the oldest memorials of our
species at present discovered in Great Britain are post-
glacial, or posterior in date to the boulder-clay. The
position of the Hoxne flint implements corresponds with
that of the Mundesley beds, from A to D, p. 268, and
the most likely stratum in which to find hereafter flint
tools is no doubt the gravel A of that section, which has
all the appearance of an old river-bed. No flint tools
have yet been observed there, but had the old alluvium of
Amiens or Abbeville occurred in the Norfolk cliffs instead
of the Valley of the Somme, and had we depended on the
waves of the sea instead of the labour of many hundred
workmen continued for twenty years, for exposing the flint

implements to view, we might have remained ignorant to this day of the fossil relics brought to light by M. Boucher de Perthes, and those who have followed up his researches.

For the present we must be content to wait and consider that we have made no investigations which entitle us to wonder that the bones or stone weapons of the era of the *Elephas meridionalis* have failed to come to light. If any such lie hid in those strata, and should hereafter be revealed to us, they would carry back the antiquity of Man to a distance of time probably more than twice as great as that which separates our era from that of the most ancient tool-bearing gravels yet discovered in Picardy or elsewhere. But even then the reader will perceive that the age of Man, though pre-glacial, would be so modern in the great geological calendar, as given at p. 5, that he would date only a very short way back into the Newer Pliocene Period.

CHAPTER XIII.

CHRONOLOGICAL RELATIONS OF THE GLACIAL PERIOD AND THE EARLIEST SIGNS OF MAN'S APPEARANCE IN EUROPE.

CHRONOLOGICAL RELATIONS OF THE CLOSE OF THE GLACIAL PERIOD AND THE EARLIEST SIGNS OF THE APPEARANCE OF MAN—SCANDINAVIA ONCE ENCRUSTED WITH ICE LIKE GREENLAND—OUTWARD MOVEMENT OF CONTINENTAL ICE IN GREENLAND—MILD CLIMATE OF GREENLAND IN THE MIOCENE PERIOD—GLACIAL STATE OF SWEDEN IN THE PLEISTOCENE PERIOD—CHRONOLOGICAL RELATIONS OF THE HUMAN AND GLACIAL PERIODS IN SWEDEN—SCOTLAND FORMERLY ENCRUSTED WITH ICE—ITS SUBSEQUENT SUBMERGENCE AND RE-ELEVATION—LATEST CHANGES PRODUCED BY GLACIERS IN SCOTLAND—REMAINS OF THE MAMMOTH AND REINDEER IN SCOTCH BOULDER-CLAY—FORMATION OF LAKES OR KAMES—RELATION OF THE HUMAN AND GLACIAL PERIODS IN SCOTLAND—RARITY OF ORGANIC REMAINS IN GLACIAL FORMATIONS.

THE chronological relations of the human and glacial periods were frequently alluded to in the last chapter, and the sections obtained near Bedford (p. 215), and at Hoxne, in Suffolk (p. 219), and a general view of the Norfolk cliffs, have taught us that the earliest signs of Man's appearance in the British Isles, hitherto detected, are of post-glacial date, in the sense of being posterior to the grand submergence of England beneath the waters of the glacial sea. But after that period, during which nearly the whole of England north of the Thames and Bristol Channel lay submerged for ages, the bottom of the sea, loaded with mud and stones melted out of floating ice, was upheaved, and glaciers filled for a second time the valleys of many mountainous regions. We may now, therefore, inquire whether the peopling of Europe by those Paleolithic races which were contemporary

T

with the *Elephas primigenius* and other mammalia now extinct, was brought about during this concluding phase of the glacial epoch.

Although it may be impossible in the present state of our knowledge to come to a positive conclusion on this head, I know of no inquiry better fitted to clear up our views respecting the geological state of the northern hemisphere at the time when the fabricators of the flint implements of the Amiens type flourished. I shall therefore now proceed to consider the chronological relations of that ancient people with the final retreat of the glaciers from the mountains of Scandinavia, Scotland, Wales, and Switzerland.

Scandinavia once Covered with Ice and a Centre of Dispersion of Erratics.

In the North of Europe, along the borders of the Baltic, an unstratified boulder formation, or till, similar to that which has been shown in the last chapter to abound in the Norfolk cliffs, is continuous for hundreds of miles east and west, and it has been long known that the erratic blocks contained in it, often of very large size, are of northern origin. Some of them have come from Norway and Sweden, others from Finland, and their present distribution implies that they were carried southwards, for a part at least of their way, by floating ice, at a time when much of the area over which they are scattered was under water. But it appears from the observations of Hoetlingk, in 1840, and those of more recent inquirers, that while many blocks have travelled to the south, others have been carried northwards, or to the shores of the Polar Sea, and others north-eastward, or to those of the White Sea. In fact, they have wandered towards all points of the compass, from the mountains of Scandinavia as a centre, and the roctilinear furrows imprinted

by them on the polished surfaces of the mountains where
the rocks are hard enough to retain such markings, radiate
in all directions, or point outwards from the highest land,
in a manner corresponding to the course of the erratics
above mentioned.[*]

Mr. Kjerulf, of Christiania, has shown that the direction
of the furrows and striæ, produced by glacial abrasion,
neither conforms to a general movement of floating ice from
the Polar regions, nor to the shape of the existing valleys,
as it would do if it had been caused by independent glaciers
generated in the higher valleys after the land had acquired
its actual shape. Their general arrangement and apparent
irregularities are, he contends, much more in accordance with
the hypothesis of there having been at one time a universal
covering of ice over the whole of Norway and Sweden, like
that now existing in Greenland, which, being annually re-
cruited by fresh falls of snow, was continually pressing outwards
and downwards to the coast and lower regions, after crossing
many of the lower ridges, and having no relations to the
minor depressions, which were all choked up with ice and
reduced to one uniform level.[†]

In support of this view, he appeals to the admirable
description of the continental ice of Greenland, published
by Dr. H. Rink, of Copenhagen, who resided three or four
years in the Danish settlements, in Baffin's Bay, on the west
coast of Greenland, between latitudes 69° and 73° N. 'In
that country, the land,' says Dr. Rink, 'may be divided into
two regions, the "inland" and the "outskirts." The "inland," which is 800 miles from west to east, and of much
greater length from north to south, is a vast unknown

---

[*] Sir R. I. Murchison, in his
'Russia and the Ural Mountains'
(1845), has indicated on a map, not
only the southern limits of the Scandinavian drift, but by arrows the

direction in which it 'proceeded ec-
centrically from a common central
region.'

[†] Zeitschrift, Geologische Gesellschaft, Berlin, 1860.

continent, buried under one continuous and colossal mass of
permanent ice, which is always moving seaward, but a small
proportion only of it in an easterly direction, since nearly
the whole descends towards Baffin's Bay.'[*] At the heads of
the fiords which intersect the coast, the ice is seen to rise
somewhat abruptly from the level of the sea beyond which
the ice of the interior rises continuously as far as the eye
can reach, and to an unknown altitude. All minor ridges
and valleys are levelled and concealed, but here and there
steep mountains protrude abruptly from the icy slope,
and a few superficial lines of stones or moraines are visible
at seasons when no recent snow has fallen.

The few glimpses that have hitherto been obtained of
this wilderness of ice reveal the fact that it is, at any rate,
enormously thick ; it rises rapidly inland from its seaward
border, often two hundred feet and more in height, to a
gently sloping plateau 2,000 feet high, and estimated to
consist entirely of ice. Professor Nordenskiöld and his
companions, who penetrated in 1870 to a distance of thirty
geographical miles into the interior, reached a height of
2,200 feet above the level of the sea, in lat. 68° 22''.
Professor Nordenskiöld describes ‚the inland ice as in
constant motion, advancing slowly but with varying velocity
towards the sea, into which it passes through a great many
large and small ice-streams. The movement of the ice
gives rise to huge chasms and clefts, which occur chiefly
where the advance is most rapid—that is to say, in the
neighbourhood of the great ice-streams[†]—and the great
thickness of this moving sheet of ice is indicated by the
enormous depth of the 'moulins,' or well-like cavities in the
ice, into which Professor Nordenskiöld saw rivulets and considerable streams of water plunge till they were lost to sight.

* Journal of Royal Geographical   Greenland, Geol. Mag., vol. ix., 1872,
Society, vol. xxiii. p. 145, 1853.   p. 305.
† Nordenskiöld, Expedition to

At the present day the surplus of ice formed in the interior which has not been melted by the summer thawing and evaporation finds an outlet in the fiords which bound the west coast of Greenland. Through these the ice is now protruded in the form of huge tongues of ice, several miles wide, which then exhibit all the phenomena of true glaciers. They frequently continue their course into the salt water, grating along the rocky bottom, which they must polish and score, at depths of hundreds and even of more than a thousand feet. At length they reach water deep enough to break off vast masses, which then float in the sea as icebergs, and are constantly drifting into Baffin's Bay.

Since the interior of Greenland consists almost entirely of this sheet of ice, it is not surprising that the surface is exceedingly free from stones and earth, such as those of which the lateral moraines of an Alpine glacier are made. Some few trains of superficial stones and rubbish have, however, been noticed by Rink; so that we must conclude that a few peaks do rise above the surrounding ice-field. At some points, where the ice of the interior of Greenland reaches the coast, Dr. Rink saw mighty springs of clayey water issuing from under the edge of the ice even in winter, showing the grinding action of the glacial mass mixed with sand, on the subjacent surface of the rocks. The débris thus shed out into the sea is peopled by a rich assemblage of molluscan and crustacean life, especially in places removed from the influx of the fresh water. Dr. Richard Brown, who visited Greenland in 1861 and again in 1867, and who has written several able papers upon the action of the Greenland ice, calculates that one of these muddy subglacial streams, when discharging itself into a long narrow fiord where the mud would collect in the bottom, would form a deposit at least twenty-five feet thick

in a century, exactly resembling the brick-earth or fossiliferous boulder-clay of Scotland.*

The 'outskirts,' where the Danish colonies are stationed,
consist of numerous islands, of which Disco Island is the
largest, in lat. 70° N., and of many peninsulas, separated by
fiords from fifty to a hundred miles long, running into the
land, and through which the ice above alluded to passes on
its way to the bay. This area is 30,000 square miles in
extent, and contains in it some mountains 4,000 feet to
5,000 feet high. The perpetual snow usually begins at the
height of 2,000 feet, below which level the land is for the
most part free from snow between June and August, and
supports a vegetation of several hundred species of flowering
plants, which ripen their seeds before the winter.† There are
even some places where phenogamous plants have been found
at an elevation of 4,500 feet, a fact which, when we reflect
on the immediate vicinity of so large and lofty a region of
continental ice in the same latitude well deserves the attention of the geologist, who should also bear in mind that
while the Danes are settled to the west in the 'outskirts,'
there exists, due east of the most southern portion of this
ice-covered continent, at the distance of about 1,200 miles,
the home of the Laplanders with their reindeer, bears,
wolves, seals, walruses, and cetacea. If, therefore, there
are geological grounds for suspecting that Scandinavia
or Scotland or Wales was ever in the same glacial condition as Greenland now is, we must not imagine that
the contemporaneous fauna and flora were everywhere poor
and stunted, or that they may not, especially at the distance
of a few hundred miles in a *southward* direction, have been
very luxuriant.

Dr. Pingel, with whom I conversed at Copenhagen in

* Brown, Quart. Geol. Journal, 1870, vol. xxvi. p. 662.
† Rink, Journ. Roy. Geog. Soc., vol. xxiii. p. 149.

1834, ascertained in 1830-32, in company with Captain
Graah, that the whole coast from lat. 60° to about 70° north
has been subsiding for the last four centuries, so that some
ancient piles driven into the beach to support the boats of
the settlers have been gradually submerged, and wooden
buildings have had to be repeatedly shifted farther inland.*

In Norway and Sweden, instead of such a movement of
subsidence, the land is slowly rising; but we have only to
suppose that formerly, when it was covered like Greenland
with continental ice, it sank at the rate of several inches or
feet in a century, and we shall be able to explain why marine
deposits are found above the level of the sea, and why these
generally overlie polished and striated surfaces of rock.

We know that Greenland was not always covered with
snow and ice, for when we examine the tertiary strata of
Disco Island (of the upper miocene period) we discover there
a multitude of fossil plants, which demonstrate that, like
many other parts of the arctic regions, it formerly enjoyed a
mild and genial climate. Among the fossils brought from
that island, lat. 70° N., Professor Heer has recognised
Sequoia Langsdorfii, a coniferous species which flourished
throughout a great part of Europe in the miocene period,
and is very closely allied to the living *Sequoia sempervirens*
of California. The same plant was found fossil by Sir
John Richardson within the arctic circle, far to the west on
the Mackenzie River, near the entrance of Bear River; also
by some Danish naturalists in Iceland to the east. The Icelandic surturbrand, or lignite, of this age has also yielded a
rich harvest of plants, more than thirty-one of them, according to Steenstrup and Heer, in a good state of preservation,
and no less than fifteen specifically identical with miocene
plants of Europe. Thirteen of the number are arborescent;
and amongst others is a tulip-tree (*Liriodendron*), with its

* Principles of Geology, 11th ed. ch. xxxi. p. 196.

fruit and characteristic leaves, a plane (*Platanus*), a walnut,
and a vine, affording unmistakeable evidence of a climate in
the parallel of the arctic circle which precludes the supposition of glaciers then existing in the neighbourhood, still less
any general crust of continental ice, like that of Greenland.[*]

As the older pliocene flora of the tertiary strata of Italy,
like the shells of the coralline crag, before adverted to,
p. 252, indicate a temperature milder than that now prevailing in Europe, though not so warm as that of the upper
miocene period, it is probable that the accumulation of snow
and glaciers on the mountains and valleys of Greenland did
not begin till after the commencement of the pliocene
period, and may not have reached its maximum until the
close of that period.

Norway and Sweden appear to have passed through all
the successive phases of glaciation which Greenland has experienced, and others which that country will one day
undergo if the milder climate which it formerly enjoyed
should ever be restored to it. There must have been first a
period of separate glaciers in Scandinavia, then a Greenlandic
state of continental ice, and thirdly, when that diminished,
a second period of enormous separate glaciers filling many a
valley now wooded with fir and birch. Lastly, under the influence of the Gulf Stream, and various changes in the height
and extent of land in the arctic circle, and possibly of astronomical changes, as pointed out by Mr. Croll,[†] a melting of
nearly all the permanent ice between latitudes 60° and 70°
north, corresponding to the parallels of the continental ice
of Greenland, has occurred, so that we have now to go
farther north than lat. 70° before we encounter any glacier

* Heer, Recherches sur la Végéta-   † See Lyell, Principles of Geology,
tion du Pays tertiaire, etc., 1861,   ch. xiii. 11th ed.
p. 178.

coming down to the sea coast. Among other signs of the
last retreat of the extinct glaciers, Kjerulf and other authors
describe large transverse moraines left in many of the Norwegian and Swedish glens.

Chronological Relations of the Human and Glacial Periods in Sweden.

We may now consider whether any, and what part, of
these changes in Scandinavia may have been witnessed by
Man.   In Sweden, in the immediate neighbourhood of
Upsala, I observed, in 1834, a ridge of stratified sand and
gravel, in the midst of which occurs a layer of marl, evidently
formed originally at the bottom of the Baltic by the slow
growth of the mussel, cockle, and other marine shells of living
species intermixed with some proper to fresh water.   The
marine shells are all of dwarfish size, like those now inhabiting the brackish waters of the Baltic; and the marl, in which
myriads of them are embedded, is now raised more than a
hundred feet above the level of the Gulf of Bothnia.   Upon
the top of this ridge (one of those called osars in Sweden)
repose several huge erratics, consisting of gneiss for the
most part unrounded, from nine to sixteen feet in diameter,
and which must have been brought into their present position
since the time when the neighbouring gulf was already characterised by its peculiar fauna.   Here, therefore, we have
proof that the transport of erratics continued to take place,
not merely when the sea was inhabited by the existing testacea, but when the north of Europe had already assumed that
remarkable feature of its physical geography which separates
the Baltic from the North Sea, and causes the Gulf of Bothnia
to have only one-fourth of the saltness belonging to the
ocean.

I cannot doubt that these large erratics of Upsala were
brought into their present position during the recent period,

not only because of their moderate elevation above the sea-
level in a country where the land is now perceptibly rising every
century, but because I observed proofs of a great oscillation of
level which had taken place at Södertelje, south of Stockholm
(about forty-five miles distant from Upsala), after the country
had been inhabited by Man. I described, in the 'Philosophical Transactions' for 1835, the section there laid open in
digging a level in 1819, which showed that a subsidence
followed by a re-elevation of land, each movement amounting
to more than sixty feet, had occurred since the time when a
rude hut had been built on the ancient shore. The wooden
frame of the hut, with a ring of hearthstones on the floor,
and much charcoal, were found, and over them marine
strata, more than sixty feet thick, containing the dwarf
variety of *Mytilus edulis*, and other brackish-water shells
of the Bothnian Gulf. Some vessels put together with
wooden pegs, of anterior date to the use of metals, were
also embedded in parts of the same marine formation, which
has since been raised, so that the upper beds are more than
sixty feet above the sea-level, the hut being thus restored to
about its original position relatively to the sea.

We have seen in the account of the Danish 'shell-
mounds,' or 'refuse-heaps,' of the recent period (p. 14), that
even at the comparatively late period of their origin the
waters of the Baltic were more salt than they are now. The
Upsala erratics may belong to nearly the same era as those
'refuse-heaps.' But were we to go back to a long antecedent
epoch, or to that of the Belgian and British caves with their
extinct animals, and the signs they afford of a state of
physical geography departing widely from the present, or to
the era of the implement-bearing alluvium of St. Acheul,
we might expect to find Scandinavia overwhelmed with
glaciers, and the country uninhabitable by Man. At a much
more remote period the same country was in the state in

which Greenland now is, overspread with one uninterrupted coating of continental ice, which has left its peculiar markings on the highest mountains. This period, probably anterior to the earliest traces yet brought to light of the human race, may have coincided with the submergence of England, and the accumulation of the boulder-clay of Norfolk and Suffolk before mentioned. It has already been stated that the syenite and some other rocks of the Norfolk till (p. 262) seem to have come from Scandinavia, and there is no era when icebergs are so likely to have floated them so far south as when the whole of Sweden and Norway were enveloped in a massive crust of ice; a state of things the existence of which is deduced from the direction of the glacial furrows, and their frequent unconformity to the shape of the minor valleys.

Glacial Period in Scotland.

Professor Agassiz, after his tour in Scotland in 1840, declared his opinion that erratic blocks had been dispersed from the Scottish mountains as from an independent centre, and that the capping of ice had once been of extraordinary thickness.[*]

Mr. Robert Chambers, after visiting Norway and Sweden, and comparing the signs of glacial action observed there with similar appearances in the Grampians, came also to the conclusion that the highlands both of Scandinavia and Scotland had once been ‘moulded in ice,’ and that the outward and downward movement and pressure of the frozen mass had not only smoothed, polished, and scratched the rocks, but had, in the course of ages, deepened and widened the valleys, and produced part of that denudation which has commonly been ascribed exclusively to aqueous action. The

* Agassiz, Proceedings Geol. Soc., 1840, and Edinb. Phil. Journ., xlix. p. 79.

glaciation of the Scotch mountains was traced by him to the
height of at least three thousand feet.[*]

Mr. T. F. Jamieson, of Ellon, in Aberdeenshire, brought
forward in 1860[†] an additional body of facts in support of
this theory. According to him the Grampians were at the
period of extreme cold enveloped 'in one great winding
sheet of snow and ice,' which reached everywhere to the
coast-line, the land being then more elevated than it is now.
He described the glacial furrows sculptured on the solid
rocks as pointing in Aberdeenshire to the south-east, those
of the valley of the Forth at Edinburgh, from west to east,
and higher up the same valley, at Stirling, from north-west
to south-east, as they should do if the ice had followed the
lines of what is now the principal drainage. The observations of Sir James Hall, Mr. Maclaren, Mr. Chambers, and
Dr. Fleming were cited by him in confirmation of this arrangement of the glacial markings, while in Sutherland and
Rossshire he showed that the glacial furrows along the north
coast point northwards, and in Argyleshire westwards, always
in accordance with the direction of the principal glens and
fiords.

Another argument was also adduced by him in proof of
the ice having exerted its mechanical force in a direction
from the higher and more inland country to the lower
region and sea coast. Isolated hills and minor prominences of rock are often polished and striated on the land
side, while they remain rough and jagged on the side
fronting the sea. This may be seen both on the east and
west coast. Mention was also made of blocks of granite
which have travelled from south to north in Aberdeenshire,
of which there would have been no examples had the erratics

* Ancient Sea Margins, Edinburgh,   1858, and January 1865.
1848. Glacial Phenomena, Edinburgh   † Quarterly Geological Journal,
New Philosophical Journal, April   vol. xvi.

been all brought by floating ice from the arctic regions when
Scotland was submerged.   It was also urged against the
doctrine of attributing the general glaciation to submergence,
that the glacial grooves radiate from a centre as in Greenland, whereas, if they had been due to ice coming from the
north, they would have been parallel to the coast-line, to
which they are now often almost at right angles.   The argument, moreover, which formerly had most weight in favour
of floating ice, namely, that it explained why so many of the
stones did not conform to the contour and direction of the
minor hills and valleys, was brought forward by Jamieson
with no small effect, in favour of the doctrine of continental
ice on the Greenlandic scale, which, after levelling up the
lesser inequalities, would occasionally flow in mighty ice-
currents, in directions often at a high angle to the smaller
ridges and glens.

Professor Archibald Geikie, in a paper published in the
transactions of the Geological Society of Glasgow in 1863,
entirely adopts Mr. Jamieson's view as to a continental ice-
sheet, and expresses his belief that wherever the boulder-clay
mantle is removed, the rock below is almost invariably found
to be smoothed and marked by ice scratches.   The same appearances are to be seen where there is no covering of clay or
gravel, and the land has that smooth flowing outline such as
may now be seen in valleys of the Alps where glaciers have
lately passed, grinding off all asperities, and in this manner
transforming small rough crags into dome-shaped hillocks,
the *roches moutonnées* of Swiss geologists.   In Scotland, he
writes, 'it would hardly be an exaggeration to say, that the
whole surface of the island, from Cape Wrath to the Solway,
is smoothed and striated.'*

If these markings are to be attributed to a covering of

* Geikie, Trans. Geol. Soc. of Glasgow, 1863, p. 16.

land-ice, as they are by Mr. Jamieson and Mr. Geikie, they
would imply that the thickness must have been enormous,
for Mr. Jamieson points out that the mountain of Schehallion
in Perthshire, 3,500 feet high, is marked near the top as
well as on its flanks by ice pressing over it from the north.*

The application to Scandinavia and Scotland of this
theory of a continental ice-sheet, makes it necessary to re-
consider the validity of the proofs formerly relied on as
establishing the submergence of a great part of Scotland
beneath the sea, at some period subsequent to the commencement of the glacial period. In all cases where marine shells
overlie till, or rest on polished and striated surfaces of rock,
the evidence of the land having been under water, and
having been since upheaved, remains unshaken; but this
special proof rarely extends to heights exceeding five hundred
feet. In the basin of the Clyde we have already seen that
recent strata occur twenty-five feet above the sea-level, with
existing species of marine testacea, and with buried canoes,
and other works of art. At the higher level of forty feet
occurs the well-known raised beach of the western coast,
which, according to Mr. Jamieson, contains, near Fort
William and on Loch Fyne and elsewhere, an assemblage of
shells implying a colder climate than that of the twenty-five
foot terrace, or that of the present sea. At still greater
elevations, older beds containing a still more arctic group of
shells have been observed at Airdrie, fourteen miles south-
east of Glasgow, 524 feet above the level of the sea.† They
were embedded in patches of gravelly sand and in stratified
clays, with the unstratified boulder-till both above and
below them in the same way as the Bridlington bed of sand
is enclosed in the boulder-clay of Holderness, and in the
overlying unstratified drift were some boulders of granite

* Jamieson, Quart. Geol. Journal, vol. xxi., 1865, p. 165.
† Student's Elements, p. 162.

which must have come from distances of sixty miles at the
least.* But while in the Bridlington sand no less than five
out of the seventy species of shells are not known as living,
every one of the long list of shells from the Scotch till
belongs to living species and even to varieties which are still
extant as inhabitants of the arctic regions, while about 60
per cent. are also British forms.†

The Scotch till is therefore to be referred to a later date
than the Holderness boulder-clay. For though the five extinct species found at Bridlington might have been absent
from Scotland, owing to some of the many causes which limit
the distribution of species, still, as the preceding deposits
in England represent so many variations of arctic sea and
climate, it might fairly be expected that some few of the
numerous extinct forms would have been detected in the
Scotch till, were it as old as the similar deposit in Yorkshire. In the north of Scotland, marine shells have been
found in deposits of the same age in Caithness and in Aberdeenshire at heights of two hundred and fifty feet, and on
the shores of the Moray Frith, as at Gamrie in Banff, at an
elevation of three hundred and fifty feet; and the stratified
sands and beds of pebbles which belong to the same formation ascend still higher—to heights of five hundred feet at
least.‡

At much greater heights, stratified masses of drift occur
in which hitherto no organic remains, whether of marine or
freshwater animals, have ever been found. It is still an un-
decided question whether the origin of all such deposits in
the Grampians can be explained without the intervention
of the sea. One of the most conspicuous examples has been

* Smith of Jordanhill, Quart.
Geol. Journal, vol. vi. p. 387, 1850.
† Jamieson, Quart. Geol. Journal,
vol. xxi. p. 174.

‡ See papers by Prestwich, Proceedings of the Geological Society,
vol. II. p. 545; and T. F. Jamieson,
Quart. Geol. Journal, vol. xvi.

described by Mr. Jamieson as resting on the flank of a hill called Meal Uaine, in Perthshire, on the east side of the valley of the Tummel, just below Killiecrankie. It consists of perfectly horizontal strata, the lowest portion of them 300 feet above the river and 600 feet above the sea. From this elevation to an altitude of nearly 1,200 feet the same series of strata is traceable, continuously, up the slope of the mountain, and some patches are seen here and there even as high as 1,550 feet above the sea. They are made up in great part of finely laminated silt, alternating with coarser materials through which stones from four to five feet in length are scattered. These large boulders, and some smaller ones, are polished on one or more sides, and marked with glacial striæ. The subjacent rocks, also, of gneiss, mica slate, and quartz, are everywhere grooved and polished as if by the passage of a glacier.* At one spot a vertical thickness of 130 feet of this series of strata is exposed to view by a mountain torrent, and in all more than 2,000 layers of clay, sand, and gravel were counted, the whole evidently accumulated under water.

The evidence of the former sojourn of the sea upon the land after the commencement of the glacial period was formerly inferred from the height to which erratic blocks derived from distant regions could be traced, besides the want of conformity in the glacial furrows to the present contours of many of the valleys. Some of these phenomena might possibly, as we have seen, be accounted for by assuming that there was once a crust of ice resembling that now covering Greenland.

The Grampians in Forfarshire and in Perthshire are from 3,000 to 4,000 feet high. To the southward lies the broad and deep valley of Strathmore, and to the south of this again rise the Sidlaw Hills to the height of 1,500 feet and upwards. On the summits of this latter chain, formed of

* Jamieson, Quart. Geol. Journal. vol. xvi. p. 360.

sandstone and shale, and at various elevations, I have observed huge angular fragments of mica-schist, some three,
and others fifteen feet in diameter, which have been conveyed
for a distance of at least fifteen miles from the nearest
Grampian rocks from which they could have been detached.
Others have been left strewed over the bottom of the large
intervening vale of Strathmore.*

It may be possible that the transportation of such blocks
has been due, not to floating ice, but to a period when
Strathmore was filled up with land-ice, a current of which
extended from the Perthshire Highlands to the summit of
the Sidlaw Hills, and the total absence of marine shells from
all deposits, stratified or unstratified, which have any connection with these erratics in Forfarshire and Perthshire,
seems to favour such a theory. But the same mode of
transport can scarcely be imagined for those fragments of
mica-schist, one of them weighing from eight to ten tons,
which were observed much farther south by Mr. Maclaren on
the Pentland Hills, near Edinburgh, at the height of 1,100
feet above the sea, the nearest mountain composed of this
formation being fifty miles distant.† On the same hills,
also, at all elevations, stratified gravels occur which, although
devoid of shells, it seems hardly possible to refer to any but
a marine origin.†

Although I am willing, therefore, to concede that the
glaciation of the Scotch mountains, at elevations exceeding
2,000 feet, may be explained by land-ice, it seems difficult
not to embrace the conclusion that a subsidence took place
not merely of 500 or 600 feet, as demonstrated by the
marine shells, but to a much greater amount, as shown by
the present position of erratics and some patches of stratified
drift. The absence of marine shells at greater heights than

* Proceedings of the Geological   † Maclaren, Geology of Fife, &c.,
Society, vol. iii. p. 311.   p. 220.

525 feet above the sea, will be treated of in a future chapter.
It may in part, perhaps, be ascribed to the action of glaciers,
which swept out marine strata from all the higher valleys,
after the re-emergence of the land.

On the whole, the evidence seems to prove that at the
time of the deposition of the till, Scotland may have been
covered by large fields of ice, radiating from an independent
local centre, and scoring the fundamental rock as far as the
sea margin, and perhaps extending beyond, as in Greenland
at the present day, driving back the salt water and producing
to a depth of several hundred feet below the ordinary sea-
level all the characteristic features of land-ice.

It may help us to conceive such local centres of glacial
action for Scotland, the north of England, and Wales, if we
consider that in our times an island situated in the southern
hemisphere corresponding in latitude to that of Yorkshire,
(55°S.), namely, South Georgia, was found by Captain Cook to
be covered with a universal and permanent sheet of ice reaching to the sea-shore in summer.[*] The cause of this is no doubt
the prevalence of floating ice in the surrounding sea, and if
the reader will consult our map (p. 327) of a supposed submergence of 100 fathoms, (600 feet) he will see by how
small a geographical change the British Isles would be converted into an archipelago, the islands of which, though the
three largest of them would exceed South Georgia in size,
would nevertheless be sufficiently isolated to be chilled by
floating icebergs,[†] which alone under the present astronomical conditions, could give rise in so low a latitude to a great
thickness of land-ice in the absence of high mountains. It
might have seemed natural when we are looking for a source
of greater cold to speculate on the elevation of land or the

* Principles of Geology, vol. i. p. 242, 11th ed., 1872.
† For the prevalence of icebergs
round the island of South Georgia, see Captain Evans's ice-chart map, Admiralty Charts, 1865, No. 1241.

increased height of adjacent mountains, such as those of the lake district of the north of England, but the present state of the southern hemisphere with its icebergs, shows clearly how a downward movement, even of a hundred fathoms, may have converted every island in the British area into an independent centre of ice-action.

Many of the arguments against adopting the doctrine of an almost indefinitely extended polar ice-cap, filling up large parts of the sea, are well treated of in the Duke of Argyll's anniversary address to the Geological Society of London, for 1873, and when we consider how easily the climate of a region, where the height of the mountains is by no means excessive, may be entirely changed by floating bergs, we see that it is not necessary to insist on the simultaneous production of glacial phenomena over the larger portion of a hemisphere, in order to explain the wide extension of the working of land-ice; for oceanic currents from an arctic or antarctic region of continental ice may sometimes take one direction and sometimes another, and produce effects, which, although originating in succession, and belonging to different phases of time during which the marine fauna remained unchanged, might appear to be the result of one and the same continental ice-sheet.

Land Fauna of the Scotch Boulder-Clay.

It must not be supposed that Scotland in its glacial period was devoid of land plants and animals. As already remarked at p. 280, the analogy with Greenland shows that an abundance of plants may exist in close proximity to a great icefield. However, nothing of the sort has yet been found, and there seem to be only three or four instances known in all Scotland of mammalia having been discovered in the boulder-clay. Mr. R. Bald, early in the present century, found a single elephant's tusk at a depth of from

fifteen to twenty feet from the surface, in unstratified
boulder-clay in the valley of the Forth;[*] and in 1817 two
other tusks and some bones of the elephant, three and a half
feet long and thirteen inches in circumference, were met
with, as we learn from the same authority, lying in an
horizontal position, seventeen feet deep in clay with marine
shells at Kilmaurs, in Ayrshire.[†] The species of shells were
not given, but Mr. Bryce, in 1865, ascertained that the formation in which the fossil remains occur is itself a fresh-
water deposit, and is covered by a bed of marine sand with
arctic shells, and then with a great mass of till with glaciated
boulders.[‡] A still more interesting fact concerning these
fossils was ascertained a few years ago by Mr. J. Young, curator
of the Glasgow Museum. Knowing that the freshwater formation from which the bones were taken contained in some
places the seed of the pond-weed *Potamogeton,* and the
aquatic Ranunculus, he washed the mud adhering to the reindeer horns of Kilmaurs, and that which filled the cracks of
the associated elephants' tusks, and detected in these fossils,
which had lain in the Glasgow Museum for half a century,
abundance of the same seeds, thus proving beyond a doubt,
that the mammoth inhabited Scotland at the time when this
freshwater formation was in course of deposition.

In another excavation through the Scotch boulder-clay,
made in digging the Clyde and Forth Junction Railway, the
antlers of a reindeer were found at Croftamie, in Dumbarton-
shire, in the basin of the river Endrick, which flows into
Loch Lomond. They were buried in twelve feet of till, and
associated with marine shells, *Cyprina islandica, Astarte
elliptica, A. compressa, Fusus antiquus, Littorina littorea,*
and a *Balanus.* The height above the level of the sea was

---

* Memoirs of the Wernerian Society,   ‡ Bryce, Quart. Geol. Journal, vol.
Edinburgh, vol. iv. p. 58.   xxi. p. 217, 1865.
† Ibid., vol. iv. p. 63.

between one hundred and one hundred and three feet. The reindeer's horn was seen by Professor Owen, who considered it to be that of a young female of the large variety, called by the Hudson's Bay trappers, the carabou.

That these three examples should be the only indications of a land fauna in Scotland is not wonderful when we reflect on the grinding and destroying power of a large mass of ice. If any drift of the age of the Cromer forest and glacial deposits of the Norfolk coast was ever formed in Scotland, they would easily have been swept away and destroyed when the country was covered by gigantic glaciers ; and indeed some thick masses of sand and gravel on the eastern coast of Aberdeenshire containing *Voluta Lamberti, Nucula Cobboldiæ,* &c., which appear to be a remnant of a Red Crag deposit, represent at present all that is known of the history of Scotland in later tertiary times.[*]

Kames or Eskers of Scotland.

Although Scotland was considerably submerged when the till was formed, the depression did not end here, but according to the opinion of the best Scottish geologists, continued until the land had sunk in places as much as 2,000 feet.[†] To this period have been referred the mounds and ridges of stratified sand and gravel, called Kames or Eskers, which lie over the till, and are especially numerous in the broad straths east of the great glen through which the Caledonian canal now passes. The marine denudation of the till appears to have furnished the material of which these kames or eskers are composed. The entire absence of any fossil remains in them makes it impossible to say whether their formation marks a genial interval in the glacial period, but there must have been much floating ice to carry about the erratic blocks

which are now strewn over the surface, resting on the mounds
of gravel as well as on the hillsides.   After the formation of
the kames there was no doubt a great resuscitation of the
glaciers, yet they were confined to separate basins.

I have elsewhere given an account of striking features in
the physical geography of Perthshire and Forfarshire,* which
I consider to belong to this period; namely a continuous
zone of boulder-clay, forming ridges and mounds from fifty
to seventy feet high (the upper part of the mounds usually
stratified), enclosing numerous lakes, some of them several
miles long, and many ponds and swamps filled with shell-
marl and peat.   This band of till, with Grampian boulders
and associated river gravel, may be traced continuously for
a distance of thirty-four miles, with a width of three and a
half miles, from near Dunkeld, by Coupar, to the south of
Blairgowrie, then through the lowest part of Strathmore,
and afterwards in a straight line through the greatest depression in the Sidlaw Hills, from Forfar to Lunan Bay.

Although no great river now takes its course through this
line of ancient lakes, moraines, and river gravel, yet it evidently marks an ancient line by which, first, a great glacier
descended from the mountains to the sea, and by which,
secondly, at a later period, the principal water drainage of the
country was effected.

Mr. James Geikie, in a paper to the Geological Magazine,†
attributes the clays with marine shells at Errol and Elie to
this later period of ice-action.   If his supposition be correct,
it implies that not much improvement in climate had taken
place, for the shells are as fully boreal in character as those
from the true lower till.   The moraines in the present high-
land valleys mark the length to which the glaciers extended
in this their period of decline.   The same writer has also

* Proceedings of the Geological   † Geological Magazine, vol. ix.
Society, vol. lii. p. 342.   p. 28.

endeavoured to show—from the evidence of the position of
the boulder-clay in Scotland due to land-ice, of erratic
blocks, and at lower levels of marine or freshwater remains,
and here and there some stratified masses intercalated in the
midst of glacial till—how complicated a series of events must
have occurred, including more than one inter-glacial period,
and requiring for their explanation several oscillations of
level and successive submergences and re-elevations of the
land, the discussion of which would lead me beyond the
limits of the present work.[*]

Relation of the Human and Glacial Periods in Scotland.

The occurrence of the mammoth and reindeer in the
Scotch boulder-clay, both of them quadrupeds known to be
contemporary with Man, favours the idea that the retreat of
the glaciers of the Grampians above-mentioned may have
coincided in time with the existence of Man in those parts
of Europe where the climate was less severe, as, for example,
in the basins of the Thames, Somme, and Seine, in which
the bones of these and other extinct or migrating mammalia
are associated with flint implements of the antique type.
In the later periods of their decline, Scotland itself may
have been inhabited by the human race, together with *Megaceros, Bos primigenius*, and other animals, whose remains
have been found in peat mosses, or the silt of ponds, and the
present river courses, together with canoes, and worked
tools of stone.

Rarity of Organic Remains in Glacial Formations.

The general dearth of fossil shells and other organic
remains in stratified drift of the glacial period, even where
the deposits seem to have been of submarine origin, is a

[*] James Geikie, Geological Magazine vols. viii. and ix.

negative character very difficult of explanation. The porous
nature of the strata, and the length of time during which
they have been permeated by rain-water, may partly account
for the destruction of organic remains, and it is also possible
that they were originally scarce, for we read of the waters of
the sea being so freshened and chilled by the melting of icebergs in some Norwegian and Icelandic fiords, that the fish
are driven away and all the mollusca killed. The moraines
of glaciers are always from the first devoid of shells, and if
transported by icebergs to a distance, and deposited where
the ice melts, may continue as barren of every indication of
life as they were when they originated.

It was formerly believed that the bottom of the sea, at
the era of extreme submergence in Scotland and Wales, was
so deep as to reach the zero of animal life, which, in part of
the Mediterranean (the Egean, for example), the late Edward
Forbes fixed, after a long series of dredgings, at 300 fathoms.
But the shells of the glacial drift of Scotland and Wales,
when they do occur, are not always those of deep seas; and,
moreover, our faith in the uninhabitable state of the ocean at
great depths has been rudely shaken, first by Sir Leopold
McClintock and Dr. Wallich finding, in 1860, living star-
fish between Greenland and Iceland at the depth of a thousand fathoms, and more recently by Dr. Wyville Thomson
having discovered mollusca, crustacea, and echinodermata[*]
existing in the Bay of Biscay, at a depth of 2,500 fathoms
(15,000 feet). Neither can we find any reason for supposing
that the extreme cold existing in the depths of arctic seas
will account for an absence of marine life.

Dr. Hooker in his antarctic voyage with Capt. Sir J. C.
Ross, when soundings were made off Victoria Land between
the parallels of 71° and 78° S., established the fact that
the bottom of the ocean was inhabited in those high latitudes

[*] Royal Soc. Proc., vol. xviii. p. 420, 1870.

at depths of from 200 to 400 fathoms by crustacea and mollusca, besides serpulæ, ophiuræ, flustræ, virgulariæ, an encrinite, and many sponges.

Some of the same series of antarctic soundings led to the belief that animal life extended to a depth of at least 550 fathoms. The bottom in those latitudes was covered with fine mud and with occasional stones derived from melting ice. It is therefore evident that the frequency of large icebergs and the proximity of elevated antarctic land entirely covered with perpetual ice are conditions by no means unfavourable to a free development of animal life in the bed of the ocean.*

If we turn our attention to the North Polar seas, we find similar conclusions borne out by the latest investigations. In a Swedish survey of the coast of Spitzbergen and the adjoining seas made in 1861 under the personal direction of Dr. Torell, no less than 150 living species of mollusca were collected, chiefly on the west and north coasts of Spitzbergen, in lat. 79° and 80° N., and the number of individuals, as well as the variety of species, was often great, especially where the bottom consisted of fine mud derived from moraines of glaciers, and from the grinding action of the land-ice on the rocks below.

Between Spitzbergen and the north of Norway, but nearer the former country, Dr. Torell and his fellow labourer Mr. Chydenius obtained, at the enormous depths of 1,000 and 1,500 *fathoms* mollusca (a *Dentalium* and *Bulla*, or *Cylichna*), a crustacean, polythalamian shells, a coral three inches long, with several red actiniæ attached to it, and a few annelids. These occurred to the west of Beeren's Island, in latitude 76° 17′ N., and longitude 13° 53′ E., in a sea where floating ice is common for ten months in the year. The temperature of the mud at the

* J. Hooker, Annals and Magazine of Natural History, 1845, p. 238.

bottom was between 32° and 33° Fahrenheit, and that of the
water at the surface 41°, and of the air 33° Fahrenheit.

In Greenland, north of Disco Island, between latitude
70° and 71° N., in a deep channel of the sea, separating the
peninsula of Noursoak from the island of Omenak, a region
where the largest icebergs come down into Baffin's Bay,
Dr. Torell dredged up, besides more than twenty other
mollusks, *Terebratella Spitzbergensis*, living at a depth of
250 fathoms. This shell I found fossil in 1835, at Uddevalla,
in the ancient post-tertiary beds, far south of its present
range. The bottom of the sea in the Omenak Channel
consisted of impalpable mud, and on the surface of some of
the floating bergs was similar mud, on which they who trod
sank knee-deep; also numerous blocks of granite and other
rocks of all sizes, most of them striated on one, two, or
more sides. Here, therefore, a deposit must be going on of
mud containing marine shells, with intermingled glaciated
pebbles and boulders.

A species of Nucula (*Leda truncata* or *Yoldia truncata*,
Brown), now living in the seas of Spitzbergen, North Greenland, and Wellington Channel, Parry Islands, was found by
Dr. Torell to be one of the most characteristic species in the
mud of those icy regions. Of old, in the glacial period, the
same shell ranged much farther south than at present, being
found embedded in the boulder-clay of the south of Norway
and Sweden, as well as of Scotland. It has been observed by
the Rev. Thomas Brown, together with many other exclusively arctic species, at Elie, in the south of Fife, in glacial
clay, at the level of high-water mark. I have myself
collected it in a fossil state in the glacial clay of Portland
and other localities in Maine in North America. It is the
shell well known as *Leda Portlandica* of Hitchcock.

In ponds and lakes in 'the outskirts' of North Greenland, in Disco Island for example, no freshwater mollusca

were ever met with by Dr. Torell, though some species of
crustacea of the genera Apus and Branchipus inhabit such
waters. This may help us to explain the want of fossils in
all glacial deposits of fluviatile or lacustrine origin. The
discoveries above referred to show that the marine glacial
beds of the Clyde (p. 266), and those of Elie in Fife, with
their arctic shells, are precisely such formations as might
be looked for as belonging to a period when Scotland was
filled with great glaciers such as that which has left its
traces in the Sidlaw Hills.

Dr. Richard Brown found that in the Jakobshavn ice-
fiord, and many others on the shores of Greenland, the bergs
grazed the bottom, in moderately deep water, to such an
extent as almost to destroy animal and vegetable life rooted
to the bottom.* He also points out that the heads of inlets,
unless very broad and open to the sea, are bare of marine
life, because (as above mentioned with regard to Norway
and Iceland) the quantity of fresh-water from the sub-glacial
stream and melting bergs makes the neighbourhood unfavourable for marine animals. Still, after exhausting all
the causes now in action by which azoic stratified beds are
in process of formation, it cannot be denied that much
difficulty remains in accounting for the absence of life in
those strata which, when the ice began to diminish, must
have been deposited in fiords as they were gradually
abandoned by the gigantic glaciers which had filled them to
the sea bottom ; but we shall again advert to this subject
when pointing out in the next chapter the irregularity of
the presence of marine shells, and their singularly capricious
distribution in the drift formed after the commencement of
the Glacial Period in districts which we can prove to have
been submerged.

* Quart. Geol. Journal, vol. xxvi., 1870, p. 686.

CHAPTER XIV.

CHRONOLOGICAL RELATIONS OF THE GLACIAL PERIOD AND THE EARLIEST SIGNS OF MAN'S APPEARANCE IN EUROPE,

Continued.

PARALLEL ROADS OF GLEN ROY FORMED IN GLACIER LAKES — PRESENT GLACIER LAKE OF SWITZERLAND — SIGNS OF EXTINCT GLACIERS IN WALES — GREAT SUBMERGENCE OF WALES DURING THE GLACIAL PERIOD PROVED BY MARINE SHELLS — STILL GREATER DEPRESSION INFERRED FROM STRATIFIED DRIFT — DRIFT OF MACCLESFIELD AND THE NORTH OF ENGLAND — SIGNS OF EXTINCT GLACIERS IN ENGLAND — MARINE DRIFT IN IRELAND — MAPS ILLUSTRATING SUCCESSIVE REVOLUTIONS IN PHYSICAL GEOGRAPHY DURING THE PLEISTOCENE PERIOD — SOUTHERNMOST EXTENT OF ERRATICS IN ENGLAND — SUCCESSIVE PERIODS OF JUNCTION AND SEPARATION OF ENGLAND, IRELAND, AND THE CONTINENT — TIME REQUIRED FOR THESE CHANGES — ANTIQUITY OF MAN CONSIDERED IN RELATION TO THE AGE OF THE EXISTING FAUNA AND FLORA.

Parallel Roads of Glen Roy in Scotland.

PERHAPS no portion of the superficial drift of Scotland can lay claim to so modern an origin on the score of the freshness of its aspect, as that which forms what are called the Parallel Roads of Glen Roy. If they do not belong to very recent times, they are at least posterior in date to the present outline of mountain and glen, and to the time when every one of the smaller burns ran in their present channels, though some of them have since been slightly deepened. The almost perfect horizontality, moreover, of the roads, one of which is continuous for about twenty miles from east to west, and twelve miles from north to south, shows that since

the era of their formation no change has taken place in the
relative levels of different parts of the district. (See
p. 311.)

Glen Roy is situated in the Western Highlands, about
ten miles N.N.E. of Fort William, near the western end of
the great glen of Scotland, or Caledonian Canal, and near
the foot of the highest of the Grampians, Ben Nevis. (See
map, p. 302.) Throughout nearly its whole length, a
distance of more than ten miles, three parallel roads or
shelves are traced along the steep sides of the mountains,
as represented in the annexed view, Plate II., by the late
Sir T. Dick Lauder, each maintaining a perfect horizontality,
and continuing at exactly the same level on the opposite
sides of the glen. I examined these 'parallel roads' in
company with Dr. Buckland in 1825. Seen at a distance,
they appear like ledges, or roads, cut artificially out of the
sides of the hills; and when we are upon them, we find
them to be pathways preserving their horizontality on the
steep hill-side, though in some places we can scarcely
recognise their existence, so uneven is their surface, and so
covered with boulders. They are from ten to sixty feet
broad, becoming, however, much wider where they cross the
bed of small rivulets.

On closer inspection, we find that these terraces are
formed of earth and loose angular stones similar to those
which the disintegrating action of the frost and rain upon
the native rock now loosens from the hillsides and washes
down to the valley beneath, together with some pebbles and
rolled fragments of rock not found in situ in the immediate
vicinity. The parallel shelves are therefore due to the
accumulation at these levels of detritus precisely similar to
that which is dispersed over the declivities of the hills above
and below. The lowest of these roads is about 850 feet
above the level of the sea, the next about 212 feet higher,

Fig. 39.

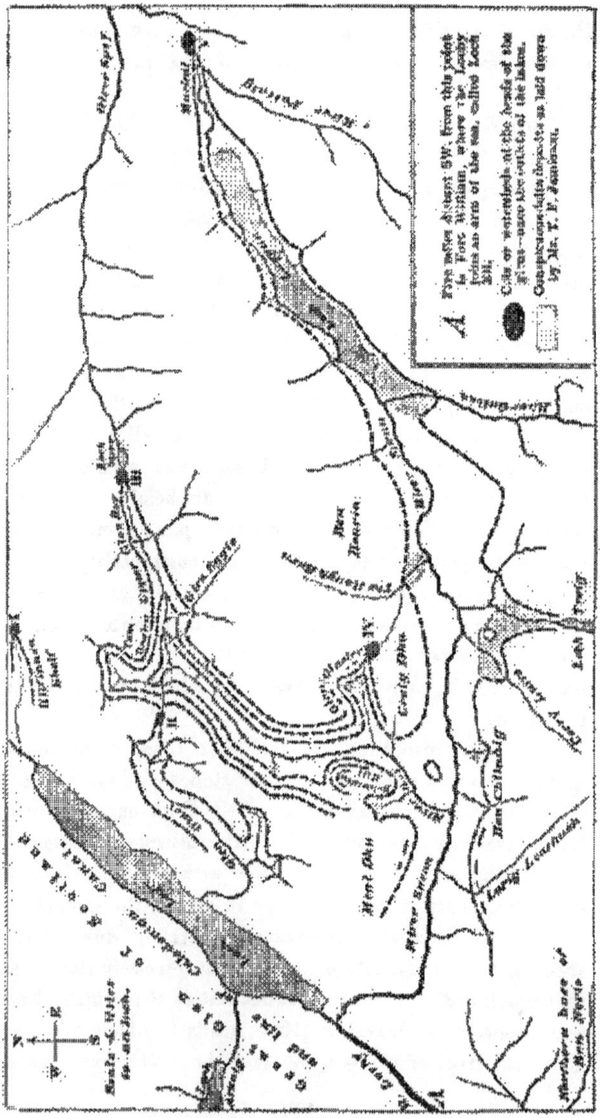

and the third 82 feet above the second. There is a fourth shelf, which occurs only in a contiguous valley called Glen Gluoy, which is twelve feet above the highest of all the Glen Roy roads, and consequently about 1,156 feet above the level of the sea.* One only, the lowest of the three roads of Glen Roy, is continued throughout Glen Spean, a large valley with which Glen Roy unites. (See Plate II. and map, fig. 30.) As the shelves, having no slope towards the sea like ordinary river terraces, are always at the same absolute height, they become continually more elevated above the river in proportion as we descend each valley; and they at length terminate very abruptly, without any obvious cause, or any change either in the shape of the ground or in the composition or hardness of the rocks.

I should exceed the limits of this work, were I to attempt to give a full description of all the geographical circumstances attending these singular terraces, or to discuss the ingenious theories which have been severally proposed to account for them by Dr Macculloch, Sir T. Dick Lauder, and Messrs. Darwin, Agassiz, Milne, and Chambers. There is one point, however, on which all are agreed, namely, that these shelves are ancient beaches, or littoral formations, accumulated round the edges of one or more sheets of water which once stood for a long time successively at the level of the several shelves.

It is well known, that wherever a lake or marine fiord exists surrounded by steep mountains subject to disintegration by frost or the action of torrents, some loose matter is washed down annually, especially during the melting of snow, and a check is given to the descent of this detritus at the point where it reaches the waters of the lake. Moreover the waves beating against the shore cut into the hillside and spread out the materials along the beach; their dis-

* Another detached shelf also occurs in Kilfinnan. (See Map, p. 302.)

persing power being aided by the ice, which often adheres
to pebbles during the winter months, and gives buoyancy
to them. The annexed diagram
illustrates the manner in which
Dr. Macculloch and Mr. Darwin
suppose 'the roads' to constitute
mere excrescences of the superficial alluvial coating which rests
upon the hill-side, and consists
chiefly of clay and sharp unrounded stones.

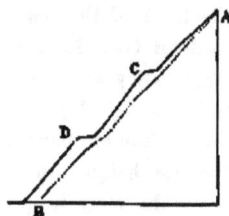

Fig. 40.

A B. Supposed original surface
of rock.
C D. Roads or shelves in the
outer alluvial covering
of the hill.

Among other proofs that the
parallel roads have really been
formed along the margin of sheets
of water, it may be mentioned,
that wherever an isolated hill rises in the middle of the glen
above the level of any particular shelf, as in Mealderry,
Plate II., a corresponding shelf is seen at the same level
passing round the hill, as would have happened if it had
once formed an island in a lake or fiord. Another very remarkable peculiarity in these terraces is this; each of them
comes in some portion of its course to a *col*, or parting ridge
between the heads of glens, the explanation of which will be
considered in the sequel.

Those writers who first advocated the doctrine that the
roads were the ancient beaches of freshwater lakes, were
unable to offer any probable hypothesis respecting the formation and subsequent removal of barriers of sufficient height
and solidity to dam up the water. To introduce any violent
convulsion for their removal was inconsistent with the uninterrupted horizontality of the roads, and with the undisturbed
aspect of those parts of the glens where the shelves come
suddenly to an end.

Mr. Agassiz and Dr. Buckland, desirous, like the defenders

PLATE II.

(To face p. 328.)

VIEW OF THE MOUTHS OF GLEN ROY AND GLEN SPEAN, BY SIR T. LAUDER DICK.

Glen Collarig.   Hill of Bohuntine.   Glen Roy.   Monliarty.   Entrance of Glen Spean.

Point of division between Glen Roy and Spean.

of the lake theory, to account for the limitation of the shelves
to certain glens, and their absence in contiguous glens, where
the rocks are of the same composition, and where the slope
and inclination of the ground are very similar, first started
the theory that these valleys were once blocked up by enormous glaciers descending from Ben Nevis, giving rise to what
are called, in Switzerland and in the Tyrol, glacier-lakes. It
will readily be conceded that this hypothesis was preferable
to any previous lacustrine theory, by accounting more easily
for the temporary existence and entire disappearance of lofty
transverse barriers, although the height required for the supposed dams of ice appeared very enormous.

Mr. T. F. Jamieson, of Ellon, when he visited Lochaber
in 1861, observed many facts highly confirmatory of the
theory of glacier-lakes, and he published an account of the
Glen Roy terraces in 1863,[*] in which he showed that
this theory affords a complete explanation of all the
most striking peculiarities. Taking for granted the gradual
retreat of the glaciers, and availing himself most ingeniously
of the peculiar arrangement of the high mountain passes in
this part of Lochaber, he was able to trace out accurately
the alternate accumulation and removal of the large masses
of ice necessary to dam the waters at each successive stage.
In the first place, he found much superficial scoring and
polishing of rocks, and accumulation of boulders at those
points where signs of glacial action ought to appear, if ice
had once dammed up the waters of the glens in which the
'roads' occur. Ben Nevis may have sent down its glaciers
from the south, and Glen Arkaig from the north, for the
mountains at the head of the last-mentioned glen are 3,000
feet high, and may, together with other tributary glens, have
helped to choke up the great Caledonian valley with ice, so

[*] Quart. Geol. Journ., vol. xix. 1863.

X

as to block up for a time the mouths of the Spean, Roy, and Gluoy. The temporary conversion of these glens into glacier-lakes is the more conceivable, because the hills at their upper ends not being lofty nor of great extent, they may not have been filled with ice at a time when great glaciers were generated in other adjoining and much higher regions.

The greatest difficulty is in conceiving how the waters could be made to stand so high in Glen Roy, as to allow the uppermost shelf to be formed. Grant a barrier of ice in the lower part of the glen, of sufficient altitude to stop the waters from flowing westward, still, what prevented them from escaping over the 'col' at the head of Glen Glaster? This 'col' coincides exactly in level, as Mr. Milne Home first ascertained, with the second or middle shelf of Glen Roy. The difficulty here stated appears now to be removed by supposing that the higher lines or roads were formed before the lower ones, and when the quantity of ice was most in excess. We must imagine that at the time when the uppermost shelf of Glen Roy was forming in a shallow lake, the lower part of that glen was filled up with ice, an offshoot from the great mass filling the valley where Fort William now stands ; and, according to Mr. Jamieson, a glacier from Loch Treig then protruded itself across Glen Spean, and rested on the flank of the hill on the opposite side, in such a manner as effectually to prevent any water from escaping over the Glen Glaster 'col,' but affording a free exit by the 'col' leading to Glen Spey. The proofs of such a glacier having actually existed at the point in question consist, he says, in numerous striæ observable in the bottom of Glen Spean, and in the presence of moraine matter in considerable abundance on the flanks of the hill extending to heights above the Glen Glaster 'col.' Protected in the shady north-east corner of Ben Nevis, it may well have survived the time

when increasing warmth had extinguished all traces of ice
in the more exposed regions of Glen Roy and the upper part
of Glen Spey. When the barrier was at its greatest height
it is possible that three different lakes may have existed
at the same time. But as the body of ice in the Great Glen
diminished, the formation of the uppermost shelf—namely,
that of Glen Gluoy—must have ceased, while the formation of
the lower ones may have continued. Finally, when all the ice
had melted except that in the Great Glen, an uninterrupted
sheet of water occupied the united areas of Glen Roy and
Glen Spean, occasioning the formation of the lowest of the
three roads—that which coincides very nearly with the level
of the Muckel Pass at the head of the Spean valley—where
there are unequivocal signs of a river having flowed out for
a considerable period.

Before the idea of glacier-lakes had been suggested by
Agassiz, another theory to account for the origin of the
parallel roads had been proposed by Mr. Darwin; namely,
that the shelves were formed when the glens were arms of
the sea, and consequently that there were never any seaward
barriers. According to this theory, the land emerged during
a slow and uniform upward movement, like that now experienced throughout a large part of Sweden and Finland;
but there were certain pauses in the upheaving process, at
which times the waters of the sea remained stationary for so
many centuries as to allow of the accumulation of an extraordinary quantity of detrital matter, and the excavation, at
many points immediately above the sea-level, of deep notches
and bare cliffs in the hard and solid rock.

But one of the most fatal objections against accepting
this explanation is the fact that the structure of the terraces
in Glen Roy is unlike that of any known marine beaches, even
those now being formed on the shores of the sheltered sea-lochs,
and the unrolled condition of the stones appears to be incom-

patible with a sea in which the daily tides must have exerted
an immense power of abrasion and rolling. On the contrary,
the appearance of the stones on the terrace is identical with
that of the fragments on the shores of a freshwater loch. At
the level of several of the shelves in Glen Roy, at points
where torrents now cut channels through the shelf as they
descend the hill-side, there are, as Mr. Jamieson has pointed
out, small delta-like extensions of the shelf, perfectly preserved, as if the materials, whether fine or coarse, had
originally settled there in a placid lake, and had not been
acted upon by tidal currents, mingling them with the sediment of other streams. These deltas are too entire to allow
us to suppose that they have at any time since their origin
been exposed to the waves of the sea. In addition to these
arguments, I may mention that in Switzerland, at present,
no testacea live in the cold waters of glacier-lakes; so that
the entire absence of fossil shells, whether marine or fresh-
water, in the stratified materials of each shelf, would be
accounted for, if the theory above mentioned be embraced.

Dr. Hooker has described some parallel terraces, very
analogous in their aspect to those of Glen Roy, as existing
in the higher valleys of the Himalaya, of which his pencil
has given us several graphic illustrations. He believes these
Indian shelves to have originated on the borders of glacier-
lakes, the barriers of which were usually formed by the ice
and moraines of lateral or tributary glaciers, which descended
into and crossed the main valley, as we have supposed in the
case of Glen Roy; but others he ascribes to the terminal
moraine of the principal glacier itself, which had retreated
during a series of milder seasons, so as to leave an interval
between the ice and the terminal moraine. This interspace
caused by the melting of ice becomes filled with water and
forms a lake, the drainage of which usually takes place by
percolation through the porous parts of the moraine, and not

by a stream overflowing that barrier.   Such a glacier-lake
Dr. Hooker actually found in existence near the head of the
Yangma valley in the Himalaya.   It was moreover partially
bounded by recently formed marginal terraces or parallel
roads, implying changes of level in the barrier of ice and
moraine matter.[*]

*Terraces now forming on the borders of the Märjelen See
in Switzerland.*

It has been sometimes objected to the hypothesis of
glacier-lakes, as applied to the case of Glen Roy, that the
shelves must have taken a very long period for their formation.   Such a lapse of time, it is said, might be consistent
with the theory of pauses or stationary periods in the rise of
the land during an intermittent upward movement, but it
is hardly compatible with the idea of so precarious and
fluctuating a barrier as a mass of ice.   But the reader will
have seen that the permanency of level in such glacier-lakes
has no necessary connection with minor changes in the height
of the supposed dam of ice.   If a glacier descending from
higher mountains through a tributary glen enters the main
valley in which there happens to be no glacier, the river is
arrested in its course and a lake is formed.   The dam may
be constantly repaired and may vary in height several hundreds of feet without affecting the level of the lake, so long
as the surplus waters escape over a 'col' or parting ridge of
rock.   The height at which the waters remain stationary is
determined solely by the elevation of the 'col,' and not by
the barrier of ice, provided the barrier is higher than the 'col.'

A most instructive example of this is afforded by the
Märjelen See, a glacier-lake in the Swiss Alps a few miles
above Brieg, which I visited, in company with my nephew,

* Hooker, Himalaya Journal, vol. i.   also profited by the author's personal
p. 212; ii. pp. 119, 121, 156. I have   explanations.

Mr. Leonard Lyell, in August 1865. The lake is caused by the great glacier of Aletsch, which, descending the main valley, blocks up the end of a tributary valley, and bars up the exit of all water. By this means a small lake is formed, which lasts for periods varying from three to five years, until some change in the internal structure of the glacier, such as occurred last June, 1872, allows the water to escape through rents or crevices, and the lake is drained in a few hours. Nothing is then left but a small stream of water flowing at the bottom of the basin, but after an interval of about a year the water accumulates again up to its old level. This level is determined not by the height of the glacier, but by a watershed or 'col' at the other end of the valley, separating the Märjelen See from a glacier which passes down the adjoining valley of Viesch. As soon as the lake has risen to the height of this 'col,' it flows over the ridge, and is thus maintained at a constant level, along which the rubbish of the mountain side accumulates, forming a counterpart of one of the parallel roads of Glen Roy.* It was after seeing this lake in 1865 that I determined to revisit Lochaber, which I did in 1869, and my nephew and I were both convinced that the glacier theory was the true explanation of these remarkable terraces.

Age of the parallel roads of Glen Roy.

But if we embrace the theory of glacier-lakes, we must be prepared to assume not only that the sea had nothing to do with the original formation of the 'parallel roads,' but that it has never, since the disappearance of the lakes, risen in any one of the glens up to the level of the lowest shelf, which is about 850 feet high; for in that case the remarkable persistency and integrity of the roads and deltas, before described, must have been impaired.

* For a diagram of this glacier-lake, see Principles, 11th ed. vol. I. p. 374.

We have seen (p. 286) that fifty miles to the south of Lochaber, the glacier formations of Lanarkshire with marine shells of arctic character have been traced to the height of 524 feet. About fifty miles to the south-east in Perthshire are those stratified clays and sands, near Killiecrankie, which were once supposed to be of submarine origin, and which in that case would imply the former submergence of what is now dry land to the extent of 1,550 feet, or several hundred feet beyond the highest of the parallel roads. Even granting that these laminated drifts may have had their origin in a glacier-lake, there are still many facts connected with the distribution of erratics and the striation of rocks in Scotland which are not easily accounted for without supposing the country to have sunk, since the era of continental ice, to a greater depth than 525 feet, the highest point to which marine shells have yet been traced.

After what was said of the pressure and abrading power of a general crust of ice, like that now covering Greenland, it is almost superfluous to say that if, as some have contended, such a state of things ever existed in Scotland, the parallel roads must have been of later date, for every trace of them would have been obliterated by the movement of such a mass of ice. It is no less clear, that as no glacier-lakes can now exist in Greenland, so there could have been none in Scotland if the mountains were covered with one great crust of ice. It may, however, be contended, that the parallel roads were produced when the general crust of ice first gave place to a period of separate glaciers, and that no period of deep submergence ever intervened in Lochaber after the time of the lakes. Even in that case, it is difficult not to suppose that the Glen Roy country participated in the downward movement above mentioned which sank part of Lanarkshire 525 feet beneath the sea, subsequently to the first great glaciation of Scotland (p. 286). But that amount of subsidence might

have occurred, and even a more considerable one, without
causing the sea to rise to the level of the lowest shelf, or to
a height of 850 feet above the present sea-level.

On the whole, therefore, I conclude that the Glen Roy
terrace-lines and those of some neighbouring valleys, were
formed on the borders of glacier-lakes, in times long subsequent to the principal glaciation of Scotland. They may
perhaps have been nearly as late, especially the lowest of the
shelves, as that portion of the pleistocene period in which
Man coexisted in Europe with the mammoth.

Glacial Formations of England and Wales.

The considerable amount of vertical movement in opposite directions, which has been suggested as affording the
most probable explanation of the position of some of the
stratified and fossiliferous drifts of Scotland, formed since
the commencement of the glacial period, will appear less
startling, if it can be shown that independent observations
lead us to infer that a geographical revolution of still greater
magnitude accompanied the successive phases of glaciation
through which the North of England and the Welsh mountains have passed.

It has long been acknowledged that Wales was once an
independent centre of the dispersion of erratic blocks. Dr.
Buckland published in 1842 his reasons for believing that
the Snowdonian mountains in Caernarvonshire were formerly
covered with glaciers, which radiated from the central heights
through the seven principal valleys of that chain, where striæ
and flutings are seen on the polished rocks directed towards
as many different points of the compass. He also described
the 'moraines' of the ancient glaciers, and the rounded
bosses of rock, called in Switzerland 'roches moutonnées.' His
views respecting the old extinct glaciers of North Wales were

subsequently confirmed by Mr. Darwin, who attributed the transport of many of the larger erratic blocks to floating ice. Much of the Welsh glacial drift had already been shown by Mr. Trimmer to have had a submarine origin, and Mr. Darwin maintained that when the land rose again to nearly its present height, glaciers filled the valleys, and 'swept them clean of all the rubbish left by the sea.' *

Professor Ramsay, in a paper read to the Geological Society in 1851, and in a later work on the glaciation of North Wales,† described three successive glacial periods, during the first of which the land was much higher than it now is, and the quantity of ice excessive; secondly, a period of submergence when the land was 2,300 feet lower than at present, and when the higher mountain tops only stood out of the sea as a cluster of low islands, which nevertheless were covered with snow; and lastly, a third period when the marine boulder drift formed in the middle period was ploughed out of the larger valleys by a second set of glaciers, smaller than those of the first period. In Wales it was certainly preceded by submergence, and the rocks had been exposed to glacial polishing and friction before they sank.

Fortunately the evidence of the sojourn of the Welsh mountains beneath the waters of the sea is not deficient, as in Scotland, in that complete demonstration which the presence of marine shells affords. The late Mr. Trimmer in 1831 discovered such shells on Moel Tryfaen, in North Wales, in drift elevated more than 1,300 feet above the level of the sea. It appeared from his observations, and those of the late Edward Forbes, corroborated by others of Professor Ramsay and Mr. Prestwich, that about twelve species of shells, including *Fusus Bamffius, F. antiquus, Venus striatula* (Forbes and Hanley), have been met with at heights of

* Philosophical Magazine, ser. 3, vol. xxi. p. 180.   † Glaciers of North Wales, 1860.

between 1,000 and 1,400 feet, in drift, through which glaciated
boulders are dispersed. In the summer of 1863 I visited
North Wales, in company with my friend the Rev. W. S.
Symonds, and we first examined some points in the neighbourhood of Snowdon, where Professor Ramsay had seen
marine shells at the height of about 1,300 feet. But at this
point we were entirely unsuccessful; and I am persuaded that,
like many of our predecessors, we should also have failed in
our search on Moel Tryfaen, had we not fortunately been
directed by Mr. R. D. Darbishire to a recent and deep
cutting in the drift made by the newly-formed Alexandra
Mining Company near the summit of that hill, probably on
the exact spot where Mr. Trimmer, in 1831, removed some
of the same gravel in search of slates.

In the cutting alluded to, we had an opportunity of
studying a mass of stratified and incoherent sand and
gravel, 35 feet thick, for the most part in thin and irregular layers, and containing here and there fragments of
shells, with a few entire ones. The beds bore every mark of
gradual and successive accumulation, some layers being composed of fine, others of coarser materials, and in the lower
beds were several large boulders, one or two of them heavier
than we could lift, of far-transported rocks glacially polished,
and scratched on more than one side. Among these
was a black mass of basalt, which Mr. Symonds believes
must have come from Ireland.* Underneath the glacial
deposit we saw the edges of the vertical slates exposed to
view; they were smoothed in places, but there was nowhere
a sufficient exposure to enable us to decide whether it was
the effect of glaciation or beach erosion.

Mr. Darbishire had already called the attention of the
workmen and overseers to the fossil shells. We received
from the men, and collected ourselves, a set of specimens,

* Records of the Rocks, 1872, p. 59.

which, though many of them were in fragments, Mr. Gwyn
Jeffreys was able to refer to twenty species, all of them now
living in the British or Northern seas. When I showed them
to Dr. Torell, he observed that, although they constituted a
Northern fauna, and bore testimony to a colder climate than
that of the present British seas, they by no means indicated
such an intensity of cold as did the assemblage of shells
brought to light on the borders of the estuaries of the
Forth and Tay, where, in the ancient glacial drift or clay of
Elie in Fife, and Errol in Perthshire, the Rev. Thomas Brown
found thirty-five shells of living species, all now inhabitants of arctic regions, such as *Leda truncata, Pecten Grœn-
landicus, Crenella lævigata* Gray, *Crenella nigra* Gray, and
others first brought by Captain Parry from the coast of
Melville Island, lat. 76° N. The same fossil fauna of Scotland
exhibits no admixture of species peculiar to the seas south of
Spitzbergen, and the individuals consist of varieties proper to
the coldest latitudes. But as the Scotch fossils occur in a
parallel of latitude 200 miles north of Moel Tryfaen, it
becomes a question whether the more southern aspect of the
Welsh fauna is due to geographical position, or to its having
originated before or after the extreme refrigeration of the
glacial period. In Massachusetts, on the east coast of North
America, it is well known that Cape Cod divides abruptly a
northern from a southern province of mollusca, and there
may have been a similar sudden change from an arctic to a
more southern fauna somewhere between Scotland and North
Wales.

We are indebted to Mr. Darbishire for having formed a
collection of no less than fifty-four species of mollusca from
the above-mentioned drift of Moel Tryfaen. A complete list
of these will be found in the ' Proceedings of the Manchester
Literary and Philosophical Society for 1863–4,' p. 177, and
corrected in vol. ii. of the ' Geological Magazine,' 1865. In

a letter to the author, dated November 13, 1863, the same naturalist observes: 'Besides *Balanus Hameri*, and traces of a sponge (*Cliona*), I have obtained shells of fifty-four species of mollusca, all of which appear to be now living in British or more northern seas, or, including three characteristically arctic varieties, fifty-seven forms of shells. Of these, eleven are well known as exclusively of the arctic division of the present seas, including—

Tellina proxima, Brown	*Natica clausa*
Astarte borealis and *A. crebricostata*	*Trophon scalariformis* and *T. Gunneri*
Leda pernula	*Dentalium abyssorum*

'Four are arctic species, which still survive within British limits :—

Astarte elliptica and *A. compressa*	*Trichotropis borealis*
Trophon clathratus = *Fusus Bamffius*	

'Of the whole list, thirty-seven species are now living in the Irish Sea, including nineteen of wide general range north and south of these islands. Amongst the latter, the more abundant are—

Tellina solidula	*Murex erinaceus*
Cardium edule and *C. echinatum*	*Nassa reticulata*
Turritella communis	*Mytilus edulis*

'Thirteen are species abundant in British seas, and ranging northwards, but not of frequent occurrence to the southward, such as—

Mya truncata	*Purpura lapillus*
Venus canina	*Buccinum undatum*
Littorina littorea	*Fusus gracilis* and *F. antiquus*
Lacuna vincta	*Mangelia turricula*

'Mr. Darbishire has carefully re-measured the height of Moel Tryfaen, and confirms Mr. Trimmer's estimate of 1,392 feet above the level of the sea. The highest level reached by the fossil shells is 1,360 feet.'

As shells are almost invariably wanting in porous drift like that of Moel Tryfaen, we naturally enquire by what

accident they can have in this instance escaped obliteration.
Mr. Darbishire suggests that an overlying yellowish-brown
sandy clay, 1 ft. 9 in. thick, which underlies the superficial
peaty soil, and covers all the beds of gravel and sand with
shells, may, by its impermeable nature, have preserved the
fossils. The antiquity of drift upraised to such a height
must be very great, and we can hardly imagine that so many
shells could have escaped being dissolved by rain-water, had
it been able to percolate freely, for countless ages, from the
peaty covering through beds of sand and pebbles, remarkably
loose in their texture.

A northern drift, similar to that of Moel Tryfaen, has been
observed in England, in Lancashire, Cheshire, Derbyshire,
Shropshire, Staffordshire, and Worcestershire, and though it
is rare to find marine shells at heights exceeding two or three
hundred feet, yet a few instances have been noticed. At
Mottram-in-Longdendale, east of Manchester, Mr. J. F.
Bateman discovered Till at a height of 568 feet above the
sea-level containing marine shells, *Turritella communis*,
Purpura lapillus, Cardium edule, and others, among which
Trophon clathratum(= *Fusus Bamffius*), though surviving
in North British seas, indicates a cold climate.[*] At Macclesfield also, about three miles east of the town, in Vale Royal
on the Buxton Road, Mr. Prestwich, in 1862, found drift
with marine shells at an elevation of between 1,100 and 1,200
feet above the level of the sea, and I was able, in 1871, to
verify the fact and to collect some of the mollusca. A similar
deposit was discovered by Mr. Sainter, six miles to the south
of Macclesfield, at the same level. The marine fauna of
these beds is extremely similar to that of Moel Tryfaen,
almost every species being identical; it has a strong northern
character, with, however, one remarkable exception, namely,
that unmistakeable fragments occur of *Cytherea chione*, a

[*] Binney, Proc. Manchester Phil. Soc., No. 3, 1862-3, p. 15.

characteristic member of the Spanish marine fauna which is
not now found north of Carnarvon Bay. The height at
which the Vale Royal shells occur acquires importance when
viewed in connection with the proofs of the still greater
elevation of the old bed of the sea at Moel Tryfaen. The
two localities are about eighty miles distant from each other
in a straight line, and the Vale Royal shelly drift is near the
watershed of the centre of England. Intermediate between
these points there are areas varying greatly in height above
the sea, composed of every description of rock, sometimes
covered with drift, but often free from it, and where proofs
of marine submergence are entirely wanting, these have been
surveyed with such care, that, but for the occasional patches
before mentioned, in which the shelly remains occur, a geologist who relied on negative evidence might have confidently
affirmed that the land had not been covered during the formation of the drift by salt water. But we are now certain
of this fact, since the largest part of the area in question is
very much lower than many of the highest sites reached
by the shelly drifts. It becomes therefore most difficult to
decide what amount of negative evidence would entitle us to
infer an uninterrupted prevalence of land during the glacial
period, more especially as Mr. Brown has shown how much
the power of icebergs running aground in the fiords of
Greenland may destroy the signs of animal life on the sea-
bottom.[*]

The mountains of the English lake district, afford equally
unequivocal vestiges of ice-action not only in the form of
polished and grooved surfaces, but also of those rounded
bosses before mentioned, as being so abundant in the Alpine
valleys of Switzerland, where glaciers exist, or have existed.
Mr. Hull has published a faithful account of these phenomena,
and has given a representation of some of the English 'roches

* Quart. Geol. Journal, vol. xxvi, p. 668, 1870.

moutonnées,' which precisely resemble hundreds of dome-shaped protuberances in North Wales, Sweden, and North America.*

The marks of glaciation on the rocks, and the transportation of erratics from Cumberland to the eastward, have been traced by Professor Phillips over a large part of Yorkshire, extending to a height of 1,500 feet above the sea.

Fig. 41.

Dome-shaped rocks, or 'roches moutonnées,' in the valley of the Rotha, near Ambleside, from a drawing by E. Hull, F.G.S.†

In treating of the former glacial sea on the East of England, allusion has already been made (p. 267) to the evidence for the existence of an enormous mass of land-ice in the Lake District and the high ground of the north-west of Yorkshire. In an important paper, lately published,‡ Mr. R. H. Tiddeman has given a summary of the work of previous observers and added his own observations and conclusions on the action of moving ice. From this paper we learn that the phenomena presented in the North of England are strikingly similar to those in Scotland, which had already

* Hull, Edinburgh New Phil. Jour., July 1860.
† Edinburgh New Philosophical Journal, vol. xi. pl. I. p. 31, 1860.
‡ Quart. Geol. Journal, vol. xxviii., 1872, p. 471.

been attributed to the action of land-ice. The peculiar local characteristics of the Till, the rude way in which the lines of glacial striæ often cross the present drainage slopes, and the considerable height to which these markings ascend on the hill side, lead to the inference that a thick sheet of moving ice must have enveloped the country. One of the exits of this ice will have been by St. George's Channel and over the low ground of Anglesey, that island, as Professor Ramsay pointed out some years ago, being glaciated from the N.N.E. and strewn with boulders from the Lake Country. Mr. Cumming also showed in 1846[*] that the Isle of Man had been glaciated from the E.N.E., and that it was covered with boulders from the Lake Mountains and the Solway Firth. After the period of severest glaciation passed away there seems to have been a fluctuation in the level of the land and in the climate analogous to that which took place in Scotland. Along the Somersetshire coast are mounds of rolled shingle resting on boulder-clay, and there are local deposits containing recent shells of semi-boreal character at inconsiderable elevations.

Signs of Ice-action and Submergence in Ireland during the Glacial Period.

In Ireland we encounter the same difficulty as in Scotland, in determining how much of the glaciation of the higher mountains should be referred to land glaciers, and how much to floating ice, during submergence. The signs of glacial action have been traced by Professor Jukes[†] to elevations of 2,500 feet in the Killarney district, and to great heights in other mountainous regions. Marine shells have rarely been met with higher than 300 feet above the sea, and that

[*] Quart. Geol. Journal, vol. ii. p. 340, 1846.
[†] Reports of Irish Geol. Survey, 1859, maps, sheet 184.

chiefly in gravel, clay, and sand in Wicklow and Wexford,
but in one instance near the Three Rock mountain, S.E. of
Dublin, marine drift, containing fragments of *Cyprina*
and glacial shells, reaches a height of from 1,000 to 1,200
feet. More than eighty species have been obtained from
these marine drifts. The great elevation of these shells, and
the still greater height to which the rocks in the mountainous
regions of Ireland have been smoothed and striated, has led
geologists to the opinion that that island was in great part
submerged during a portion of the glacial period, and the
wide extent of drift spread over large areas in Ireland seems
to indicate that the land above water formed an archipelago,
as represented in the maps, figs. 42, p. 325, and 43, p. 327.

Speaking of the Wexford drift, the late Professor E.
Forbes states that Sir H. James found in it, together with
many of the usual glacial shells, several species which are
characteristic of the crag; among others the reversed variety
of *Fusus antiquus*, called *F. contrarius*, and the extinct
species *Nucula Cobboldiæ*, and *Turritella incrassata*.[*]
Perhaps a portion of this drift of the south of Ireland may
belong to the close of the newer pliocene period, and may
be of a somewhat older date than the shells of the Clyde,
alluded to at p. 266.

The scarcity of mammalian remains in the Irish drift
favours the theory of its marine origin. In the superficial
deposits of the whole island, I have only met with three
recorded examples of *Elephas primigenius*; the first in the
south, near Dungarvan,[†] about the year 1770; the second in
a cave near the same town, where bones of the mammoth,
reindeer, horse, and of two species of bear (*Ursus Arctos,*
and *Ursus spelæus?*), were found;[‡] and the third in the
centre of the island near Belturbet, in the county of Cavan.

---

[*] Forbes' Memoirs of Survey, &c., Waterford, p. 88.
vol. i. p. 377.   [‡] E. Brenan and Dr. Carte,
[†] Smith's History of the County Dublin, 1859.

Perhaps the conversion into land of the eastern portion of the bed of the glacial sea may have preceded in time the elevation of the Irish drift. Ireland may have continued for a longer time in the state of an archipelago, and was therefore for a much shorter time inhabited by the large extinct pleistocene pachyderms which coexisted with the fabricators of the St. Acheul flint hatchets, but could not immigrate into Ireland till after its union with England and the Continent.

Although the course taken by the Irish erratics in general is such that their transportation seems to have been due to floating ice or coast-ice, yet some granite blocks have travelled from south to north, as recorded by Sir R. Griffiths, namely, those of the Ox Mountains in Sligo; a fact from which Mr. Jamieson infers that those mountains formed at one time a centre of dispersion. In the same part of Ireland, the general direction in which the boulders have travelled is everywhere from north-west to south-east, a course directly at right angles to the prevailing trend of the present mountain ridges.

Maps illustrating successive Revolutions in Physical Geography during the Pleistocene Period.

The late Mr. Trimmer, before referred to, endeavoured to assist our speculations as to the successive revolutions in physical geography, through which the British Islands have passed since the commencement of the glacial period, by four ' sketch maps ' as he termed them, in the first of which he gave an ideal restoration of the original Continental period, called by him the first elephantine period, or that of the forest of Cromer, before described (p. 254). He was not aware that the prevailing elephant of that era (*E. meridionalis*) was distinct from the mammoth. At this era he conceived Ireland and England to have been united with each other and with France, but much of the area represented as land in the map, fig. 44, p. 328, was supposed to be under water. His second map, of the great submergence of the glacial period, was not essentially different from our map, fig. 42, p. 325. His third map expressed a period of partial re-elevation, when Ireland was reunited to Scotland and the north of England; England being still separated from France. This restoration appears to me to rest on insufficient data, being constructed to suit the supposed area over which the gigantic Irish deer, or *Megaceros*, migrated from east to west, also to explain an assumed submergence of the district called the Wealden, in the south-east of England, which had remained land during the grand glacial submergence.

The fourth map is a return to nearly the same continental conditions as the first—Ireland, England, and the Continent being united. This he called the second elephantine period; and it would coincide very closely with that part of the pleistocene era in which Man coexisted with the mammoth, and when, according to Mr. Trimmer's hypothesis previously indicated by Mr. Godwin-Austen, the Thames was a tributary of the Rhine.[*]

These geographical speculations were indulged in ten years after Edward Forbes had published his bold generalisations on the geological changes which accompanied the successive establishment of the Scandinavian, Germanic, and other living floras and faunas in the British Islands, and, like the theories of his predecessor, were the results of much reflection on a vast body of geological facts. It is by repeated efforts of this kind, made by geologists who are prepared for the partial failure of some of their first attempts, that we shall ultimately arrive at a knowledge of the long series of geographical revolutions which have followed each other since the beginning of the pleistocene period.

* Joshua Trimmer. Quarterly Geo-
logical Journal, vol. ix. plate xiii.   1853; and Godwin-Austen, ibid. map,
plate vii. vol vii. p. 134, 1851.

The map, fig. 42, p. 325, will give some idea of the great extent of land which would be submerged, were we to infer, as many geologists have done, from the joint evidence of marine shells, erratics, glacial striæ and stratified drift at great heights, that Scotland was, during part of the glacial period, 2,000 feet below its present level, and other parts of the British Isles, 1,300 feet. A subsidence to this amount can be demonstrated in the case of North Wales and Ireland by marine shells (see above, p. 313). As to central England, or the country north of the Thames and Bristol Channel, marine shells of the glacial period sometimes reach as high as 1,200 feet, as near Macclesfield, (see above, p. 317). A subsidence of 600 feet, as will be seen by the map, fig. 43, given at p. 327, would reduce the whole of the British Isles to an archipelago of very small islands, with the exception of parts of Scotland, and the north of England and Wales, where four islands of considerable dimensions would still remain.

The map does not indicate a state of things supposed to have prevailed at any one moment of the past: it simply represents the effects of a downward movement of a hundred fathoms, or 600 English feet, assumed to be uniform over the whole of the British Isles. It shows the very different state of the physical geography of the area in question, when contrasted with the results of an opposite movement, or one of upheaval, to an equal amount, of which Sir Henry de la Beche had already given us a picture, in his excellent treatise called 'Theoretical Researches.'[*] His map I have reproduced (fig. 44, p. 328), after making some important corrections in it.

If we are surprised when looking at the map, fig. 43, at the vast expanse of sea which so moderate a subsidence as 600 feet would cause, we shall probably be still more

[*] Also repeated in De la Beche's Geological Observer.

Fig. 42.

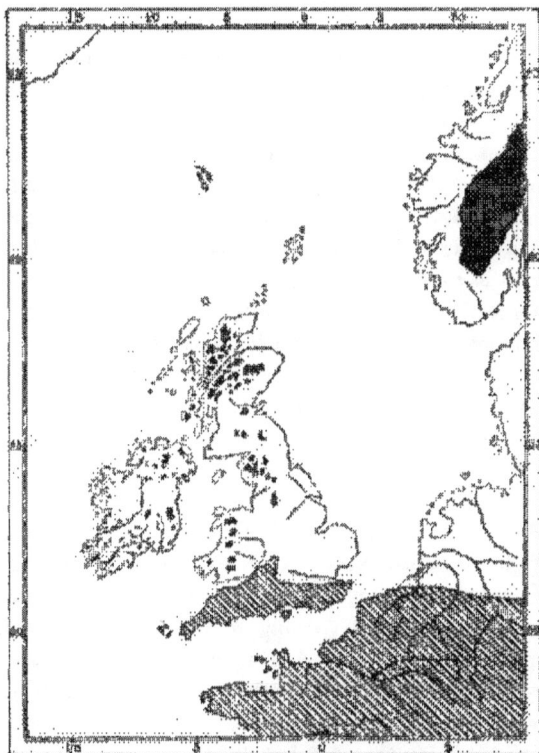

MAP OF THE BRITISH ISLES AND PART OF THE NORTH-WEST OF EUROPE, SHOWING THE GREAT AMOUNT OF SUPPOSED SUBMERGENCE OF LAND BENEATH THE SEA DURING PART OF THE GLACIAL PERIOD.

The submergence of Scotland is to the extent of 2,000 feet, and of other parts of the British Isles, 1,300.

In the map, the dark shade expresses the land which alone remained above water. The area shaded by diagonal lines is that which cannot be shown to have been under water at the period of floating ice by the evidence of erratics, or by marine shells of northern species.* How far the several parts of the submerged area were simultaneously or successively laid under water, in the course of the glacial period, cannot, in the present state of our knowledge, be determined.

* Since this map was constructed it has been shown that probably some parts of the northern coast of Devon and southern shores of Sussex may have been submerged.

astonished to perceive, in fig. 44, that a rise of the same
number of feet would unite all the British Isles, including
the Hebrides, Orkneys, and Shetlands, with one another and
the continent, and lay dry the sea now separating Great
Britain from Sweden and Denmark.

It appears from soundings made during various Admiralty
surveys, that the gained land thus brought above the level of
the sea, instead of presenting a system of hills and valleys
corresponding with those usually characterising the interior
of most of our island, would form a nearly level terrace, or
gently inclined plane, sloping outwards like those terraces of
denudation and deposition which I have elsewhere described
as occurring on the coasts of Sicily and the Morea.*

It seems that, during former and perhaps repeated oscillations of level undergone by the British Isles, the sea has
had time to eat back the cliffs for miles in many places;
while in others the detritus derived from wasting cliffs
drifted along the shores, together with the sediment brought
down by rivers and swept by currents into submarine valleys,
has exerted a levelling power, filling up such depressions as
may have pre-existed. Owing to this twofold action few
marked inequalities of level have been left on the sea-bottom,
the ' silver-pits ' off the mouth of the Humber offering a
rare exception to the general rule, and even there the narrow
depression is less than 300 feet in depth. Beyond the 100
fathom line, the submarine slope surrounding the British
coast is so much steeper that a second elevation of equal
amount (or of 600 feet) would add but slightly to the area
of gained land; in other words, the 100 and 200 fathom
lines run very near each other.†

The naturalist would have been entitled to assume the
former union, within the pleistocene period, of all the British

---

* Student's Elements of Geology, p. 80.
† De la Beche, Geological Researches, p. 191.

CHAP. XIV. IN BRITISH PHYSICAL GEOGRAPHY. 327

Fig. 43.

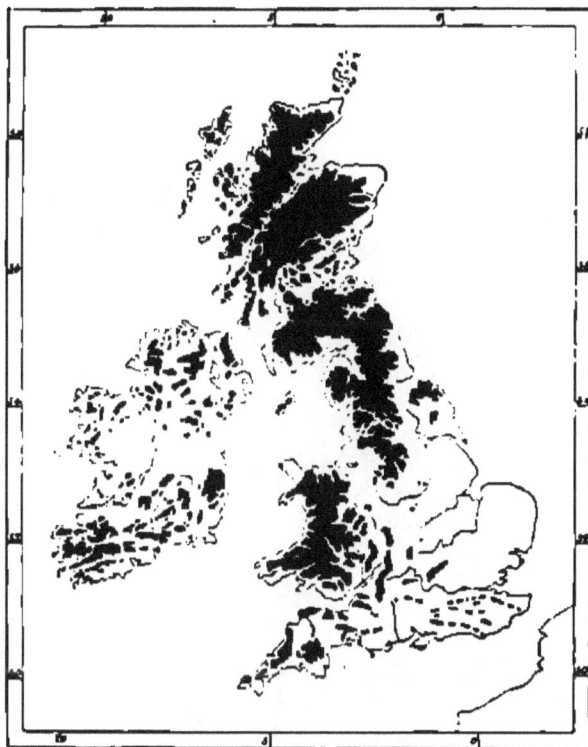

MAP SHOWING WHAT PARTS OF THE BRITISH ISLANDS WOULD REMAIN
ABOVE WATER AFTER A SUBSIDENCE OF THE AREA TO THE EXTENT
OF 500 FEET.

The authorities to whom I am indebted for the information contained in this
map are—for

SCOTLAND,—A. Geikie, Esq., F.G.S. and T. F. Jamieson, Esq., of Ellon, Aberdeenshire.

ENGLAND, — For the counties of Yorkshire, Lancashire, and Durham — Col.
Sir Henry James, R.E.

Dorsetshire, Hampshire, and Isle of Wight—H. W. Bristow, Esq.

Gloucestershire, Somersetshire, and part of Devon—R. Etheridge, Esq.

Kent and Sussex—Frederick Drew, Esq.

Isle of Man—W. Whitaker, Esq.

IRELAND,—Reduced from a contour map constructed by Lieut. Larcom, R.E.
in 1837, for the Railway Commissioners.

Fig. 41.

MAP OF PART OF THE NORTH-WEST OF EUROPE, INCLUDING THE BRITISH
ISLES, SHOWING THE EXTENT OF SEA WHICH WOULD BECOME LAND IF
THERE WERE A GENERAL RISE OF THE AREA TO THE EXTENT OF 600
FEET.

The darker shade expresses what is now land, the lighter shade the space
intervening between the present coast line and the 100 fathom line, which
would be converted by such a movement into land.

The original of this map will be found in Sir H. de la Beche's 'Theoretical
Researches,' p. 190, 1834, but several important corrections have been introduced into it from recently published Admiralty Surveys.

Mr. John Murray, C.E., has published, in the Proceedings of the Institute of
Civil Engineers for 1860–61, a most instructive map of the soundings in the
North Sea, showing certain zones of equal depth, derived from the British Admiralty surveys and from other sources. He has shown that there is a channel
exceeding 100 fathoms in depth, which extends from the North Sea into the

Isles with each other and with the continent, as expressed in
the map, fig. 44, even if there had been no geological facts in
favour of such a junction. For in no other way would he be
able to account for the identity of the fauna and flora found
throughout these lands. Had they been separated ever since
the miocene period, like Madeira, Porto Santo, and the
Desertas, constituting the small Madeiran Archipelago, we
might have expected to discover a difference in the species
of land-shells, not only when Ireland was compared to England, but when different islands of the Hebrides were contrasted one with another, and each of them with England.
It would not, however, be necessary, in order to effect the
complete fusion of the animals and plants which we witness,
to assume that all parts of the area formed continuous land
at one and the same moment of time, but merely that the
several portions were so joined within the pleistocene era
as to allow the animals and plants to migrate freely in
succession from one district to another.

Southernmost Extent of Erratics in England.

In reference to that portion of the south of England
which is marked by diagonal lines in the map at p. 325, the

Baltic, and which was not noticed in Sir H. De la Beche's map of 1834 (*Theoretical Researches*, p. 190). This channel, at the point marked *b* in our map,
near the entrance of the Baltic, is no less than 430 fathoms, or 2,580 feet, deep.
The introduction of these straits, separating Scandinavia from the British area,
constitutes a striking geographical feature in the corrected map. The outline
also of the west coasts of Sweden and part of Norway, as they would appear
after the assumed upheaval, has been corrected by reference to the most
modern surveys.

Secondly. Mr. Murray has also called my attention to the new Admiralty
chart of the west coast of Ireland, which shows that the 100-fathom line
approaches much nearer the west coast than it was represented to do in the old
charts—so much so, as to diminish by 80 miles the westward trend of the land
in our map; for the Porcupine Bank, instead of being joined to the mainland,
as in Sir H. De la Beche's map of 1834, and in fig. 43 of my two former editions,
now forms an island in latitude 53° 50′ N., off the coast of Connemara.

Lastly. In the narrowest part of the channel which separates Ireland from
Scotland, at the point marked *a* in map, fig. 44, there is a small space exceeding
100 fathoms in depth.

theory of its having been an area of dry land during the
period of great submergence and floating-ice does not depend
merely on negative evidence, such as the absence of the
northern drift or boulder clay on its surface; but we have
also, in favour of the same conclusion, the remarkable fact
of the presence of erratic blocks on the southern coast of
Sussex, implying the existence there of an ancient coast-line
at a period when the cold must have been at its height.

These blocks are to be seen in greatest number at Pagham
and Selsea, fifteen miles south of Chichester, in lat. 50° 40′ N.
They consist of fragments of granite, syenite, and greenstone, as well as of Devonian and Silurian rocks, some of
them of large size. I measured one of granite at Pagham, twenty-seven feet in circumference. They are not of
northern origin, but must have come from the coast of
Normandy or Brittany, or from land which may once have
existed to the south-west, in what is now the English Channel.

They were probably drifted into their present site by coast
ice, and the yellow clay and gravel in which they are embedded are a littoral formation, as shown by the shells.
Beneath the gravel containing these large erratics, is a blue
mud in which skeletons of *Elephas antiquus*, and other
mammalia, have been observed. Still lower occurs a sandy
loam, from which Mr. R. G. Austen * has collected thirty-
eight species of marine shells, all recent, but forming an
assemblage differing as a whole from that now inhabiting
the English Channel. The presence among them of *Lutraria
rugosa* and *Pecten polymorphus*, not known to range
farther north in the actual seas than the coast of Portugal,
indicates a somewhat warmer temperature at the time when
they flourished. Subsequently, there must have been great
cold when the Selsea erratics were drifted into their present
position, and this cold doubtless coincided in time with a low

* Geological Quarterly Journal, vol. xiii. p. 40.

temperature farther north. These transported rocks of Sussex
are somewhat older than a sea-beach with recent marine
shells which at Brighton is covered by chalk rubble, called
the 'elephant-bed,' which I cannot describe in this place, but
I allude to it as one of many geological proofs of the former
existence of a seashore in this region, and of ancient cliffs
bounding the channel between France and England, all of
older date than the close of the glacial period.

In order to form a connected view of the most simple
series of changes in physical geography which can possibly
account for the phenomena of the glacial period, and the
period of the establishment of the present provinces of animals
and plants, the following geographical states of the British
and adjoining areas may be enumerated.

First, a continental period, towards the close of which the
forest of Cromer flourished (p. 254) : when the land was at
least 500 feet above its present level, perhaps much higher,
and its extent probably greater than that given in the map,
fig. 44, p. 328.   The remains of *Hippopotamus major*, and
Rhinoceros etruscus, found in beds of this period, seem to
indicate a climate somewhat milder than that now prevailing
in Great Britain.

Secondly, a period of submergence, by which the land
north of the Thames and Bristol Channel, and that of Ireland,
was gradually reduced to such an archipelago as is pictured
in map, fig. 43, p. 327 ; and finally to such a general prevalence of sea as is seen in map, fig. 42, p. 325.   This was
the period of great submergence and of floating ice, when
the Scandinavian flora, which occupied the lower grounds
during the first continental period, may have obtained exclusive possession of the only lands not covered with perpetual
snow.

Thirdly, a second continental period, when the bed of the
glacial sea, with its marine shells and erratic blocks, was laid

dry, and when the quantity of land equalled that of the first
period, and therefore probably exceeded that represented in
the map, fig. 44, p. 328. During this period there were glaciers
in the higher mountains of Scotland and Wales, and the
Welsh glaciers, as we have seen, pushed before them and
cleared out the marine drift with which some valleys had
been filled during the period of submergence. The parallel
roads of Glen Roy are referable to some part of the same era.

As a reason for presuming that the land which in map,
fig. 44, p. 328, is only represented as 600 feet above its present
level, was during part of this period much higher, Professor
Ramsay has suggested that, as the previous depression far
exceeded a hundred fathoms (amounting in Wales to 1,400
feet, as shown by marine shells, and to 2,300, by stratified
drift), it is not improbable that the upward movement was on
a corresponding scale.

In passing from the period of chief submergence to this
second continental condition of things, we may conceive a
gradual change first from that of map 42 to map 43, then
from the latter phase to that of map 44, and finally to still
greater accessions of land. During this last period the
passage of the Germanic flora into the British area took place,
and the Scandinavian plants, together with northern insects,
birds, and quadrupeds, retreated into the higher grounds.
It was during this second continental period that Palæolithic
Man probably inhabited Europe together with the mammoth
and woolly rhinoceros, or with the *Elephas antiquus*, *Rhinoceros hemitœchus*, and *Hippopotamus major*.

Fourthly, the next and last change comprised the breaking up of the land of the British area once more into
numerous islands, ending in the present geographical condition of things. There were probably many oscillations of
level during this last conversion of continuous land into
islands, and such movements in opposite directions would

account for the occurrence of marine shells at moderate
heights above the level of the sea, notwithstanding a
general lowering of the land. To the close of this era belong the marine deposits of the Clyde and the Carses of the
Tay and Forth, before alluded to, pp. 51, 59, which mark a
slight elevation. During this period a gradual amelioration
of temperature took place, from the cold of the glacial period
to the climate of historical times.

In a memoir by Professor E. Forbes, before cited, he
observes that the land of passage by which the plants and
animals migrated into Ireland consisted of the upraised
marine drift which had previously formed the bottom of the
glacial sea. Portions of this drift extend to the eastern shores
of Wicklow and Wexford, others are found in the Isle of Man
full of arctic shells, others on the British coast opposite
Ireland. The freshwater marl, containing numerous skeletons
of the great deer, or *Megaceros*, overlie in the Isle of Man
that marine glacial drift. Professor Forbes also remarks
that the subsequent disjunction of Ireland from England, or
the formation of the St. George's Channel, which is less than
400 feet in its greatest depth, preceded the opening of the
Straits of Dover, or the final separation of England from the
Continent. This he inferred from the present distribution of
species both in the animal and vegetable kingdoms. Thus
for example, there are twice as many reptiles in Belgium as
in England, and the number inhabiting England is twice
that found in Ireland. Yet the Irish species are all common
to England, and all the English to Belgium. It is therefore assumed that the migration of species westward having
been the work of time, there was not a sufficient lapse of
ages to complete the fusion of the continental and British
reptilian fauna, before France was separated from England
and England from Ireland.

For the same reason there are also a great number of birds

of short flight, and small quadrupeds, inhabiting England which do not cross to Ireland, the St. George's Channel seeming to have arrested them in their westward course.*

The depth of St. George's Channel in the narrower parts is only 360 feet, and the English Channel between Dover and Calais less than 200, and rarely anywhere more than 300 feet ; so that vertical movements of slight amount compared to some of those previously considered, with the aid of denuding operations or the waste of sea cliffs, and the scouring out of the channel, might in time effect the insulation of the lands above alluded to.

Time required for successive Changes in Physical Geography in the Pleistocene Period.

The time which it would require to bring about such changes of level, according to the average rate assumed at p. 64, however vast, will not be found to exceed that which would best explain the successive fluctuations in terrestrial temperature, the glaciation of solid rocks, the transportation of erratics above and below the sea level, the height of arctic shells above the sea, and last, not least, the migration of the existing species of animals and plants into their actual stations, and the extinction of some conspicuous forms which flourished during the pleistocene ages. When we duly consider all these changes which have taken place since the beginning of the glacial epoch, or since the Forest of Cromer and the *Elephas meridionalis* flourished, we shall find that the phenomena become more and more intelligible in proportion to the slowness of the rate of elevation and depression which we assume.

The submergence of Wales to the extent of 1,400 feet, as

* E. Forbes, 'Fauna and Flora of British Isles; Memoirs of Geological Survey,' vol. I. p. 344, 1846.

proved by glacial shells, would require 56,000 years, at the
rate of 2½ feet per century; but taking Professor Ramsay's
estimate of 800 feet more, as stated at p. 313, that depression
being implied by the position of some of the stratified drift,
we must demand an additional period of 32,000 years,
amounting in all to 88,000; and the same time would be
required for the re-elevation of the tract to its present height.
But if the land rose in the second continental period as much
as 600 feet above its present level, as in map, p. 328, this
600 feet, first of rising and then of sinking, would require
48,000 years more; the whole of the grand oscillation, comprising the submergence and re-emergence, having taken
about 224,000 years for its completion; and this, even if
there were no pause or stationary period, when the downward
movement ceased, and before it was converted into an upward
one.

I am aware that it may be objected that the average rate
here proposed is a purely arbitrary and conjectural one,
because, at the North Cape, it is supposed that there has been
a rise of about five feet in a century, and at Spitzbergen,
according to Mr. Lamont, a still faster upheaval during the
last 400 years.[*]   But, granting that in these and some exceptional cases (none of them as yet very well established)
the rising or sinking has, for a time, been accelerated, I do
not believe the average rate of motion to exceed that above
proposed. Mr. Darwin, I find, considers that such a mean
rate of upheaval would be as high as we could assume for
the west coast of South America, where we have more evidence
of sudden changes of level than anywhere else. He has not,
however, attempted to estimate the probable rate of secular
elevation in that or any other region.

I see no reason for supposing that any part of the revolutions in physical geography, to which the maps above

[*] Seasons with the Sea-Horses, p. 202.

described have reference, indicate any catastrophes greater
than those which the present generation has witnessed. If
man was in existence when the Cromer forest was becoming
submerged, he would have felt no more alarm than the
Danish settlers on the east coast of Baffin's Bay, when they
found the poles, which they had driven into the beach to
secure their boats, had subsided below their original level.

Already, perhaps, the melting ice has thrown down clay,
sand, and boulders upon those poles, a counterpart in
mineral character, though not in age, of the boulder-clay
which overlies the forest-bed on the Norfolk cliffs.

If we reflect on the long series of events of the pleistocene
and recent periods contemplated in this chapter, it will be
remarked that the time assigned to the first appearance of
Man, so far as our geological inquiries have yet gone, is
extremely modern in relation to the age of the existing
fauna and flora, or even to the time when most of the living
species of animals and plants attained their actual geographical distribution.   At the same time it will also be seen,
that if the advent of Man in Europe occurred before the
close of the second continental period, and antecedently to
the separation of Ireland from England, and of England
from the continent, the event would be sufficiently remote
to cause the historical period to appear quite insignificant
in duration, when compared to the antiquity of the human
race.

CHAPTER XV.

EXTINCT GLACIERS OF THE ALPS AND THEIR CHRONOLOGICAL
RELATION TO THE HUMAN PERIOD.

EXTINCT GLACIERS OF SWITZERLAND—ALPINE ERRATIC BLOCKS ON
THE JURA—TRANSPORTED BY GLACIERS AND NOT BY FLOATING ICE—
EXTINCT GLACIERS OF THE ITALIAN SIDE OF THE ALPS—THEORY OF
THE ORIGIN OF LAKE-BASINS BY THE EROSIVE ACTION OF GLACIERS
CONSIDERED—SUCCESSIVE PHASES IN THE DEVELOPMENT OF GLACIAL
ACTION IN THE ALPS—LACUSTRINE FORMATIONS OF INTERGLACIAL
AGE—PROBABLE RELATION OF THESE TO THE EARLIEST KNOWN DATE
OF MAN—COLD PERIOD IN SICILY AND SYRIA.

Extinct Glaciers of Switzerland.

WE have seen in the preceding chapters that the mountains
of Scandinavia, Scotland, and North Wales have served,
during the glacial period, as so many independent centres
for the dispersion of erratic blocks, just as at present the
ice-covered continent of North Greenland is sending down
ice in all directions to the coast, and filling Baffin's Bay
with floating bergs, many of them laden with fragments of
rocks.

Another great European centre of ice-action during the
pleistocene period was the Alps of Switzerland, and I shall
now proceed to consider the chronological relations of the
extinct Alpine glaciers to those of more northern countries
previously treated of.

The Alps lie far south of the limits of the northern drift
described in the foregoing pages, being situated between the
44th and 47th degrees of north latitude. On the flanks of

these mountains, and on the sub-Alpine ranges of hills or plains adjoining them, those appearances which have been so often alluded to, as distinguishing or accompanying the drift, between the 50th and 70th parallels of north latitude, suddenly reappear and assume, in a southern region, a truly arctic development. Where the Alps are highest, the largest erratic blocks have been sent forth; as, for example, from the regions of Mont Blanc and Monte Rosa, into the adjoining parts of Switzerland and Italy; while in districts where the great chain sinks in altitude, as in Carinthia, Carniola, and elsewhere, no such rocky fragments, or a few only and of smaller bulk, have been detached and transported to a distance.

In the year 1821, M. Venetz first announced his opinion that the Alpine glaciers must formerly have extended far beyond their present limits, and must since have varied from century to century, so that between the 11th and 15th centuries they were less advanced than now, and began in the 17th and 18th centuries to push forward again, covering roads formerly open, and overwhelming forests of ancient growth. The proofs appealed to by him in confirmation of this doctrine were afterwards acknowledged by M. Charpentier, who strengthened them by new observations and arguments, and declared, in 1836, his conviction that the glaciers of the Alps must once have reached as far as the Jura, and have carried thither their moraines across the great valley of Switzerland. M. Agassiz, after several excursions in the Alps with M. Charpentier, and after devoting himself some years to the study of glaciers, published, in 1840, an admirable description of them and of the marks which attest the former action of great masses of ice over the entire surface of the Alps and the surrounding country.*

* Agassiz, Études sur les Glaciers et Système Glaciaire. For a summary of the action of glaciers and the phenomena produced by them, see 'Principles of Geology,' 11th ed. vol. i. p. 359 et seq., and 'Student's Elements of Geology,' p. 143 et seq.

Moraines, erratics, polished surfaces, domes, striæ, and perched blocks are all observed in the Alps at great heights above the present glaciers, and far below their actual extremities; also in the great valley of Switzerland, fifty miles broad; and almost everywhere on the Jura, a chain which lies to the north of this valley. The average height of the Jura is about one-third that of the Alps, and it is now entirely destitute of glaciers; yet it presents almost everywhere moraines, and polished and grooved surfaces of rocks. The erratics, moreover, which cover it present a phenomenon which has astonished and perplexed the geologist for more than half a century. No conclusion can be more incontestable than that these angular blocks of granite, gneiss, and other crystalline formations, came from the Alps, and that they have been brought for a distance of fifty miles and upwards across one of the widest and deepest valleys of the world; so that they are now lodged on the hills and valleys of a chain composed of limestone and other formations, altogether distinct from those of the Alps. Their great size and angularity, after a journey of so many leagues, has justly excited wonder, for hundreds of them are as large as cottages; and one in particular, composed of gneiss, celebrated under the name of Pierre à Bot, rests on the side of a hill about 900 feet above the lake of Neufchatel, and is no less than forty feet in diameter. But there are some far-transported masses of granite and gneiss which are still larger, and which have been found to contain 50,000 and 60,000 cubic feet of stone; and one limestone block at Devens, near Bex, which has travelled thirty miles, contains 161,000 cubic feet, its angles being sharp and unworn.

Von Buch, Escher, and Studer inferred, from an examination of the mineral composition of the boulders, that those resting on the Jura, opposite the lakes of Geneva and Neufchatel, have come from the region of Mont Blanc and the

Valais, as if they had followed the course of the Rhone, to the lake of Geneva, and had then pursued their way uninterruptedly in a northerly direction.

Alpine Erratic Blocks on the Jura, and in Switzerland generally, due to Glaciers and not to Floating Ice.

M. Charpentier, who conceived the Alps in the period of greatest cold to have been higher by several thousand feet than they are now, had already suggested that the Alpine glaciers once reached continuously to the Jura, conveying thither the large erratics in question.* M. Agassiz, on the other hand, instead of introducing distinct and separate glaciers, imagined that the whole valley of Switzerland might have been filled with ice, and that one great sheet of it extended from the Alps to the Jura, the two chains being of the same height as now relatively to each other. To this idea it was objected that the difference of altitude, when distributed over a space of fifty miles, would give an inclination of two degrees only, or far less than that of any known glacier. In spite of this difficulty, the hypothesis has since received the support of Professor James Forbes, in his very able work on the Alps, published in 1843.

In 1841, I advanced, jointly with Mr. Darwin,† the theory that the erratics might have been transferred by floating ice to the Jura, at the time when the greater part of that chain, and the whole of the Swiss valley to the south, was under the sea. We pointed out that if at that period the Alps had attained only half their present altitude, they would yet have constituted a chain as lofty as the Chilian Andes, which, in a latitude corresponding to Switzerland, now send down glaciers to the head of every sound, from which ice-

* D'Archiac, Histoire des Progrès,   † See Elements of Geology, 2nd
&c., tom. ii. p. 249.   ed., 1841.

bergs, covered with blocks of granite, are floated seaward.
Opposite that part of Chili where the glaciers abound, is
situated the island of Chiloe, one hundred miles in length,
with a breadth of thirty miles, running parallel to the continent. The channel which separates it from the mainland
is of considerable depth, and twenty-five miles broad. Parts
of its surface, like the adjacent coast of Chili, are overspread
with recent marine shells, showing an upheaval of the land
during a very modern period ; and beneath these shells is a
boulder deposit, in which Mr. Darwin found large blocks of
granite and syenite, which had evidently come from the
Andes.

A continuance in future of the elevatory movement, now
observed to be going on in this region of the Chilian Andes
and of Chiloe, might cause the former chain to rival the Alps
in altitude, and give to Chiloe a height equal to that of the
Jura. The same rise might dry up the channel between
Chiloe and the mainland, so that it would then represent
the great valley of Switzerland.

Sir Roderick I. Murchison, after making several important geological surveys of the Alps, proposed, in 1849, a
theory agreeing essentially with that suggested by Mr. Darwin and myself, viz. that the erratics were transported to the
Jura, at a time when the great strath of Switzerland, and
many valleys receding far into the Alps, were under water.
He thought it impossible that the glacial detritus of the
Rhone could ever have been carried to the lake of Geneva
and beyond it by a glacier, or that so vast a body of ice
issuing from one narrow valley could have spread its erratics
over the low country of the Cantons of Vaud, Friburg, Berne,
and Soleure, as well as the slopes of the Jura, comprising a
region of about a hundred miles in breadth from south-west
to north-east, as laid down in the map of Charpentier. He
therefore imagined the granitic blocks to have been trans-

ported to the Jura by ice-floats when the intermediate country
was submerged.* It may be remarked that this theory, provided the water be assumed to have been salt or brackish,
demands quite as great an oscillation in the level of the land
as that on which Charpentier had speculated, the only difference being that the one hypothesis requires us to begin with
a subsidence of 2,500 or 3,000 feet, and the other with an
elevation to the same amount. We should also remember
that the crests or watersheds of the Alps and Jura are about
eighty miles apart, and if once we suppose them to have been
in movement during the glacial period, it is very probable
that the movements at such a distance may not have been
strictly uniform. If so, the Alps may have been relatively
somewhat higher, which would greatly have facilitated the
extension of Alpine glaciers to the flanks of the less elevated
chain.

Five years before the publication of the memoir last
mentioned, M. Guyot had brought forward a great body of
new facts in support of the original doctrine of Charpentier,
that the Alpine glaciers once reached as far as the Jura, and
that they had deposited thereon a portion of their moraines.†
The scope of his observations and argument was laid with
great clearness before the British public in 1852 by Mr.
Charles Maclaren, who had himself visited Switzerland for
the sake of forming an independent opinion on a theoretical
question of so much interest, and on which so many eminent
men of science had come to such opposite conclusions.‡

M. Guyot had endeavoured to show that the Alpine
erratics, instead of being scattered at random over the Jura
and the great plain of Switzerland, are arranged in a certain
determinate order, strictly analogous to that which ought to

* Quarterly Geological Journal,   Naturelles de Neufchâtel. 1845.
1850. vol. vi. p. 65.   ‡ Edinburgh New Philosophical
† Bulletin de la Société des Sciences   Magazine, October 1852.

prevail if they had once constituted the lateral, medial, and
terminal moraines of great glaciers. The rocks chiefly relied
on as evidence of this distribution consist of three varieties of
granite, besides gneiss, chlorite-slate, euphotide, serpentine,
and a peculiar kind of conglomerate, all of them mineral
compounds, foreign alike to the great strath between the
Alps and Jura, and to the structure of the Jura itself. In
these two regions, limestones, sandstones, and clays of the
secondary and tertiary formations alone crop out at the surface, so that the travelled fragments of Alpine origin can
easily be distinguished, and in some cases the precise localities pointed out from whence they must have come.

The accompanying map or diagram, slightly altered from
one given by Mr. Maclaren, will enable the reader more
fully to appreciate the line of argument relied on by M.
Guyot. The dotted area is that over which the Alpine
fragments were spread by the supposed extinct glacier of the
Rhone. The site of the present reduced glacier of that name
is shown at A. From that point, the boulders may first be
traced to B, or Martigny, where the valley takes an abrupt
turn at right angles to its former course. Here the blocks
belonging to the right side of the river, or derived from c, d, e,
have not crossed over to the left side at B, as they should
have done had they been transported by floating ice, but
continue to keep to the side to which they belonged, assuming that they once formed part of a right lateral moraine of
a great extinct glacier. That glacier, after arriving at the
lower end of the long narrow valley of the upper Rhone at
F, filled the lake of Geneva, F, I, with ice. From F, as from
a great vomitory, it then radiated in all directions, bearing
along with it the moraines with which it was loaded, and
spreading them out on all sides over the great plain. But
the principal icy mass moved straight onwards in a direct
line towards the hill of Chasseron, G (precisely opposite F),

where the Alpine erratics attain their maximum of height
on the Jura, that is to say, 2,015 English feet above the
level of the Lake of Neufchatel, or 3,450 feet above the

Fig. 45

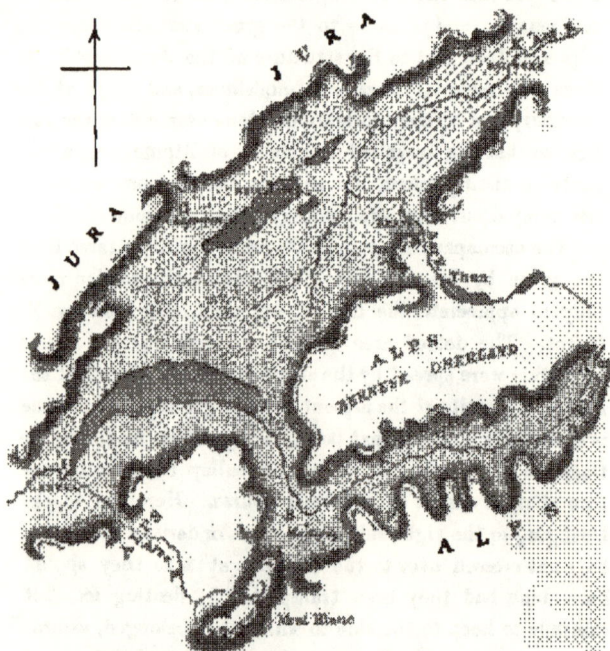

MAP SHOWING THE SUPPOSED COURSE OF THE ANCIENT AND NOW EX-
TINCT GLACIER OF THE RHONE, AND THE DISTRIBUTION OF THE
ERRATIC BLOCKS AND DRIFT CONVEYED BY IT TO THE GREAT VALLEY
OF SWITZERLAND AND THE JURA.

sea.   The granite blocks which have ascended to this
eminence, a, came from the east shoulder of Mont Blanc, A,
having travelled in the direction B, F, G.

When these and the accompanying blocks resting on the
south-eastern declivity of the Jura are traced from their

culminating point, *o*, in opposite directions, whether westward towards Geneva, or eastward towards Soleure, they are
found to decline in height from the middle of the arc *o*
towards the two extremities *i* and *x*, both of which are at a
lower level than *o* by about 1,500 feet. In other words,
the ice of the extinct glacier, having mounted up on the
sloping flanks of the Jura in the line of greatest pressure
to its highest elevation, began to decline laterally with a
gentle inclination, till it reached two points distant from
each other no less than 100 miles.

In further confirmation of this theory, M. Guyot observed
that fragments derived from the right bank of the great
valley of the Rhone, *c*, *d*, *e*, are found on the right side of
the great Swiss basin or strath, as at *l* and *m*, while those
derived from the left bank, *p*, *h*, occur on the left side of
the basin, or on the Jura, between *o* and *i*; and those again
derived from places farthest up on the left bank and nearest
the source of the Rhone, as *n o*, occupy the middle of the
great basin, constituting, between *m* and *x*, what M. Guyot
calls the frontal or terminal moraine of the eastern prolon-
gation of the old glacier.

It is evident that the above-described restriction of
certain fragments of peculiar lithological character to that
bank of the Rhone where the parent rocks are alone met
with, and the linear arrangement of the blocks in corresponding order on the opposite side of the great plain of
Switzerland, are facts which harmonise singularly well with
the theory of glaciers, while they are much less reconcilable
with that of floating ice. Against the latter hypothesis, all
the arguments which Charpentier originally brought forward
in opposition to the first popular doctrine of a grand
débâcle, or sudden flood, rushing down from the Alps to the
Jura, might be revived. Had there ever been such a rush
of muddy water, said he, the blocks carried down the basins

of the principal Swiss rivers, such as the Rhone, Aar, Reuss,
and Limmat, would all have been mingled confusedly
together instead of having each remained in separate and
distinct areas as they do and should do according to the
glacial hypothesis.

M. Morlot presented me in 1857 with an unpublished
map of Switzerland in which he had embodied the results of
his own observations, and those of MM. Guyot, Escher, and
others, marking out by distinct colours the limits of the
ice-transported detritus proper to each of the great river-
basins. The arrangement of the drift and erratics thus
depicted accords perfectly well with Charpentier's views, and
is quite irreconcilable with the supposition of the scattered
blocks having been dispersed by floating ice when Switzerland was submerged.

As opposed to the latter hypothesis, I may also state that
nowhere as yet have any marine shells or other fossils than
those of a terrestrial character, such as the bones of the
mammoth, and a few other mammalia, and some coniferous
wood, been detected in those drifts, though they are often
many hundreds of feet in thickness.

A glance at M. Morlot's map, above mentioned,* will
show that the two largest areas, indicated by a single colour,
are those over which the Rhone and the Rhine are supposed
to have spread out in ancient times their enormous moraines.
One of these glaciers only, that of the Rhone, has been exhibited in our diagram, fig. 45, p. 344. The distinct character
of the drift in the two cases is such as it would be if two
colossal glaciers should now come down from the higher Alps
through the valleys traversed by those rivers, leaving their
moraines in the low country. The space occupied by the glacial
drift of the Rhine is equal in dimensions to, or rather exceeds,
that of the Rhone, and its course is not interfered with in the

* See map, Geological Quarterly Journal, vol. xviii. pl. 18, p. 185.

least degree by the Lake of Constance, forty-five miles long, any more than is the dispersion of the erratics of the Rhone by the Lake of Geneva, about fifty miles in length. The angular and other blocks have in both instances travelled on precisely as if those lakes had no existence, or as if, which was no doubt the case, they had been filled with solid ice.

During my visit to Switzerland in 1857, I made excursions, in company with several distinguished geologists, for the sake of testing the relative merits of the two rival theories above referred to, and examined parts of the Jura above Neufchatel in company with M. Desor, the country round Soleure with M. Langen, the southern side of the great strath near Lausanne with M. Morlot, the basin of the Aar, around Berne, with M. Escher von der Linth; and having satisfied myself that all the facts which I saw north of the Alps were in accordance with M. Guyot's views, I crossed to the Italian side of the great chain, and became convinced that the same theory was equally applicable to the ancient moraines of the plains of the Po.

M. Escher pointed out to me at Trogen in Appenzel, on the left bank of the Rhine, fragments of a rock of a peculiar mineralogical character, commonly called the granite of Pontelyas, the natural position of which is well known near Trons, a hundred miles from Trogen, on the left bank of the Rhine, about thirty miles from the source of that river. All the blocks of this peculiar granite keep to the left bank, even where the valley turns almost at right angles to its former course near Mayenfeld below Chur, making a sharp bend, resembling that of the valley of the Rhone at Martigny. The granite blocks, where they are traced to the low country, still keep to the left side of the Lake of Constance. That they should not have crossed over to the opposite river-

bank below Chur is highly improbable, if, rejecting the aid
of land-ice, we appeal to floating ice as the transporting
power.

In M. Morlot's map, already cited, we behold between
the areas occupied by the glacial drift of the Rhine and
Rhone three smaller yet not inconsiderable spaces, distinguished by distinct colours, indicating the peculiar detritus
brought down by the three great rivers, the Aar, Reuss, and
Limmat. The ancient glacier of the first of these, the Aar,
has traversed the lakes of Brienz and Thun, and has borne
angular, polished, and striated blocks of limestone and other
rocks as far as Berne, and somewhat below that city. The
Reuss has also stamped the lithological character of its own
mountainous region upon the lower part of its hydrographical basin by covering it with its peculiar Alpine drift.
In like manner the old extinct glacier of the Limmat, during
its gradual retreat, has left monuments of its course in the
lake of Zurich in the shape of terminal moraines, one of
which has almost divided that great sheet of water into two
lakes.

The ice-work done by the extinct glaciers, as contrasted
with that performed by their dwarfed representatives of the
present day, is in due proportion to the relative volume of
the supposed glaciers, whether we measure them by the
distances to which they have carried erratic blocks, or the
areas which they have strewed over with drift, or the hard
surfaces of rock and number of boulders which they have
polished and striated. Instead of a length of five, ten, or
twenty miles, and a thickness of 200, 300, or at the utmost
600 feet, those giants of the olden time must have been
from 50 to 150 miles long, and between 1,000 and 3,000
feet deep. In like manner the glaciation, although identical
in kind, is on so small a scale in the existing Alpine glaciers

as at first sight to disappoint a Swedish, Scotch, Welsh, or
North American geologist. When I visited the terminal
moraine of the glacier of the Rhone in 1859, and tried to
estimate the number of angular or rounded pebbles and
blocks which exhibited glacial polishing or scratches as
compared to those bearing no such markings, I found that
several thousand had to be reckoned before I arrived at the
first, which was so striated or polished as to differ from the
stones of an ordinary torrent-bed. Even in the moraines
of the glaciers of Zermatt, Viesch, and others, in which
fragments of limestone and serpentine are abundant (rocks
which most readily receive and most faithfully retain the
signs of glaciation), I found, for one which displayed such
indications, several hundreds entirely free from them. Of
the most opposite character were the results obtained by me
from a similar scrutiny of the boulders and pebbles of the
terminal moraine of one of the old extinct glaciers, namely,
that of the Rhone in the suburbs of Soleure. Thus at the
point κ, in the map, fig. 45, p. 344, I observed a mass of
unstratified clay or mud, through which a variety of angular
and rubbed stones were scattered, and a marked proportion
of the whole were polished and scratched, and the clay
rendered so compact, as if by the incumbent pressure of a
great mass of ice, that it has been found necessary to blow
it up with gunpowder in making railway cuttings through
part of it. A marble rock of the age of our Portland stone,
on which this old moraine rests, has its surface polished like
a looking-glass, displaying beautiful sections of fossil shells
of the genera Nerinæa and Pteroceras, while occasionally,
besides finer striæ, there are deep rectilinear grooves, agreeing in direction with the course in which the extinct glacier
would have moved according to the theory of M. Guyot,
before explained.

Extinct Glaciers of the Italian Side of the Alps.

To select another example from the opposite or southern side of the Alps. It will be seen in the elaborate map, executed in 1862 by Signor Gabriel de Mortillet, of the ancient glaciers of the Italian flank of the Alps, that the old moraines descend in narrow strips from the snow-covered ridges, through the principal valleys, to the great basin of the Po, on reaching which they expand and cover large circular or oval areas. Each of these groups of detritus is observed (see map, p. 351) to contain exclusively the wreck of such rocks as occur *in situ* on the Alpine heights of the hydrographical basins to which the moraines respectively belong.

I had an opportunity of verifying this fact, in company with Signor Gastaldi as my guide, by examining the erratics and boulder formation between Susa and Turin, on the banks of the Dora Riparia, which brings down the waters from Mont Cenis, and from the Alps SW. of it. I there observed striated fragments of dolomite and gypsum, which had come down from Mont Cenis, and had travelled as far as Avigliana; also masses of serpentine, brought from less remote points, some of them apparently exceeding in dimensions the largest erratics of Switzerland. I afterwards visited, in company with Signori Gastaldi and Michelotti, a still grander display of the work of a colossal glacier of the olden time, twenty miles NE. of Turin, the moraine of which descended from the two highest of the Alps, Mont Blanc and Monte Rosa, and, after passing through the valley of Aosta, issued from a narrow defile above Ivrea (see map, fig. 46). From this vomitory, the old glacier poured into the plains of the Po that wonderful accumulation of mud, gravel, boulders, and large erratica, which extends for fifteen miles from above Ivrea to below Caluso, and which, when

seen in profile from Turin, have the aspect of a chain of
hills.   In many countries, indeed, they might rank as an important range of hills, for where they join the mountains

Fig. 46.

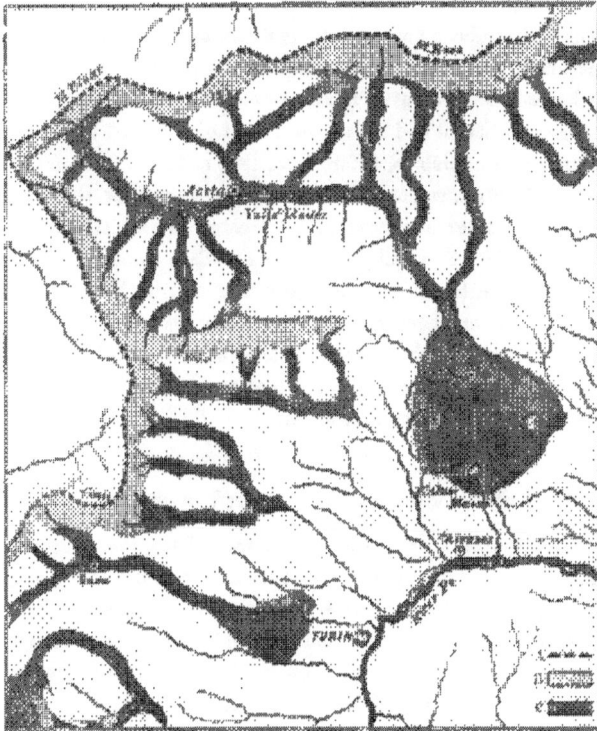

they are more than 1,500 feet high, and retain more than
half that height for a great part of their course, rising very
abruptly from the plain, often with a slope of from 20° to
30°. This glacial drift reposes near the mountains on ancient
metamorphic rocks, and farther from them on marine pliocene strata. Portions of the ridges of till and stratified
matter have been cut up into mounds and hillocks by the
action of the river, the Dora Baltea, and there are numerous
lakes, so that the entire moraine much resembles, except in
its greater height and width, the line of glacial drift of Perthshire and Forfarshire, before described, p. 294. Its complicated structure can only be explained by supposing that the
ancient glacier advanced and retreated several times, and
left large lateral moraines, the more modern mounds within
the limits of the older ones, and masses of till thrown down
upon the re-arranged and stratified materials of the first set
of moraines. Such appearances accord well with the hypothesis of the successive phases of glacial action in Switzerland,
to which I shall presently advert.

Contorted Strata of Glacial Drift south of Ivrea.

At Mazzé near Caluso (see map, p. 351), the southern
extremity of this great moraine has recently been cut
through in making a tunnel for the railway which runs from
Turin to Ivrea. In the fine section thus exposed Signor
Gastaldi and I had an opportunity of observing the internal
structure of the glacial formation. In close juxtaposition to
a great mass of till with striated boulders, we saw stratified
beds of alternating gravel, sand, and loam, which were so
sharply bent that many of them had been twice pierced
through in the same vertical cutting. Whether they had
been thus folded by the mechanical power of an advancing
glacier, which had pushed before it a heap of stratified matter,

as the glacier of Zermatt has been sometimes known to shove
forward blocks of stone through the walls of houses, or
whether the melting of masses of ice, once interstratified with
sand and gravel, had given rise to flexures, in the manner
before suggested, pp. 183 and 265 ; it is at least satisfactory
to have detected this new proof of a close connection between
ice-action and contorted stratification, such as has been
described as so common in the Norfolk cliffs, p. 262, and
which is also very often seen in Scotland and North America,
where stratified gravel overlies till. I have little doubt that
if the marine pliocene strata, which underlie a great part of
the moraine below Ivrea, were exposed to view in a vertical
section, those fundamental strata would be found not to participate in the least degree in the plications of the sands and
gravels of the overlying glacial drift.

To return to the marks of glaciation : in the moraine at
Mazzé, there are many large blocks of protogene, and large
and small ones of limestone and serpentine, which have been
brought down from Monte Rosa, through the gorge of Ivrea,
after having travelled for a distance of fifty miles. Confining
my attention to a part of the moraine, where pieces of limestone and serpentine were very numerous, I found that no less
than one-third of the whole number bore unequivocal signs
of glacial action ; a state of things which seems to bear some
relation to the vast volume and pressure of the ice which
once constituted the extinct glacier and to the distance which
the stones had travelled. When I separated the pebbles of
quartz, which were never striated, and those of granite, mica
schist, and diorite, which do not often exhibit glacial markings, and confined my attention to the serpentine alone, I
found no less than nineteen in twenty of the whole number
polished and scratched ; whereas in the terminal moraines of
some modern glaciers, where the materials have travelled not

A A

more than ten or fifteen, instead of fifty miles, scarce one in twenty even of the serpentine pebbles exhibit glacial polish and striation.

Theory of the Origin of Lake-basins by the erosive Action of Glaciers, considered.

Geologists are all agreed that the last series of movements to which the Alps owe their present form and internal structure occurred after the deposition of the miocene strata; and it has been usual to refer the origin of the numerous lake-basins of Alpine and sub-Alpine regions, both in Switzerland and Northern Italy, to the same movements; for it seemed not unnatural to suppose, that forces capable of modifying the configuration of the greatest European chain, by uplifting some of its component tertiary strata (those of marine origin of the miocene period) several thousand feet above their former level, and throwing them into vertical and contorted positions, must also have given rise to many superficial inequalities, in some of which large bodies of water would collect. M. Desor, in a memoir on the Swiss and Italian lakes, suggested that they may have escaped being obliterated by sedimentary deposition, by having been filled with ice during the whole of the glacial period.

Subsequently to the retreat of the great glaciers, we know that the lake-basins have been to a certain extent encroached upon and turned into land by river deltas; one of which, that of the Rhone, at the head of the lake of Geneva, is no less than twelve miles long and several miles broad; besides which there are many torrents on the borders of the same lake, forming smaller deltas.

M. Gabriel de Mortillet, after a careful study of the glacial formations of the Alps, agreed with his predecessors, that the great lakes had existed before the glacial period, but came to the opinion, in 1859, that they had all been first filled up with alluvial matter, and then re-excavated by the action of ice, which during the epoch of intense cold had by its weight and force of propulsion scooped out the loose and incoherent alluvial strata, even where they had accumulated to a thickness of 2,000 feet. Besides this erosion, the ice had carried the whole mass of mud and stones up the inclined planes, from the central depths to the lower outlets of the lakes, and sometimes far beyond them. As some of these rock-basins are 500, others more than 2,000 feet deep, having their bottoms in some cases 500, in others 1,000 feet below the level of the sea, and having areas from twenty to fifty miles in length and from four to twelve in breadth, we may well be startled at the boldness of this hypothesis.

The following are the facts and train of reasoning which induced M. de Mortillet to embrace these views. At the lower ends of the great Italian lakes, such as Maggiore, Como, Garda, and others, there are vast moraines which are proved by their contents to have come from the upper Alpine valleys above the lakes. Such moraines often repose on an older stratified alluvium, made up of rounded and worn pebbles of precisely the same rocks as those forming the moraines, but not derived from them, being small in size, never angular, polished or striated, and the whole having evidently come from a great distance. These older alluvial strata must, according to M. de Mortillet, be of pre-glacial date, and could not have been carried past the sites of the lakes, unless each basin had previously been filled and levelled up with mud, sand, and gravel, so that the river channel was continuous from the upper to the lower extremity of each basin.

Professor Ramsay, in 1859, brought forward an able theory to account for the origin of lake-basins by the action of ice. After acquiring an intimate knowledge of the glacial phenomena of the British Isles, he had taught, many years before, that small tarns and shallow rock-basins, such as we see in many mountain regions, owe their origin to glaciers which erode the softer rocks, leaving the harder ones standing out in relief and comparatively unabraded. Following up this idea after he had visited Switzerland, and without any communication with M. de Mortillet or cognisance of his views, he suggested that the lake-basins were not of pre-glacial date, but had been scooped out by ice during the glacial period, the excavation having for the most part been effected in miocene sandstone, provincially called, on account of its softness, ' molasse.' By this theory he dispensed with the necessity of filling up pre-existing cavities with stratified alluvium, in the manner proposed by M. de Mortillet.

It is no doubt true, as Professor Ramsay remarks, that heavy masses of ice, creeping for ages over a land-surface (whether this comprise hills, plateaus, and valleys, as in the case of Greenland, before described (p. 276), or be confined to the bottoms of great valleys, as now in the higher Alps), must often, by their grinding action, produce depressions, in consequence of the different degrees of resistance offered by rocks of unequal hardness. Thus, for example, where quartzose beds of mica schist alternate with clay-slate, or where trap-dykes cut through sandstone or slate—these and innumerable other common associations of dissimilar stony compounds must give rise to a very unequal amount of erosion, and consequently to lake-basins on a small scale. But the larger the size of any lake, the more certain it will be to contain within it rocks of every degree of hardness, toughness, and softness; and if we find a gradual deepening from the head towards the central parts, and a shallowing again from the middle to the lower end, as in several of the great Swiss and Italian lakes, which are thirty or forty miles in length, we require a power capable of acting with a considerable degree of uniformity on these masses of varying powers of resistance.

It has been ascertained experimentally, that in a glacier, as in a river, the rate of motion is accelerated or lessened, according to the greater or less slope of the ground; also, that the lower strata of ice, like those of running water, move more slowly than those above them. In the Lago Maggiore, which is more than 2,600 feet deep (797 metres), the ice, says Professor Ramsay, had to descend a slope of about 3° for the first twenty-five miles, and then to *ascend* for the last twelve miles (from the deepest part towards the outlet), at an angle of 5°. It is for those who are conversant with the dynamics of glacier motion to divine whether, in such a case, the discharge of ice would not be entirely effected by the superior and faster-moving strata, and whether the lowest would not be motionless or nearly so, and would therefore exert very little, if any, friction on the bottom.

Unlike the great question of atmospheric denudation, the erosive action of glaciers is free from the complication due to chemical action, and remains simply a piece of rough mechanical work. The final product is the fine mud or flour of rock which is suspended in the water of the stream that flows from the foot of the glacier, and which, no doubt, as we see in Greenland at the present time, is mainly derived from the wearing away of the underlying rock. But it is possible to overrate the amount of denudation implied by this muddy overflow. It must not be forgotten that the rocky fragments which are showered down on to a glacier from the hillsides above, and from lateral and medial moraines, must to a considerable extent fall through crevasses, and so reach the bottom of the glacier. It is these masses which, by their friction on the underlying floor, produce the flour of rock; and there can be little doubt, that from being already decomposed by rain and frost they will suffer more in the

crushing and grinding process than will the rocky floor,
which has long been worn down to a smooth surface, and is
not exposed to atmospheric changes. A large share of the
detritus, therefore, issuing from the foot of an ordinary
glacier must be derived from the transported fragments, and
cannot be adduced as a proof of erosion.

In any case denudation by glaciers must be so slow a
process as to require periods of time, even geologically
speaking, of great duration. And in a mountainous region
it would be very rash to assume that during the occupation
of a valley or lake-basin by ice there were no changes of
level such as may also have had a share in the formation of
lake-basins.

Professor Ramsay has objected to the theory of the
origin of lakes through variations of level in the land, that,
supposing the oscillations to have taken place in the middle
of such a lake as Lago Maggiore, the most favourable
position in order to convert the basin into a river valley, it
would require an elevation of the Alps and a depression of
the plain of Lombardy to an extent for which we have not
sufficient evidence.* But this objection rests upon an
assumption which appears to me quite unnecessary, namely,
that the oscillation must have extended at the same angle
from the confines of the lake up to the crest of the Alps.
On the contrary, the analogy of all anticlinals would lead
us to expect that the upheaval took place along a curved
line, which would involve an elevation of the central Alps
no greater than that of many known mountain ranges. Our
experience in the lifetime of the present generation, of
changes in land in New Zealand during great earthquakes is
entirely opposed to the notion that the movements, whether
upward or downward, are uniform throughout areas of indefinite extent. Moreover it is just in such mountain chains

* Philosophical Magazine, April 1865.

as the Alps that unequal movements in very limited areas are most strikingly exemplified. The huge contortions and vertical position of the older rocks sufficiently show how in very small areas the greatest inequalities of movement have taken place, and it is perfectly unwarrantable to suppose that those movements which carried up miocene strata to a height of 9,000 feet and gave them a dip of many degrees, were entirely suspended during the immense lapse of time during which the glaciers of the Alps swept through the valleys of Switzerland and Piedmont. In the countless post-miocene ages which preceded the glacial period there was ample time for the slow erosion by water of all the principal hydrographical basins of the Alps, and the sites of all the great lakes coincide, as Professor Ramsay truly says, with these great lines of drainage. The lake-cavities do not lie in synclinal troughs, following the strike and folding of the strata, but often, as the same geologist remarks, cross them at high angles; nor are they due to rents or fissures, although these, with other accidents connected with the disturbing movements of the Alps, may sometimes have determined originally the direction of the valleys. The conformity of the lake-basins to the principal watercourses is explicable if we assume them to have resulted from inequalities in the upward and downward movements of the whole country in pleistocene times, after the valleys were eroded.

We know that in Sweden the rate of the rise of the land is far from uniform, being only a few inches in a century near Stockholm, while north of it, and beyond Gefle, it amounts to as many feet in the same number of years. Let us suppose, with Charpentier, that the Alps gained in height several thousand feet at the time when the intense cold of the glacial period was coming on. This gradual rise would be an era of aqueous erosion, and of the deepening, widening, and lengthening of the valleys. It is very improbable that the elevation would be everywhere identical in quantity, but if it was never in excess in the outskirts as compared to the central region or crest of the chain, it would not give rise to lakes. When, however, the period of upheaval was followed by one of gradual subsidence, the movement not being everywhere strictly uniform, lake-basins would be formed wherever the rate of depression was in excess in the upper country.

We have no certainty that such movements may not now be in progress in the Alps; for if they are as slow as we have assumed, they would be as insensible to the inhabitants as is the upheaval of Scandinavia, or the subsidence of Greenland, to the Swedes and Danes who dwell there. They only know of the progress of such geographical revolutions because a slight change of level becomes manifest on the margin of the sea. The lines of elevation or depression above supposed might leave no clear geological traces of their action on the high ridges and table-lands separating the valleys of the principal rivers; it is only when they cross such valleys that the disturbance caused in the course of thousands of years in the drainage becomes apparent. If there were no ice the sinking of the land might not give rise to lakes. To accomplish this, in the absence of ice, it is necessary that the rate of depression should be sufficiently fast to make it impossible for the depositing power of the river to keep pace with it, or, in other words, to fill up the incipient cavity, as fast as it begins to form. Such levelling operations, once complete, the running water, aided by sand and pebbles, will gradually cut a gorge through the newly raised rock, so as to prevent it from forming a barrier. But if a great glacier fill the lower part of the valley, all the conditions of the problem are altered. Instead of the mud, sand, and stones drifted down from the higher regions

being left behind in the incipient basin, they all travel
onwards in the shape of moraines on the top of the ice,
passing over and beyond the new depression, so that when,
at the end of fifty or a thousand centuries, the glacier melts,
a large and deep basin representing the difference in the
movement of two adjoining mountain areas—namely, the
central and the circumferential—is for the first time rendered
visible.

The gravest objection to the hypothesis of glacial erosion
on a stupendous scale, and unaided by changes of level, is
afforded by the entire absence of lakes of the first magnitude
in several areas where they ought to exist if the enormous
glaciers which once occupied those spaces had possessed the
deep excavating power ascribed to them. Thus in the area
laid down on the map, p. 351, or that covered by the ancient
moraine of the Dora Baltea, we see the monuments of a
colossal glacier derived from Mont Blanc and Monte Rosa,
which descended from points nearly a hundred miles distant,
and then emerging from the narrow gorge above Ivrea,
deployed upon the plains of the Po, advancing over a floor
of marine pliocene strata of no greater solidity than the
miocene sandstone and conglomerate in which the lake-
basins of Geneva, Zurich, and some others are situated.
Why did this glacier fail to scoop out a deep and wide basin
rivalling in size the lakes of Maggiore or Como, instead of
merely giving rise to a few ponds above Ivrea, which may
have been due to ice action? There is one lake, it is true,
that of Candia, near the southern extremity of the moraine,
which is larger; but even this, as will be seen by the map,
p. 351, is of quite subordinate importance, and whether it
is situated in a rock-basin or is simply caused by a dam of
moraine matter, has not yet been fully made out.

Professor Gastaldi of Turin has recently communicated
to me the results of his close examination of the Pied-

montese Alps, which have led him to adopt Professor Ramsay's theory of the erosion of lake-basins exclusively by ice. His most important observations relate to the structure of the great Alpine valleys and the question why lakes exist at the outlets of some of them and not of others. He first remarks that the valleys of the Dora Riparia, the Stura, and the Dora Baltea have very narrow mouths, the latter being little more than half a mile wide at its opening into the valley of the Po; while the Stura at Lanza passes through a very narrow opening, only leaving room for the torrent and a road. He suggests as an explanation of this fact, that a belt of hornblende rock, diorite, syenite, or serpentine, crosses the mouths of these valleys, traversing the country nearly in a straight line from Monte Viso on the south-west, to the St. Gothard pass on the north-east; and that these rocks are those which longest resist the action of water, whether in the liquid or solid form, whereas calcareous, felspathic, granitic, and porphyritic rocks decompose much more quickly. I quite agree with Professor Gastaldi that the difference in the hardness of the rocks has greatly influenced the position of the lake-basins. But we must bear in mind, what he himself points out, that these hornblendic rocks would also offer greater resistance to subaerial denudation than granite, gneiss, and the calcareous schists.

Assuming, as all geologists will be ready to admit, that the largest and deepest valleys will be made in the rocks which yield most to atmospheric and aqueous erosion, then, if there have been changes of level in the lower part of the valley, such as those above mentioned (at a time perhaps when the Alps, according to Charpentier's hypothesis, were several thousand feet higher than at present), it will follow that it will be precisely where these greatest valleys were hollowed out, whether by rivers or by a glacier, or both, that the great lakes, on the melting of the ice, will be situated. But if lake-basins are to be attributed solely to glacial action, it would be very difficult to explain their absence, not only in Alpine districts already alluded to, but in the mountains of the Caucasus, where not only sub-Alpine sheets of water, but even mountain-tarns are absent, although there are glaciers equal or superior in dimensions to those of Switzerland. M. Favre points out that moraines of great height and huge erratics justify the assertion that the present glaciers are only the shadows of their former selves; on the other hand we know that there exist in Cashmere several lakes which cannot be attributed to glacial erosion, but which have shifted their position during changes of level accompanying the earthquakes which have repeatedly convulsed that region in the course of the last 2,000 years, and caused the submergence and silting-up in freshwater strata of the buried temples of Cashmere.

The proofs of ice-action in all places where large glaciers have existed, are so manifest to the eye that I believe it needs some caution lest we should attribute to ice great physical changes upon which it has, so to speak, only left the final stamp. The only region which I have myself seen where a great change of level has taken place in the course of my own life is that called the 'sunk country' around New Madrid on the Mississippi. This area, which was permanently submerged in the earthquake of 1811-12, is said by Flint, the geographer, to extend for a distance of seventy miles north and south, and thirty east and west, and the vertical movement, independent of the opening here and there of huge chasms, amounts in places to eight feet.* Now, as the higher country drained by the Mississippi bears the marks of having been once glaciated, it is geologically possible that a great glacier may one day descend again into this great valley, and pass over the sunk country, so that ice may occupy the lowest part of the plain. In this case there

* Principles of Geology, 11th ed. vol. ii. p. 107.

can be no doubt that ice-erosion would efface all previous superficial markings, and might destroy all traces of the submerged forests which now bear evidence to a change of level. Unless history, therefore, preserved a record of the changes of 1811-12, geologists would be unable to assign to the two distinct and independent causes, namely, subterranean heat and the action of ice on the surface, their appropriate amount of efficacy in producing a shallow depression.

On the whole, therefore, it appears to me that the safest conclusion at which we can arrive in the present state of our knowledge is, that there is an intimate connection between the glacial period and a predominance of lakes, in producing which the action of ice is threefold : First—by its direct power in scooping out shallow basins where the rocks are of unequal hardness, giving rise to what are commonly called mountain-tarns, many of which are in rock-basins. This is an operation which can by no means be confined to the land, for it must extend to below the level of high water a thousand feet and more, in such friths as have been described as filled with ice in Greenland (see above, p. 277). Secondly—The ice will act indirectly by preventing cavities caused by inequalities of subsidence or elevation from becoming the receptacles first of water, and then of sediment, by which the cavities would be levelled up and the lakes obliterated. Thirdly—The ice is also an indirect cause of lakes, by heaping up mounds of moraino matter, which, by damming up streams, give rise to ponds and even to sheets of water several miles in diameter. The comparative scarcity, therefore, of lakes of pleistocene date in tropical countries, and very generally south of the fortieth and fiftieth parallels of latitude, may be accounted for by the absence of glacial action in such regions.

Yet when we have conceded all that is due to the action of ice, directly and indirectly, still when it becomes a question

of larger and deeper lakes, like those of Switzerland, or the
north of Italy, or inland freshwater seas, like those of Canada,
it will probably be found that those movements by which
changes of level in the earth's crust are gradually brought
about, have exerted a dominant influence in producing lakes
during those lengthened periods in which glaciers or vast
ice-fields have filled the principal valleys or covered continental areas.

Successive Phases of Glacial Action in the Alps, and Interglacial Formations.

According to the geological observations of M. Morlot,
the following successive phases in the development of ice-
action in the Alps are plainly recognisable.

1st. There was a period when the ice was in its greatest
excess, as described at p. 345 et seq. when the glacier of the
Rhone not only reached the Jura, but climbed to the height
of 2,015 feet above the lake of Neufchatel, and 4,800 above
the sea, at which time the Alpine ice actually entered
the French territory at some points, penetrating by certain
gorges, as through the defile of the Fort de l'Ecluse among
others.

2nd. To this succeeded a prolonged retreat of the great
glaciers, when they evacuated not only the Jura and the low
country between that chain and the Alps, but retired some
way back into the Alpine valleys. M. Morlot, supposes their
diminution in volume to have accompanied a general subsidence of the country, to the extent of at least 1,000 feet.
The geological formations of the second period consist of
stratified masses of sand and gravel, called the 'ancient
alluvium' by MM. Necker and Favre, corresponding to the
'older or lower diluvium' of some writers. Their origin is
evidently due to the action of rivers, swollen by the melting
of ice, by which the materials of parts of the old moraines

were re-arranged and stratified, and left usually at considerable heights above the level of the present valley plains.

3rd. The glaciers again advanced and became of gigantic dimensions, though they fell far short of those of the first period. That of the Rhone, for example, did not again reach the Jura, though it filled the lake of Geneva and formed enormous moraines on its borders, and in many parts of the valley between the Alps and Jura.

4th. A second retreat of the glaciers took place when they gradually shrunk nearly into their present limits, accompanied by another accumulation of stratified gravels, which form in many places a series of terraces above the level of the alluvial plains of the existing rivers.

In the gorge of the Dranse, near Thonon, M. Morlot discovered two of these glacial formations in direct superposition, namely, at the bottom of the section, a mass of compact till or boulder-clay twelve feet thick, including striated boulders of Alpine limestone, and covered by regularly stratified ancient alluvium 150 feet thick, made up of rounded pebbles in horizontal beds. This mass is in its turn overlaid by a second formation of unstratified boulder-clay, with erratic blocks and striated pebbles, which constituted the left lateral moraine of the great glacier of the Rhone, when it advanced for the second time to the lake of Geneva. At a short distance from the above section, terraces composed of stratified alluvium are seen at the heights of 20, 50, 100, and 150 feet above the lake of Geneva, which, by their position, can be shown to be posterior in date to the upper boulder-clay, and therefore belong to the fourth period, or that of the last retreat of the great glaciers. In the deposits of this fourth period, the remains of the mammoth have been discovered, as at Morges, for example, on the lake of Geneva. The conical delta of the Tinière, mentioned at p. 29 as containing at different depths monuments of the Roman as well as of the antecedent bronze and stone ages, is the work of alluvial deposition going on when the terrace of 50 feet was in progress. This modern delta is supposed by M. Morlot to have required 10,000 years for its accumulation. At the height of 150 feet above the lake, following up the course of the same torrent, we come to a more ancient delta, about ten times as large, which is therefore supposed to be the monument of about ten times as many centuries, or 100,000 years, all referable to the fourth period mentioned in the preceding page, or that which followed the last retreat of the great glaciers.*

If the lower flattened cone of Tinière be referred in great part to the age of the oldest lake-dwellings, the higher one might, perhaps, correspond to the pleistocene period of St. Acheul, or the era when Man and the *Elephas primigenius* flourished together; but no human remains or works of art have as yet been found in deposits of this age, or in any alluvium containing the bones of extinct mammalia in Switzerland.

We have on the borders of the lake of Zurich other interglacial formations containing a mammalian fauna, such as elsewhere is associated with human remains or flint implements of Paleolithic type. The first of these, that of Utznach, is a delta formed at the head of the ancient and once more extensive lake of Zurich, the decrease in area of which may have been caused by the deepening of its outlet, or, as I think not improbable, by the upheaval of the higher country round its head, as compared to the lower region where its outlet is now situated. The argillaceous and lignite-bearing strata at Utznach, situated about 350 feet above the present level of the lake, are more than 100 feet in thickness, and rest unconformably on highly inclined, and sometimes vertical,

* Morlot, Terrain quaternaire du   Vaudoise des Sciences Naturelles,
Bassin de Léman. Bulletin de Société   No. 44.

miocene molasse. These clays are covered conformably by
stratified sand and gravel sixty feet thick, partly consolidated,
in which the pebbles are of rocks belonging to the upper
valleys of the Limmat and its tributaries, all of them small
and not glacially striated, and wholly without admixture of
large angular stones. On the top of all repose very large
erratic blocks, affording clear evidence that the colossal
glacier which once filled the valley of the Limmat covered
the old littoral deposit. The great age of the lignite is partly
indicated by the bones of *Elephas antiquus* found in it.

I visited Utznach in company with M. Escher von der
Linth in 1857, and during the same year examined the lignite
of Dürnten, many miles further down on the right bank of
the lake, in company with Professor Heer and M. Marcou.
The beds there are of the same age and within a few feet of
the same height above the level of the lake. They might
easily have been overlooked or confounded with the general
glacial drift of the neighbourhood, had not the bed of lignite,
which is from five to twelve feet thick, been worked for fuel,
during which operation many organic remains came to light.
Among these are the teeth of *Elephas antiquus*, determined
by Dr. Falconer, and *Rhinoceros leptorhinus?* (*R. megarhinus*, Cuvier), the wild bull and red deer (*Bos primigenius*,
Boj., and *Cervus Elaphus*, L.), the last two determined by
Professor Rütimeyer. In the same beds I found many fresh-
water shells of the genera *Paludina*, *Limnea*, &c., all of
living species. The plants named by Professor Heer are
also recent, and agree singularly with those of the Cromer
buried forest, before described (p. 256).

Among them are the Scotch and spruce firs, *Pinus sylvestris* and *Pinus Abies*, and the buckbean, or *Menyanthes
trifoliata*, &c., besides the common birch and other European
plants.

Overlying this lignite are, first, as at Utznach, stratified

gravel, not of glacial origin, about thirty feet thick ; and,
secondly, highest of all, huge angular erratic blocks, clearly
indicating the presence of a great glacier, posterior in date
to all the organic remains above enumerated.

At Wetzikon, near the Pfaffikon Lake, a deposit occurs in
which the plants are exactly similar to those of Utznach and
Dürnten, and Professor Heer considers it to be of the same
age, but it is not only covered by a glacial formation, it also
rests upon boulder-clay containing glaciated erratics. We
have, therefore, here the evidence wanting at Utznach, which
establishes the interglacial age of these deposits. ' The beds
of lignite of Unterwetzikon,' writes Professor Heer, ' attain a
thickness of from 13 to 30 feet, and are covered like those of
Dürnten by strata of sand and pebbles ; they themselves rest
upon a light-coloured clay, which contains some freshwater
mollusks, the rolled pebbles reappear under the clay, and
with them calcareous striated blocks, a block of the granite
of Pontaigles, and an erratic block six feet in diameter.'[*]

There are other freshwater formations with lignite at
Morschweil, between St. Gall and Rorschaach, and at Kalt-
brunnen and Buchberg, all probably of about the same age
and containing the same assemblage of fossil plants. The
thickness of the beds of lignite in these deposits implies a
long period, lasting probably several thousand years, when
there was little or no ice in the Alps, and a vegetation similar
to that of the more temperate parts of Europe. The pebble-
beds of Dürnten, composed of well-rounded shingle which are
sorted according to their size, imply the action of rivers, not
of ice, and the fossil animals, the *Elephas antiquus, Rhinoceros leptorhinus, Bos primigenius*, and *Cervus Elaphus* are
all species which in some parts of Europe have been found
accompanying Palæolithic implements. And although no

* Heer, Urwelt der Schweiz, 1865, p. 487.

B B

human remains or works of art have yet been found in
working the Swiss lignite, it would be rash to speculate on
the non-existence of the human race in the region where
these interglacial deposits accumulated on the margin of the
lakes.

It must be confessed that in the present state of our
knowledge, any attempt to compare the chronological relations of the periods of upheaval and subsidence of areas so
widely separated as are the mountains of Scandinavia, the
Alps, and the British Isles, or the times of the advance and
retreat of glaciers in those several regions, and the greater
or less intensity of cold, must be looked upon as very conjectural. Charpentier, after pointing out the gigantic size
of the ancient Alpine glaciers, compared to those of our
time, declared his opinion that the Alps wore formerly 3,000
feet higher than now, and it is natural to connect greater
altitudes with periods of extreme cold, and for the same
reason to suppose that the deposits which have been called
interglacial and which contain living species of freshwater
shells and of plants indicating a temperate climate such as
those on the borders of the lake of Zurich, should be connected with the lowering of the great central chain of Europe,
when the snow would be greatly diminished in quantity.

We may presume with confidence that when the Alps
were highest and the Alpine glaciers most developed, filling
all the great lakes of northern Italy, and loading the plains
of Piedmont and Lombardy with ice, the waters of the
Mediterranean were chilled and of a lower average temperature than now.   Such a period of refrigeration is required
by the conchologist to account for the prevalence of northern
shells in the Sicilian seas about the close of the newer
pliocene or commencement of the pleistocene period. For
such shells as *Cyprina islandica*, *Panopæa Norvegica*
(= *P. Bivonæ, Philippi*), *Leda pygmæa*, Müust, and some

others, enumerated among the fossils of the latest tertiary formations of Sicily by Philippi and Edward Forbes, point unequivocally to a former more severe climate. Dr. Hooker also, in his journey to Syria, in the autumn of 1860, found the moraines of extinct glaciers, on which the whole of the ancient cedars of Lebanon grow, to descend 4,000 feet below the summit of that chain. The temperature of Syria is now so much milder, that there is no longer perpetual snow even on the summit of Lebanon, the height of which was ascertained to be 10,200 feet above the Mediterranean.[*]

Such monuments of a cold climate in latitudes so far south as Syria and the north of Sicily, between 33° and 38° north, may be confidently referred to an early part of the glacier period, or to times long anterior, not only to those of the men and extinct mammalia of Abbeville and Amiens, but also to those temperate periods indicated by the interglacial formations of Switzerland.

[*] Hooker, Natural History Review, No. 5, January 1862, p. 11.

CHAPTER XVI.

NATURE, ORIGIN, AND AGE OF THE EUROPEAN LOESS.

NATURE, ORIGIN, AND AGE OF THE LOESS OF THE RHINE AND
DANUBE—IMPALPABLE MUD PRODUCED BY THE GRINDING ACTION OF
GLACIERS—DISPERSION OF THIS MUD AT THE PERIOD OF THE RETREAT
OF THE GREAT ALPINE GLACIERS—CONTINUITY OF THE LOESS FROM
SWITZERLAND TO THE LOW COUNTRIES—CHARACTERISTIC ORGANIC
REMAINS NOT LACUSTRINE—ALPINE GRAVEL IN THE VALLEY OF THE
RHINE COVERED BY LOESS—GEOGRAPHICAL DISTRIBUTION OF THE
LOESS AND ITS HEIGHT ABOVE THE SEA—FOSSIL MAMMALIA—LOESS OF
THE DANUBE—HIMALAYAN MUD OF THE PLAINS OF THE GANGES COM-
PARED TO EUROPEAN LOESS—OSCILLATIONS IN THE LEVEL OF THE ALPS
AND LOWER COUNTRY REQUIRED TO EXPLAIN THE FORMATION AND DENU-
DATION OF THE LOESS—MORE RAPID MOVEMENT OF THE INLAND COUNTRY.

Nature and Origin of the Loess.

INTIMATELY connected with the subjects treated of in
the last chapter, is the nature, origin, and age of certain loamy deposits, commonly called loess, which form a
marked feature in the superficial formations of the basins of
the Rhine, Danube, and some other large rivers draining the
Alps, and which extend down the Rhine into the Low
Countries, and were once perhaps continuous with others of
like composition in the north of France.

Some skilful geologists, peculiarly well acquainted with
the physical geography of Europe, have styled the loess
the most difficult geological problem, although belonging
to the period of existing land-shells, and the highest
and newest by position of all the great formations; and
in a work dealing with the evidences of the antiquity of
man, I the more willingly devote a chapter to this Alpine

mud, because, as we have seen at p. 239, it has yielded near
Maestricht a human jaw at a depth of 19 feet; and M. Boué
found in it, near Strasburg, the bones of nearly half a
skeleton, at a depth which he calculated at 80 feet, allowing
for the subsequent denudation of the terraces which have
been shaped out of it.

In every country, and at all geological periods, rivers have
been depositing fine loam on their inundated plains in the
manner explained above at p. 36, where the Nile mud was
spoken of. This mud of the plains of Egypt, according
to Professor Bischoff's chemical analysis, agrees closely
in composition with the loess of the Rhine.* I have
also shown (p. 237), when speaking of the fossil man of
Natchez, how identical in mineral character, and in the
genera of its terrestrial and amphibious shells, is the ancient
fluviatile loam of the Mississippi with the loess of the
Rhine. Thus we find that loam presenting the same aspect
has originated at different times and in distinct hydrographical basins; but there can be no doubt that the loess
of the Alps may be traced to the advance and retreat of
those gigantic glaciers which originated during the glacial
period, when the Alps were a great centre of dispersion,
not only of erratics, as we have seen in the last chapter, and
of gravel, which was carried further than the erratics, but
also of very fine mud, which was transported to still greater
distances and in greater volume down the principal river-
courses between the mountains and the sea.

Mud produced by Glaciers.

They who have visited Switzerland are aware that every
torrent which issues from an icy cavern at the extremity of a
glacier is densely charged with an impalpable powder, produced by the grinding action to which the subjacent floor of

* Chemical and Physical Geology, vol. i. p. 132.

rock and the stones and sand frozen into the ice are exposed
in the manner before described (p. 357). We may therefore
readily conceive that a much greater volume of fine sediment
was swept along by rivers swollen by melting ice at the time
of the retreat of the gigantic glaciers of the olden time. The
fact that a large proportion of this mud, instead of being
carried to the ocean, where it might have formed a delta on
the coast, or have been dispersed far and wide by the tides
and currents, has accumulated in inland valleys, will be found
to be an additional proof of the former occurrence of those
grand oscillations in the level of the Alps and parts of the
adjoining continent which were required to explain the
alternate advance and retreat of the glaciers, and the superposition of more than one boulder-clay and stratified alluvium
as before mentioned (p. 366).

The position of the loess between Basle and Bonn is such
as to imply that the great valley of the Rhine had already
acquired its present shape, and in some places perhaps more
than its actual depth and width, previously to the time when
it was gradually filled up to a great extent with fine loam.
The greater part of this loam has been since removed, so that
a fringe only of the deposit is now left on the flanks of the
boundary hills, or occasionally some outliers in the middle of
the great plain of the Rhine where it expands in width.

These outliers are sometimes on such a scale as to admit of
minor hills and valleys having been shaped out of them by
the action of rain and small streamlets, as near Freiburg in
the Brisgau and other districts.

Fossil Shells of the Loess.

The loess is generally devoid of fossils, although in many
places they are abundant, consisting of land-shells, all of
living species, and comprising no small part of the entire

molluscous fauna now inhabiting the same region. The
three shells most frequently met with are those represented
in the annexed figures. The *Succinea* is not strictly
aquatic, but lives in damp places, and may be seen in full
activity far from rivers, in meadows, where the grass is wet
with rain or dew; but shells of the genera *Limnea, Planorbis,
Paludina, Cyclas*, and others, requiring to be constantly in
the water, are extremely exceptional in the loess, occurring
only at the bottom of the deposit, where it begins to alternate
with ancient river gravel, on which it usually reposes.

This underlying gravel consists, in the valley of the Rhine,

Fig. 47.   Fig. 48.   Fig. 49.

Succinea elongata   *Pupa muscorum.*   *Helix hispida,* Lin.; *H. plebeia,* Drap.
(*oblonga*).

for the most part, of pebbles and boulders of Alpine origin,
showing that there was a time when the rivers had power to
convey coarse materials for hundreds of miles northwards
from Switzerland, towards the sea; whereas, at a later period,
an entire change was brought about in the physical geography
of the same district, so that the same river deposited nothing
but fine mud, which accumulated to a thickness of 800 feet
or more above the original alluvial plain.

But although most of the fundamental gravel was derived
from the Alps, there has been observed in the neighbourhood
of the principal mountain chains bordering the great valley,
such as the Black Forest, Vosges, and Odenwald, an admixture of detritus characteristic of those several chains.
We cannot doubt, therefore, that as some of these mountains,
especially the Vosges, had, during the glacial period, their
own glaciers, a part of the fine mud of their moraines must
have been mingled with loess of Alpine origin; although the

principal mass of the latter must have come from Switzerland,
and can in fact be traced continuously from Basle to Belgium.

Geographical Distribution of the Loess.

It was stated in the last chapter, p. 347, that at the time of
the greatest extension of the Swiss glaciers, the Lake of
Constance, and all the other great lakes, were filled with ice,
so that gravel and mud could pass freely from the upper
Alpine valley of the Rhine, to the lower region between Basle
and the sea, the great lake intercepting no part of the
moraines, whether fine or coarse. On the other hand, the Aar,
with its great tributaries the Limmat and the Reuss, does not
join the Rhine till after it issues from the lake of Constance :
and by their channels a large part of the Alpine gravel and
mud could always have passed without obstruction into the
lower country, even after the ice of the great lake had melted.

It will give the reader some idea of the manner in which
the Rhenish loess occurs, if he is told that some of the earlier
scientific observers imagined it to have been formed in a vast
lake which occupied the valley of the Rhine from Basle to
Mayence, sending up arms or branches into what are now the
valleys of the Main, Neckar, and other large rivers. They
placed the barrier of this imaginary lake in the narrow and
picturesque gorge of the Rhine between Bingen and Coblentz :
and when it was objected that the lateral valley of the Lahn,
communicating with that gorge, had also been filled with loess,
they were compelled to transfer the great dam fartherdown, and
to place it below Bonn. Strictly speaking, it must be placed
much farther north, or in the 51st parallel of latitude, where
the limits of the loess have been traced out by MM. Omalius
D'Halloy, Dumont, and others, running east and west by
Cologne, Juliers, Louvain, Oudenarde, and Courtray, in
Belgium, to Cassel, near Dunkirk, in France. This boundary

line may not indicate the original seaward extent of the
formation, as it may have stretched still farther north, and its
present abrupt termination may only show how far it was
cut back at some former period by the denuding action of
the sea.

Even if the embedded fossil shells of the loess had
been lacustrine, instead of being, as we have seen, terrestrial
and amphibious, the vast height and width of the required
barrier would have been fatal to the theory of a lake : for the
loess is met with in great force at an elevation of no less than
1,600 feet above the sea, covering the Kaiserstuhl, a volcanic
mountain which stands in the middle of the great valley of the
Rhine, near Freiburg in Brisgau. The extent to which the
valley has there been the receptacle of fine mud afterwards
removed is most remarkable.

M d'Archiac, when speaking of the loess, observes that it
envelopes Hainault, Brabant, and Limburg like a mantle,
everywhere uniform and homogeneous in character, filling up
the lower depressions of the Ardennes, and passing thence
into the north of France, though not crossing into England.
In France, he adds, it is found on high plateaus, 600 feet
above some of the rivers, such as the Marne; but as we go
southwards and eastwards of the basin of the Seine, it diminishes in quantity, and finally thins out in those directions.[*]
It may even be a question whether the ' *limon des plateaux*,'
or upland loam of the Somme valley, before alluded to,[†] may
not be a part of the same formation. As to the higher and
lower level gravels of that valley, which, like that of the Seine,
contain no foreign rocks, we have seen that they are each of
them covered by deposits of loess or inundation-mud belonging respectively to the periods of the gravels, whereas the
upland loam is of much older date, more widely spread, and

* D'Archiac, Histoire des Progrès, vol. ii. pp. 169, 170.
† No. 4, fig. 11, p. 153.

occupying positions often independent of the present lines of
drainage. To restore in imagination the geographical outline
of Picardy, to which rivers charged with so much homogeneous
loam, and running at such heights, may once have belonged,
is now impossible.

In the valley of the Rhine, as I before observed, the main
body of the loess, instead of having been formed at successively lower and lower levels, as in the case of the basin of the
Somme, was deposited in a wide and deep pre-existing basin,
or strath, bounded by lofty mountain chains, such as the Black
Forest, Vosges, and Odenwald. In some places the loam
accumulated to such a depth as first to fill the valley and
then to spread over the adjoining tablelands, as in the case
of the Lower Eifel, where it encircled some of the modern
volcanic cones of loose pumice and ashes. In these instances it does not appear to me that the volcanoes were in
eruption during the time of the deposition of the loess, as
some geologists have supposed. The interstratification of
loam and volcanic ejectamenta was probably occasioned by
the fluviatile mud having gradually enveloped the cones of
loose scoriæ after they were completely formed. I am the
more inclined to embrace this view after having seen the
junction of granite and loess on the steep slopes of some of
the mountains bounding the great plain of the Rhine on its
right bank in the Berg-strasse. Thus between Darmstadt
and Heidelberg vertical sections are seen of loess 200 feet
thick, at various heights above the river, some of them at
elevations of 800 feet and upwards. In one of these may be
seen, resting on the hillside of Melibocus in the Odenwald,
the usual yellow loam free from pebbles at its contact with a
steep slope of granite, but divided into horizontal layers for a
short distance from the line of junction. In these layers,
which abut against the granite, a mixture of mica and of
unrounded grains of quartz and felspar occur, evidently

derived from the disintegration of the crystalline rock, which
must have decomposed in the atmosphere before the mud
had reached this height. Entire shells of *Helix*, *Pupa*, and
Succinea, of the usual living species, are embedded in the
granite mixture. As, therefore, we are sure that the valley
bounded by steep hills of granite existed before the tranquil
accumulation of this vast body of loess, so we may explain
the alternation of ash and mud by the action of the air and
water on the hills of volcanic ejectamenta during the deposition of the mud.

During the re-excavation of the basin of the Rhine, successive deposits of loess of newer origin were formed at various
heights ; and it is often difficult to distinguish their relative
ages, especially as fossils are often entirely wanting, and the
mineral composition of the formation is so uniform.

The loess in Belgium is variable in thickness, usually
ranging from ten to thirty feet. It caps some of the highest
hills or tableland around Brussels at the height of 300 feet
above the sea. In such places it usually rests on gravel and
rarely contains shells, but when they occur, they are of recent
species. I found the *Succinea elongata*, before mentioned,
p. 375, and *Helix hispida* in the Belgian loess at Neerepen,
between Tongres and Hasselt, where M. Bosquet had previously obtained remains of an elephant referred to *Elephas
primigenius*. This pachyderm and *Rhinoceros tichorhinus* are
cited as characterising the loess in various parts of the valley
of the Rhine. Several perfect skeletons of the marmot have
been disinterred from the loess of Aix-la-Chapelle. But
much remains to be done in determining the species of
mammalia of this formation, and the relative altitudes above
the valley-plain at which they occur.

If we ascend the basin of the Neckar, we find that it is
filled with loess of great thickness, far above its junction
with the Rhine. At Canstadt, near Stuttgart, loess resem-

bling that of the Rhine contains many fossil bones, especially
those of *Elephas primigenius*, together with some of *Rhinoceros tichorhinus*, the species having been determined by
Dr. Falconer. At this place the loess is covered by a thick
bed of travertin, used as a building stone, the product of a
mineral spring. In the travertin are many fossil plants, all
recent except two, an oak and poplar, the leaves of which
Professor Heer has not been able fully to identify with any
living varieties.

Below the loess of Canstadt, in which bones of the mammoth are so abundant, is a bed of gravel, evidently an old
river channel, now many feet above the level of the Neckar,
the valley having there been excavated to some depth below
its ancient channel so as to lie in the underlying red
sandstone or keuper. Although the loess, when traced from
the valley of the Rhine into that of the Neckar, or into any
other of its tributaries, often undergoes some slight alteration
in its character, yet there is so much identity of composition
as to suggest the idea, that the mud of the main river passed
far up the tributary valleys, just as that of the Mississippi,
during floods, flows far up the Ohio, carrying its mud with
it into the basin of that river. But the uniformity of colour
and mineral composition does not extend indefinitely into
the higher parts of every basin. In that of the Neckar, for
example, near Tübingen, I found the fluviatile loam or
brick-earth, enclosing the usual helices and succinem, together with the bones of the mammoth, very distinct in
colour and composition from ordinary Rhenish loess, and
such as no one could confound with Alpine mud. It is
mottled with red and green, like the New Red Sandstone or
Keuper, from which it has clearly been derived.

Such examples, however, merely show that where a basin
is so limited in size that the detritus is derived chiefly or
exclusively from one formation, the prevailing rock will

impart its colour and composition in a very decided manner
to the loam: whereas, in the basin of a great river which
has many tributaries, the loam will consist of a mixture of
almost every variety of rock, and will therefore exhibit an
average result nearly the same in all countries. Thus, the
loam which fills to a great depth the wide Valley of the
Saone, which is bounded on the west side by an escarpment
of inferior oolite, and by the chain of the Jura on the east,
is very like the loess found in the continuation of the same
great basin after the junction of the Rhone, by which a large
supply of Alpine mud has been added and intermixed.

In the higher parts of the basin of the Danube, loess, of
the same character as that of the Rhine, and which I believe
to be chiefly of Alpine origin, attains a far greater elevation
above the sea than any deposits of Rhenish loess; but the
loam which, according to Mr. Stur, fills valleys on the north
slope of the Carpathians, almost up to the watershed between Galicia and Hungary, may be derived from a distinct
source.

Himalayan Mud of the Ganges compared to European Loess.

In India, where, as in the Alps, we have abundant evidence of extinct glaciers, we find a deposit which may be
regarded as a counterpart of the European loess.

The vast plains of Bengal are overspread with Himalayan
mud, which, as we ascend the Ganges, extends inland for
1,200 miles from the sea, continuing very homogeneous on
the whole, though becoming more sandy as it nears the hills.
They who sail down the river during a season of inundation,
see nothing but a sheet of water in every direction, except
here and there where the tops of trees emerge above its level.
To what depth the mud extends is not known, but it resem-

bles the loess in being generally devoid of stratification, and
of shells, though containing occasionally land-shells in abundance, as well as calcareous concretions, called kunkur,
similar to the nodules of carbonate of lime sometimes observed to form layers in the Rhenish loess, and to the
nodules known as 'race' in the English brick-earth. I am
told by Colonel Strachey and Dr. Hooker, that below Calcutta,
in the Hooghly, when the flood subsides, the Gangetic mud
may be seen in river cliffs eighty feet high, in which no
organic remains were detected, a remark which I found
to hold equally in regard to the recent mud of the Mississippi.

Dr. Wallich, while confirming these observations, informs
me that at certain points in Bengal, farther inland, he met
with land-shells in the banks of the great river. Borings
have been made at Calcutta, beginning not many feet above
the sea-level, to the depth of 300 and 400 feet; and wherever
organic remains were found in the strata pierced through,
they were of a fluviatile or terrestrial character, implying,
that during a long and gradual subsidence of the country,
the sediment thrown down by the Ganges and Burrampooter
had accumulated at a sufficient rate to prevent the sea from
invading that region.

At the bottom of the borings, after passing through much
fine loam, beds of pebbles, sand, and boulders were reached,
such as might belong to an ancient river channel: and the
bones of a crocodile, and the shell of a freshwater tortoise
embedded in it, were met with, at the depth of four hundred
feet from the surface. No pebbles are now brought down
within a great distance of this point, so that the country
must once have had a totally different character, and may
have had its valleys, hills, and rivers, before all was reduced
to one common level by the accumulation upon it of fine
Himalayan mud. If the latter were removed during a

gradual re-elevation of the country, many old hydrographical
basins might reappear, and portions of the loam might alone
remain in terraces, on the flanks of hills, or on platforms, attesting the vast extent, in ancient times, of the muddy enve-
lope. A similar succession of events has, in all likelihood,
occurred in Europe during the deposition and denudation of
the loess of the pleistocene period, which, as we have seen
in a former chapter, was long enough to allow of the gradual
development of almost any amount of such physical changes.

Oscillations of Level required to explain the Accumulation and Denudation of the Loess.

A theory which attempts to account for the position
of the European loess cannot be satisfactory unless it be
equally applicable to the basins of the Rhine and Danube.
So far as relates to the source of so much homogeneous
loam, there are many large tributaries of the Danube which,
during the glacial period, may have carried an ample supply
of moraine-mud from the Alps to that river; and in regard
to grand oscillations in the level of the land, it is obvious
that the same movements, both downward and upward, of
the great mountain-chain would be attended with analogous
effects, whether the great rivers flowed northwards or eastwards. In each case fine loam would be accumulated during
subsidence, and removed during the upheaval of the land.
Changes, therefore, of level, analogous to those on which we
have been led to speculate when endeavouring to solve the
various problems presented by the glacial phenomena, are
equally available to account for the nature and geological
distribution of the loess. But we must suppose that the
amount of depression and re-elevation in the central region
was considerably in excess of that experienced in the lower
countries, or those nearer the sea, and that the rate of sub-

sidence in the latter was never so considerable as to cause
submergence or the admission of the sea into the interior of
the continent, by the valleys of the principal rivers.

We have already assumed that the Alps were loftier than
now, when they were the source of those gigantic glaciers
which reached the flanks of the Jura. At that time gravel was
borne to the greatest distances from the central mountains
through the main valleys, which had a somewhat steeper slope
than now, and the quantity of river-ice must at that time
have aided in the transportation of pebbles and boulders.
To this state of things gradually succeeded another of an
opposite character, when the fall of the rivers from the
mountains to the sea became less and less, while the Alps
were slowly sinking and the first retreat of the great glaciers
was taking place. Suppose the depression to have been at
the rate of five feet in a century in the mountains, and only
as many inches in the same time nearer the coast, still, in
such areas as the eye could survey at once, comprising a
small part only of Switzerland or of the basin of the Rhine,
the movement might appear to be uniform, and the pre-
existing valleys and heights might seem to remain relatively
to each other as before.

Such inequality in the rate of rising or sinking, when we
contemplate large continental spaces, is quite consistent with
what we know of the course of nature in our own times, as
well as at remote geological epochs. Thus in Sweden, as
before stated, the rise of land now in progress is nearly uniform as we proceed from north to south for moderate distances,
but it greatly diminishes southwards if we compare areas
hundreds of miles apart.*

To cite an example of high geological antiquity, M. Hébert
has demonstrated that, during the oolitic and cretaceous

* Principles of Geology, see Index, rise of land in Sweden.

periods, similar inequalities in the vertical movements of
the earth's crust were experienced in Switzerland and France.
By his own observations and those of M. Lory he has proved
that the area of the Alps was rising and emerging from
beneath the ocean towards the close of the oolitic epoch, and
was above water at the commencement of the cretaceous era ;
while, on the other hand, the area of the Jura, about one hundred miles to the north, was slowly sinking at the close of the
oolitic period, and had become submerged at the commencement of the cretaceous. Yet these oscillations of level were
accomplished without any perceptible derangement in the
strata, so that the lower cretaceous or neocomian beds were
deposited conformably on the oolitic.[*]

Taking for granted, then, that the depression was more
rapid in the more elevated region, the great rivers would
lose, century after century, some portion of their velocity or
carrying power, and would leave behind them on their
alluvial plains more and more of the moraine-mud or flour
of rock with which they were charged, till at length, in the
course of thousands or some tens of thousands of years, a
large part of the main valleys would begin to resemble the
plains of Egypt, where nothing but mud is deposited during
the flood season. The thickness of loam containing shells
of land and amphibious mollusca might in this way accumulate to any extent, so that the waters might overflow some
of the heights originally bounding the valley, and deposits
of ' platform mud,' as it has been termed in France, might
be extensively formed. At length, whenever a re-elevation
of the Alps at the time of the second extension of the glaciers
took place, there would be renewed denudation and removal
of such loess ; and if, as some geologists believe, there has
been more than one oscillation of level in the Alps since the

* Bulletin de la Société Géologique de France, 2 series, tom. xvi. p. 496, 1859.

commencement of the glacial period, the changes would be
proportionally more complicated, and terraces of gravel
covered with loess might be formed at different heights, and
at different periods. If such changes of level are assumed
as being connected with the formation of the great valleys,
it would be the more rash to deny the probability of the
larger and deeper lakes being the result of the joint operations of ice-action and subterranean movements, as we have
suggested in the last chapter.

CHAPTER XVII.

POST-GLACIAL DISLOCATIONS AND FOLDINGS OF CRETACEOUS AND
DRIFT STRATA IN THE ISLAND OF MÖEN, IN DENMARK.

GEOLOGICAL STRUCTURE OF THE ISLAND OF MÖEN—GREAT DIS-
TURBANCES OF THE CHALK POSTERIOR IN DATE TO THE GLACIAL
DRIFT—M. PUGGAARD'S SECTIONS OF THE CLIFFS OF MÜEN—FLEXURES
AND FAULTS COMMON TO THE CHALK AND GLACIAL DRIFT—DIFFERENT
DIRECTION OF THE LINES OF SUCCESSIVE MOVEMENT, FRACTURE, AND
FLEXURE—UNDISTURBED CONDITION OF THE ROCKS IN THE ADJOINING
DANISH ISLANDS—UNEQUAL MOVEMENTS OF UPHEAVAL IN FINMARK—
EARTHQUAKE OF NEW ZEALAND IN 1855 — PREDOMINANCE IN ALL
AGES OF UNIFORM CONTINENTAL MOVEMENTS OVER THOSE BY WHICH
THE ROCKS ARE LOCALLY CONVULSED.

IN the preceding chapters I have endeavoured to show that
the study of the successive phases of the glacial period
in Europe, and the enduring marks which they have left on
many of the solid rocks and on the character of the superficial drift, are of great assistance in enabling us to appreciate
the vast lapse of ages which are comprised in the pleistocene
epoch. They enlarge at the same time our conception of the
antiquity and present geographical distribution of the living
species of animals and plants, and throw light on the chronological relations of these species to the earliest date yet
ascertained for the existence of the human race. That date,
it will be seen, is very remote if compared to the times of history
and tradition, yet very modern if contrasted with the length
of time during which all the living testacea, and even many
of the mammalia, have inhabited the globe.

In order to render my account of the phenomena of the
glacial epoch more complete, I shall describe in this chapter

c c 2

some other changes in physical geography, and in the internal structure of the earth's crust, which have happened
in the pleistocene period, because they differ in kind from
any previously alluded to, and are of a class which were
thought by the earlier geologists to belong exclusively to
epochs anterior to the origin of the existing fauna and flora.
Of this nature are those faults and violent local dislocations
of the rocks, and those sharp bendings and foldings of the
strata, which we so often behold in mountain chains, and
sometimes in low countries also, especially where the rock
formations are of ancient date.

*Post-glacial Dislocations and Foldings of cretaceous and
drift Strata in the Island of Möen, Denmark.*

A striking illustration of such convulsions of pleistocene
date may be seen in the Danish island of Möen, which
is situated about fifty miles south of Copenhagen. The
island is about sixty miles in circumference, and consists of
white chalk, several hundred feet thick, overlaid by boulder-
clay and sand, or glacial drift which is made up of several
subdivisions, some unstratified and others stratified, the whole
having a mean thickness of sixty feet, but sometimes attaining nearly twice that thickness. In one of the oldest members
of the formation, fossil marine shells of existing species have
been found.

Throughout the greater part of Möen, the strata of the
drift are undisturbed and horizontal, as are those of the
subjacent chalk; but on the north-eastern coast they have
been, throughout a certain area, bent, folded, and shifted,
together with the beds of the underlying cretaceous formation. Within this area they have been even more deranged
than is the English chalk with flints along the central axis

of the Isle of Wight in Hampshire, or of Purbeck in Dorsetshire. The whole displacement of the chalk is evidently
posterior in date to the origin of the drift, since the beds of
the latter are horizontal where the fundamental chalk is horizontal, and inclined, curved, or vertical where the chalk displays signs of similar derangement. Although I had come
to these conclusions respecting the structure of Möen in
1835, after devoting several days in company with Dr. Forchhammer to its examination,[*] I should have hesitated to cite
the spot as exemplifying convulsions on so grand a scale, of
such extremely modern date, had not the island been since
thoroughly investigated by a most able and reliable authority,
the Danish geologist, Professor Puggaard, who has published
a series of detailed sections of the cliffs.

These cliffs extend through the north-eastern coast of the
island, called Möens Klint,[†] where the chalk precipices are
bold and picturesque, being 300 and 400 feet high, with tall
beech-trees growing on their summits, and covered here and
there at their base with huge taluses of fallen drift, verdant
with wild shrubs and grass, by which the monotony of a
continuous range of white chalk cliffs is relieved.

In the low part of the island, at A, fig. 50 (p. 390), or the
southern extremity of the line of section above alluded to,
the drift is horizontal, but when we reach B, a change,
both in the height of the cliffs and in the inclination of the
strata, begins to be perceptible, and the chalk No. 1 soon
makes its appearance from beneath the overlying members
of the drift Nos. 2, 3, 4, and 5.

This chalk, with its layers of flints, is so like that of
England as to require no description. The incumbent drift

* Lyell, Geological Transactions,   Möen, Bern, 1851 ; and Bulletin de la
2nd series, vol. ii. p. 243.   Société Géologique de France, 1851.
† Puggaard, Geologie der Insel

consists of the following subdivisions, beginning with the lowest :

No. 2. Stratified loam and sand, five feet thick, containing at one spot, near the base of the cliff at s, fig. 51, *Cardium edule*, *Tellina solidula*, and *Turritella*, with fragments of other shells. Between No. 2 and the chalk No. 1, there usually intervenes a breccia of broken chalk flints.

No. 3. Unstratified blue clay or till, with small pebbles

Fig. 50.

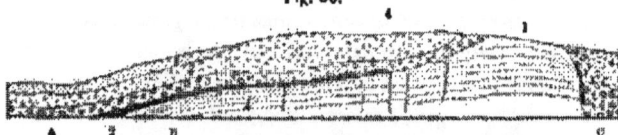

Southern extremity of Möens Klint (Puggaard).

a Horizontal drift.
B Chalk and overlying drift beginning to rise.
c First flexure and fault.   Height of cliff at this point, 180 feet.

Fig. 51.

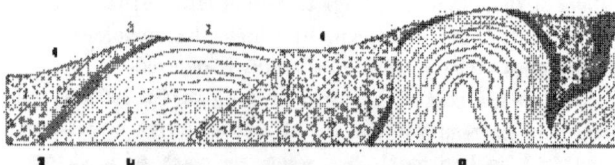

Section of Möens Klint (Puggaard), continued from fig. 50.

s Fossil shells of recent species in the drift at this point.
o Greatest height near o, 200 feet.

and fragments of Scandinavian rocks occasionally scattered through it, twenty feet thick.

No. 4. A second unstratified mass of yellow and more sandy clay forty feet thick, with pebbles and angular polished and striated blocks of granite and other Scandinavian rocks, transported from a distance.

No. 5. Stratified sands and gravel, with occasionally large

erratic blocks; the whole varying from forty to a hundred feet in thickness, but this only in a few spots.

The angularity of many of the blocks in Nos. 3 and 4, and the glaciated surfaces of others, and the transportation from a distance attested by their crystalline nature, proves them to belong to the northern drift or glacial period.

It will be seen that the four subdivisions 2, 3, 4, and 5, begin to rise at B, fig. 50, and that at c, where the cliff is 180 feet high, there is a sharp flexure shared equally by the chalk and the incumbent drift. Between D and o, fig. 51, we observe a great fracture in the rocks with synclinal and anticlinal folds, exhibited in cliffs nearly 300 feet high, the drift beds participating in all the bendings of the chalk; that is to say, the three lower members of the drift, including No. 2, which, at the point s in this diagram, contains the shells of recent species before alluded to.

Near the northern end of the Möens Klint, at a place called 'Taler,' more than 300 feet high, are seen similar folds, so sharp that there is an appearance of four distinct alternations of the glacial and cretaceous formations in vertical or highly inclined beds; the chalk at one point bending over, so that the position of all the beds is reversed.

But the most wonderful shiftings and faultings of the beds are observable in the Dronningestol, part of the same cliff, 400 feet in vertical height, where, as shown in fig. 52 (p. 392), the drift is thoroughly entangled and mixed up with the dislocated chalk.

If we follow the lines of fault, we may see, says M. Puggaard, along the planes of contact of the shifted beds, the marks of polishing and rubbing, which the chalk flints have undergone, as have many stones in the gravel of the drift, and some of these have also been forced into the soft chalk. The manner in which the top of some of the arches of bent chalk have been cut off in this and several adjoining sections,

attests the great denudation which accompanied the disturbances, portions of the bent strata having been removed,
probably while they were emerging from beneath the sea.

Fig. 52.

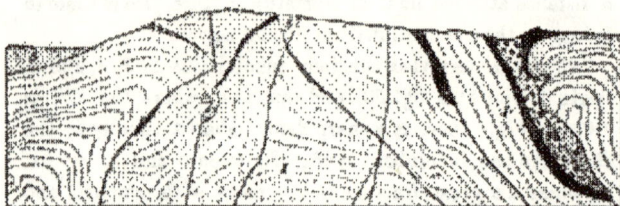

Post-glacial disturbances of vertical, folded, and shifted strata of chalk and drift,
in the Dronningestol, Möen, height 400 feet (Puggaard).

1 Chalk, with flints.
2 Marine stratified loam, lowest member of glacial formation.
3 Blue clay or till, with erratic blocks unstratified.
4 Yellow sandy till, with pebbles and glaciated boulders.
5 Stratified sand and gravel with erratics.

M. Puggaard has deduced the following conclusions from
his study of the cliffs.

1st. The white chalk, when it was still in a horizontal
position, but after it had suffered considerable denudation,
subsided gradually, so that the lower beds of drift No. 2,
with their littoral shells, were superimposed on the chalk
in a shallow sea.

2nd. The overlying unstratified boulder-clays 3 and 4 were
thrown down in deeper water by the aid of floating ice coming
from the north.

3rd. Irregular subsidences then began, and occasionally
partial failures of support, causing the bending and sometimes the engulfment of overlying masses both of the chalk
and drift, and causing the various dislocations above described
and depicted. The downward movement continued till it
exceeded 400 feet, for upon the surface even of No. 5, in
some parts of the island, lie huge erratics twenty feet or

more in diameter, which imply that they were carried by ice
in a sea of sufficient depth to float large icebergs. But
these big erratics, says Puggaard, never enter into the fissures
as they would have done had they been of date anterior to
the convulsions.

4th. After this subsidence, the re-elevation and partial
denudation of the cretaceous and glacial beds took place
during a general upward movement, like that now experienced
in parts of Sweden and Norway.

In regard to the lines of movement in Möen, M. Puggaard
believes, after an elaborate comparison of the cliffs with the
interior of the island, that they took at least three distinct
directions at as many successive eras, all of post-glacial date ;
the first line running from E.S.E. to W.N.W., with lines of
fracture at right angles to them ; the second running from
S.S.E. to N.N.W., also with fractures in a transverse direction ; and lastly, a sinking in a N. and S. direction, with other
subsidences of contemporaneous date running at right angles,
or E. and W.

When we approach the north-west end of Möens Klint, or
the range of coast above described, the strata begin to be less
bent and broken, and, after travelling for a short distance
beyond, we find the chalk and overlying drift in the same
horizontal position as at the southern end of the Möens Klint.
What makes these convulsions the more striking is the fact
that in the other adjoining Danish islands, as well as in a
large part of Möen itself, both the secondary and tertiary
formations are quite undisturbed.

It is impossible to behold such effects of reiterated local
movements, all of post-tertiary date, without reflecting that,
but for the accidental presence of the stratified drift, all of
which might easily, where there has been so much denudation,
have been missing, even if it had once existed, we might
have referred the verticality and flexures and faults of the

rocks to an ancient period, such as the era between the chalk
with flints and the Maestricht chalk, or to the time of the
latter formation, or to the Eocene, or Miocene, or older
Pliocene eras, even the last of them, long prior to the commencement of the glacial epoch. Hence we may be permitted
to suspect that in some other regions, where we have no such
means at our command for testing the exact date of certain
movements, the time of their occurrence may be far more
modern than we usually suppose. In this way some apparent
anomalies in the position of erratic blocks, seen occasionally
at great heights above the parent rocks from which they
have been detached, might be explained, as well as the irregular direction of certain glacial furrows like those described
by Professor Keilhau and Mr. Hörbye on the mountains of
the Dovrefield in lat. 62° N., where the striation and friction
is said to be independent of the present shape and slope of
the mountains.* Although even in such cases it remains to
be proved whether a general crust of continental ice, like
that of Greenland (see above, p. 276), would not account for
the deviation of the furrows and striæ from the normal directions which they ought to have followed had they been due
to separate glaciers filling the existing valleys.

It appears that in general the upward movements in Scandinavia, which have raised sea-beaches containing marine
shells of recent species to the height of several hundred feet,
have been tolerably uniform over very wide spaces; yet a
remarkable exception to this rule was observed by M. Bravais,
at Altenfiord, in Finmark, between lat. 70° and 71° N. An
ancient water-level, indicated by a sandy deposit forming a
terrace, and by marks of the erosion of the waves, can be
followed for thirty miles from south to north along the
borders of a fiord rising gradually from a height of eighty-

* Observations sur les Phénomènes d'Érosion en Norwège, 1857.

five feet to an elevation of 220 feet above the sea, or at the
rate of about four feet in a mile.[*]

To pass to another and very remote part of the world, we
have witnessed, so late as January 1855, in the northern
island of New Zealand, a sudden and permanent rise of land
on the northern shores of Cook's straits, which at one point,
called Muko-muka, was so unequal as to amount to nine feet
vertically, while it declined gradually from this maximum of
upheaval in a distance of about twenty-three miles north-
west of the greatest rise, to a point where no change of level
was perceptible. Mr. Edward Roberts, of the Royal Engineers,
employed by the British Government at the time of the
shock in executing public works on the coast, ascertained
that the extreme upheaval of certain ancient rocks followed
a line of fault running at least ninety miles from south to
north into the interior; and, what is of great geological
interest, immediately to the east of this fault, the country,
consisting of tertiary strata, remained unmoved or stationary;
a fact well established by the position of a line of nullipores
marking the sea-level before the earthquake, both on the
surface of the tertiary and palæozoic rocks.[†]

The repetition of such unequal movements, especially if
they recurred at intervals along the same lines of fracture,
would in the course of ages cause the strata to dip at a high
angle in one direction, while towards the opposite point of
the compass they would terminate abruptly in a steep escarpment.

But it is probable that the multiplication of such movements in the post-tertiary period has rarely been so great as
to produce results like those above described in Möen, for

[*] Proceedings of the Geological
Society, 1816, vol. iv. p. 94.
[†] Bulletin de la Société Géologique
de France, vol. xiii. p. 660, 1856,

where I have described the facts communicated to me by Messrs. Roberts
and Walter Mantell.

the principal movements in any given period seem to be of that more uniform kind spoken of at p. 384, by which the topography of limited districts and the position of the strata are not visibly altered except in their height relatively to the sea. Were it otherwise, we should not find conformable strata of all ages, including the primary fossiliferous of shallow-water origin, which must have remained horizontal throughout vast areas during downward movements of several thousand feet, going on at the period of their accumulation. Still less should we find the same primary strata, such as the Carboniferous, Devonian, or Silurian, still remaining horizontal over thousands of square leagues, as in parts of North America and Russia, having escaped dislocation and flexure throughout the entire series of epochs which separate palæozoic from recent times. Not that they have been motionless, for they have undergone so much denudation, and of such a kind, as can only be explained by supposing the strata to have been subjected to great oscillations of level, and exposed in some cases repeatedly to the destroying and planing action of the waves of the sea.

It seems probable that the successive convulsions in Möen were contemporary with those upward and downward movements of the glacial period which were described in the thirteenth and some of the following chapters, and that they ended before the upper beds of No. 5, p. 392, with its large erratic blocks, were deposited, as some of those beds occurring in the disturbed parts of Möen appear to have escaped the convulsions to which Nos. 2, 3, and 4 were subjected. If this be so, the whole derangement, although pleistocene, may have been anterior to the human epoch, or rather to the earliest date to which the existence of man has as yet been traced back.

CHAPTER XVIII.

THE GLACIAL PERIOD IN NORTH AMERICA.

POST-GLACIAL STRATA CONTAINING REMAINS OF MASTODON GIGANTEUS
IN NORTH AMERICA—SCARCITY OF MARINE SHELLS IN GLACIAL DRIFT
OF CANADA AND THE UNITED STATES—GREATER SOUTHERN EXTENSION
OF ICE-ACTION IN NORTH AMERICA THAN IN EUROPE—TRAINS OF
ERRATIC BLOCKS OF VAST SIZE IN BERKSHIRE, MASSACHUSETTS—
DESCRIPTION OF THEIR LINEAR ARRANGEMENT AND POINTS OF DE-
PARTURE—THEIR TRANSPORTATION REFERRED TO FLOATING AND
COAST ICE—GENERAL REMARKS ON THE CAUSES OF FORMER CHANGES
OF CLIMATE AT SUCCESSIVE GEOLOGICAL EPOCHS—DEVELOPMENT OF EX-
TREME COLD ON THE OPPOSITE SIDES OF THE ATLANTIC IN THE GLACIAL
PERIOD NOT STRICTLY SIMULTANEOUS.

ON the North American Continent, between the arctic
circle and the 42nd parallel of latitude, we meet with
signs of ice-action on a scale as grand if not grander than in
Europe; and there also the excess of cold appears to have been
first felt, at the close of the tertiary, and to have continued
throughout a large portion of the pleistocene period.

The general absence of organic remains in the North
American glacial formation, makes it as difficult as in Europe
to determine what mammalia lived on the continent at the
time of the most intense refrigeration, or when extensive
areas were becoming strewed over with glacial drift and
erratic blocks, but it is certain that a large proboscidean now
extinct, the *Mastodon giganteus*, Cuv., together with many
other quadrupeds, some of them now living and others
extinct, played a conspicuous part in the post-glacial era.
By its frequency as a fossil species, this pachyderm represents
the European *Elephas primigenius*, although the latter also

occurs fossil in the United States and Canada, and abounds, as I learn from Sir John Richardson, in latitudes farther north than those to which the mastodon has been traced.

In the State of New York, the mastodon is not unfrequently met with in bogs and lacustrine deposits formed in hollows in the drift, and therefore, in a geological position, much resembling that of recent peat and shell-marl in the British Isles, Denmark, or the Valley of the Somme, as before described. Sometimes entire skeletons have been discovered within a few feet of the surface, in peaty earth at the bottom of small ponds, which the agriculturists had drained. The accompanying shells in these cases belong to freshwater genera, such as *Lymnea*, *Physa*, *Planorbis*, *Cyclas*, and others. They differ from European species, but are the same as those now proper to ponds and lakes in the same parts of America.

I have elsewhere given an account of several of these localities which I visited in 1842,[*] and can state that they certainly have a more modern aspect than almost all the European deposits in which remains of the mammoth occur, although a few instances are cited of *Elephas primigenius* having been dug out of peat in Great Britain. Thus I was shown a mammoth's tooth in the museum at Torquay, in Devonshire, which was dredged up in Torbay, and there is good reason to believe that it was torn out of the submerged peat or forest known to exist there. A more elevated part of the same peaty formation constitutes the bottom of the valley in which Tor Abbey stands. This individual elephant must certainly have been of more modern date than his fellows found fossil in the gravel of the Brixham cave, before described (p. 102), for he flourished when the physical geography of Devonshire, unlike that of the cave period, was almost identical with that now established. The Hon.

[*] Travels in North America, vol. i. p. 55, London, 1846.

W. Stanley also found, in 1849, two perfect heads of the
mammoth in a bed of peat in Holyhead harbour. The tusks
and molars lay two feet below the surface in the peat which
was covered by the stiff blue clay.* No doubt between the
period when the mammoth was most abundant, and that
when it died out, there must have elapsed a long interval
of ages when it was growing more and more scarce ; and we
may expect to find occasional stragglers buried in deposits
long subsequent in date to others, until at last we may
succeed in tracing a passage from the pleistocene to the
recent fauna, by geological monuments, which will fill up
the gap before alluded to (p. 189) as separating the era of
the flint tools of Amiens and Abbeville from that of the peat
of the Valley of the Somme.

How far the lacustrine strata of North America, above
mentioned, may help to lessen this hiatus, and whether some
individuals of the *Mastodon giganteus* may have come down
to the confines of the historical period, is a question not so
easily answered as might at first sight be supposed. A geologist might naturally imagine that the fluviatile formation of
Goat Island, seen at the falls of Niagara, and at several
points below the falls,† was very modern, seeing that the
fossil shells contained in it are all of species now inhabiting
the waters of the Niagara, and seeing also that the deposit is
more modern than the glacial drift of the same locality. In
fact, the old river bed, in which bones of the mastodon occur,
holds the same position relatively to the boulder formation as
the strata of shell-marl and boggy-earth, with bones of mastodon, so frequent in the State of New York, bear to the glacial
drift, and all may be of contemporaneous date. But in the
case of the valley of the Niagara, we happen to have a measure

* Principles of Geology, 11th ed.,   † Travels in North America, by the
vol. i. p. 559.   Author, vol. i. ch. ii., and vol. ii. ch.
  xix.

of time, which is wanting in the other localities, namely,
the test afforded by the recession of the falls, an operation
still in progress, by which the deep ravine of the Niagara,
seven miles long, between Queenstown and Goat Island, has
been hollowed out. This ravine is not only post-glacial, but
also posterior in date to the fluviatile or mastodon-bearing
beds. The individual, therefore, found fossil near Goat Island
flourished before the gradual excavation of the deep and long
chasm, and we must reckon its antiquity, not by thousands,
but by tens of thousands of years, if I have correctly estimated
the minimum of time which was required for the erosion of
that great ravine.[*]

The stories widely circulated of bones of the mastodon
having been observed with their surfaces pierced as if by
arrow-heads, or bearing the marks of wounds inflicted
by some stone implement, must in future be more carefully
inquired into, for we can scarcely doubt that the mastodon
in North America lived down to a period when the mammoth
coexisted with Man in Europe. But I need say no more on
this subject, having already (p. 237) explained my views in
regard to the evidence of the antiquity of Man in North
America, when treating of the human bone discovered at
Natchez, on the Mississippi.

In Canada and the United States, we experience the same
difficulty as in Europe, when we attempt to distinguish
between glacial formations of submarine and those of supra-
marine origin. In the New World, as in Scotland and
England, marine shells of this era have rarely been traced
higher than five hundred feet above the sea ; and while in
England isolated instances are now known as at Moel Tryfaen
and Macclesfield, where they ascend to 1,200 feet and upwards,
seven hundred feet seems to be the maximum to which, in

* Principles of Geology, 9th ed. p. 3; and Travels in North America,
vol. i. p. 32, 1845.

America, they are at present known to ascend. In the same countries, erratic blocks have travelled from N. to S., following the same direction as the glacial furrows and striæ imprinted almost everywhere on the solid rocks underlying the drift. Their direction rarely deviates more than fifteen degrees E. or W. of the meridian, and they may have been conveyed by icebergs.

There are, nevertheless, in the United States, as in Europe, several groups of mountains which have acted as independent centres for the dispersion of erratics, as, for example, the White Mountains, latitude 44° N., the highest of which, Mount Washington, rises to about 6,300 feet above the sea; and according to Professor Hitchcock, some of the loftiest of the hills of Massachusetts once sent down their glaciers into the surrounding lower country.

Great southern Extension of Trains of Erratic Blocks in Berkshire, Massachusetts, U. S., lat. 42° N.

Having treated so fully in this volume of the events of the glacial period, I am unwilling to conclude without laying before the reader the evidence displayed in North America, of ice-action in latitudes farther south, by about ten degrees than any seen on an equal scale in Europe. This extension southwards of glacial phenomena, in regions where there are no snow-covered mountains like the Alps to explain the exception, nor any hills of more than moderate elevation, constitutes a feature of the western as compared to the eastern side of the Atlantic, and must be taken into account when we speculate on the causes of the refrigeration of the northern hemisphere during the pleistocene period.

In 1852, accompanied by Mr. James Hall, State geologist of New York, and author of many able and well-known works on geology and paleontology, I examined the glacial drift

D D

and erratics of the county of Berkshire, Massachusetts, and those of the adjoining parts of the State of New York, a district about 130 miles inland from the Atlantic coast, and situated due west of Boston, in lat. 42° 25′ north. This latitude corresponds in Europe to that of the north of Portugal. Here numerous detached fragments of rock are seen, having a linear arrangement or being continuous in long parallel trains, running nearly in straight lines over hill and dale for distances of five, ten, and twenty miles, and sometimes greater distances. Seven of the more conspicuous of these trains, from 1 to 7 inclusive, are laid down in the accompanying map or ground plan, fig. 53.* It will be remarked that they run in a N.W. and S.E. direction, or transversely to the ranges of hills A, B, and C, which run N.N.E. and S.S.W. The crests of these chains are about 800 feet in height above the intervening valleys. The blocks of the northernmost train, No. 7, are of limestone, derived from the calcareous chain B; those of the two trains next to the south, Nos. 6 and 5, are composed exclusively in the first part of their course of a green chloritic rock of great toughness, but after they have passed the ridge B, a mixture of calcareous blocks is observed. After traversing the valley for a distance of six miles, these two trains pass through depressions or gaps in the range C, as they had previously done in crossing the range B, showing that the dispersion of the erratics bears some relation to the actual inequalities of the surface, although the course of the same blocks is perfectly independent of the more leading features of the geography of the country, or those by which the present lines of drainage are determined. The greater number of the green chloritic fragments in trains 5 and 6 have evidently come from the ridge A,

* This ground plan, and a farther account of the Berkshire erratics, was given in an abstract of a lecture delivered by me to the Royal Institution of Great Britain, April 27, 1855, and published in their Proceedings.

Fig. 53.

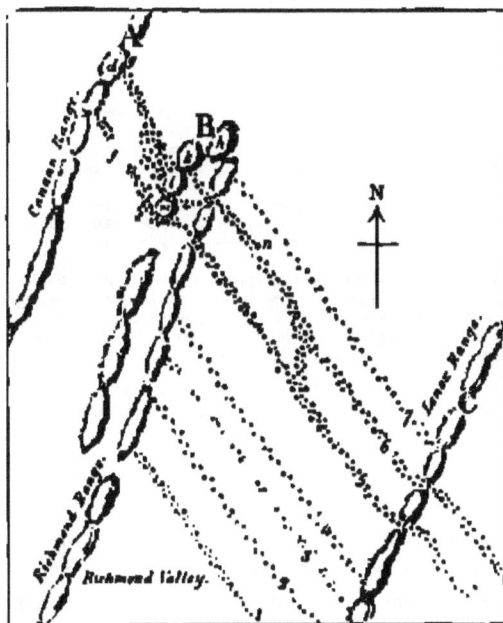

MAP SHOWING THE RELATIVE POSITION AND DIRECTION OF SEVEN TRAINS OF
ERRATIC BLOCKS IN BERKSHIRE, MASSACHUSETTS, AND IN PART OF THE STATE
OF NEW YORK.

*Distance in a straight line, between the mountain ranges A and c, about
eight miles.*

A Canaan range, in the State of New York. The crest consists of green
chloritic rock.

B Richmond range, the western division of which consists in Merriman's
Mount of the same green rock as A, but in a more schistose form, while the
eastern division is composed of slaty limestone.

C The Lenox range, consisting in part of mica-schist, and in some districts
of crystalline limestone.

d Knob in the range A, from which most of the No. 6 is supposed to have
been derived.

e Supposed starting point of the train No. 6 in the range A.

f Hiatus of 175 yards, or space without blocks.

g Sherman's House.

h Perry's Peak.

k Flat Rock.

l Merriman's Mount.

m Dupey's Mount.
n Largest block of train, No. 6. See figs. 54 and 55, p. 405.
p Point of divergence of part of the train No. 6, where a branch is sent off to No. 5.

No. 1 The most southerly train examined by Messrs Hall and Lyell, between Stockbridge and Richmond, composed of blocks of black slate, blue limestone, and some of the green Canaan rock, with here and there a boulder of white quartz.

No. 2 Train composed chiefly of large limestone masses, some of them divided into two or more fragments, by natural joints.

No. 3 Train composed of blocks of limestone and the green Canaan rock; passes south of the Richmond station on the Albany and Boston railway; is less defined than Nos. 1 and 2.

No. 4 Train chiefly of limestone blocks, some of them thirty feet in diameter, running to the north-west of the Richmond Station, and passing south of the Methodist Meeting-house, where it is intersected by a railway cutting.

No. 5 South train of Dr. Reid, composed entirely of large blocks of the green chloritic Canaan rock; passes north of the old Richmond Meeting-house, and is three-quarters of a mile north of the preceding train (No. 4).

No. 6 The great or principal train (north train of Dr. Reid), composed of very large blocks of the Canaan rock, diverges at p, and unites by a branch with train No. 5.

No. 7 A well-defined train of limestone blocks, with a few of the Canaan rock, traced from the Richmond to the slope of the Lenox range.

and a large proportion of the whole from its highest summit, d, where the crest of the ridge has been worn into those dome-shaped masses called 'roches moutonnées,' already alluded to (p. 319), and where several fragments having this shape, some of them thirty feet long, are seen in situ, others only slightly removed from their original position, as if they had been just ready to set out on their travels. Although smooth and rounded on their tops, they are angular on their lower parts, where their outline has been derived from the natural joints of the rock. Had these blocks been conveyed from d by glaciers, they would have radiated in all directions from a centre, whereas not one even of the smaller ones is found to the westward of A, though a very slight force would have made them roll down to the base of that ridge, which is very steep on its western declivity. It is clear, therefore, that the propelling power, whatever it may have been, acted exclusively in a south-easterly direction. Professor Hall and I observed one of the green blocks, twenty-four feet long,

poised upon another about nineteen feet in length. The
largest of all on the west flank of *m*, or Dupcy's Mount,
called the Alderman, is above ninety feet in diameter, and

Fig. 54.

Erratic dome-shaped block of compact chloritic rock (*a* map,
fig. 53), near the Richmond Meeting-house, Berkshire, Massachusetts,
lat. 42° 25′ N. Length, fifty-two feet; width, forty feet; height
above the soil, fifteen feet.

Fig. 55.

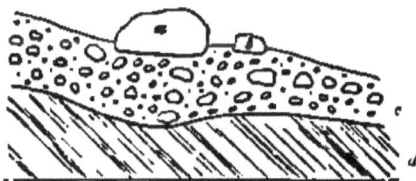

Section showing position of the block, fig. 54.
a The large block, fifty-two feet by forty. Fig. 54 and *a* map, p. 403.
b Fragment detached from the same, twenty feet by fourteen.
c Unstratified drift with boulders.
d Silurian limestone in inclined stratification.

nearly three hundred feet in circumference. We counted at
some points between forty and fifty blocks visible at once,
the smallest of them larger than a camel.

The annexed drawing represents one of the best known of

train No. 6, being that marked n on the map, p. 403. According to our measurement it is fifty-two feet long by forty
in width, its height above the drift in which it is partially
buried being fifteen feet. At the distance of several yards
occurs a smaller block, three or four feet in height, twenty
feet long, and fourteen broad, composed of the same compact
chloritic rock, and evidently a detached fragment from the
bigger mass, to the lower and angular part of which it would
fit on exactly. This erratic n has a regularly rounded top,
worn and smooth like the roches moutonnées before mentioned, but no part of the attrition can have occurred since
it left its parent rock, the angles of the lower portion being
quite sharp and unblunted.

From railway cuttings through the drift of the neighbourhood, and other artificial excavations, we may infer that the
position of the block n, if seen in a vertical section, would
be as represented in fig. 55. The deposit c in that section,
p. 405, consists of sand, mud, gravel, and stones, for the
most part unstratified, resembling the till or boulder-clay of
Europe. It varies in thickness from ten to fifty feet, being
of greater depth in the valleys. The uppermost portion is
occasionally, though rarely, stratified. Some few of the embedded stones have flattened, polished, striated, and furrowed
sides. They consist invariably, like the seven trains above
mentioned, of kinds of rock confined to the region lying to
the N.W., none of them having come from any other quarter.
Whenever the surface of the underlying rock has been exposed by the removal of the superficial detritus, a polished
and furrowed surface is seen, like that underneath a glacier,
the direction of the furrows being from N.W. to S.E., or corresponding to the course of the large erratics.

As all the blocks, instead of being dispersed from a centre,
have been carried in one direction, and across the ridges A, B,
c, and the intervening valleys, the hypothesis of glaciers in

valleys at all similar to those now existing in that country,
is out of the question. I conceive, therefore, that the
erratics were conveyed to the places they now occupy by
coast ice, when the country was submerged beneath the
waters of a sea cooled by icebergs coming annually from
arctic regions.

Fig. 54.

d, e Masses of floating ice carrying fragments of rock.

Suppose the highest peaks of the ridges A, B, C, in the annexed diagram, to be alone above water, forming islands,
and *d e* to be masses of floating ice, which drifted across the
Canaan and Richmond valleys at a time when they were
marine channels, separating islands, or rather chains of islands,
having a N.N.E. and S.S.W. direction. A fragment of ice
such as *d*, freighted with a block A, might run aground, and
add to the heap of erratics at the N.W. base of the island
(now ridge) B, or, passing through a sound between B and
the next island of the same group, might float on till it
reached the channel between B and C. Year after year two
such exposed cliffs in the Canaan range as *d* and *e* of the
map, fig. 53, p. 403, undermined by the waves, might serve
as the points of departure of blocks composing the trains
Nos. 5 and 6. It may be objected that oceanic currents
could not always have had the same direction; this may
be true, but during a short season of the year, when the
ice was breaking up, the prevailing current may have always
run S.E.

If it be asked why the blocks of each train are not more
scattered, especially when far from their source, it may be

observed, that after passing through sounds separating
islands, they issued again from a new and narrow starting
point; moreover, we must not exaggerate the regularity of
the trains, as their width is sometimes twice as great in one
place as in another; and No. 6. sends off a branch at *p*,
which joins No. 5. There are also stragglers, or large blocks,
here and there in the spaces between the two trains. As to
the distance to which any given block would be carried, that
must have depended on a variety of circumstances, such as
the strength of the current, the direction of the wind, the
weight of the block, or the quantity and draught of the ice
attached to it. The smaller fragments would, on the whole,
have the best chance of going farthest; because, in the first
place they were more numerous, and then, being lighter,
they required less ice to float them, and would not ground so
readily on shoals, or, if stranded, would be more easily started
again on their travels. Many of the blocks, which at first
sight seem to consist of single masses, are found, when examined, to be made up of two, three, or more pieces, divided
by natural joints. In case of a second removal by ice, one
or more portions would become detached and be drifted to
different points further on. Whenever this happened, the
original size would be lessened, and the angularity of the
block previously worn by the breakers would be restored, and
this tendency to split may explain why some of the far-transported fragments remain very angular.

These various considerations may also account for the
fact that the average size of the blocks of all the seven trains
laid down on the plan, fig. 53, lessens sensibly in proportion
as we recede from the principal points of departure of particular kinds of erratics, yet not with any regularity, a huge
block now and then recurring when the rest of the train
consists of smaller ones.

All geologists acquainted with the district now under consideration are agreed that the mountain ranges A, B, and C, as well as the adjoining valleys, had assumed their actual form and position before the drift and erratics accumulated on and in them, and before the surface of the fixed rocks was polished and furrowed. I have the less hesitation in ascribing the transporting power to coast ice, because I saw, in 1852, an angular block of sandstone, eight feet in diameter, which had been brought down several miles by ice, only three years before, to the mouth of the Petitcodiac estuary, in Nova Scotia, where it joins the Bay of Fundy; and I ascertained that on the shores of the same bay, at the South Joggins, in the year 1850, much larger blocks had been removed by coast ice, and after they had floated half a mile, had been dropped in salt water by the side of a pier built for loading vessels with coal, so that it was necessary at low tide to blast these huge ice-borne rocks with gunpowder in order that the vessels might be able to draw up alongside the pier. These recent exemplifications of the vast carrying powers of ice occurred in lat. 46° N. (corresponding to that of Bordeaux), in a bay never invaded by icebergs.

I may here remark that a sheet of ice of moderate thickness, if it extend over a wide area, may suffice to buoy up the largest erratics which fall upon it. The size of these will depend, not on the intensity of the cold, but on the manner in which the rock is jointed, and the consequent dimensions of the blocks into which it splits, when falling from an undermined cliff.

The fact that no marine shells accompany the Berkshire erratics or have been detected in the associated Till raises a question which I will not attempt to discuss in this place, but which will no doubt tempt many geologists to speculate on a sheet of continental ice, like that of Greenland above alluded to (p. 277) as the possible cause of the transportation of the blocks which have been carried in a direction so indepen-

dent of that which glaciers following the existing valley
must have taken.

*Probable causes of the Glacial Climates of Europe and
America.*

When I first endeavoured in the ' Principles of Geology,'
in 1830* to explain the causes, both of the warmer and
colder climates, which have at former periods prevailed on
the globe, I referred to successive variations in the height
and position of the land, and its extent relatively to the sea
in polar and equatorial latitudes—also to fluctuations in the
course of oceanic currents, the Gulfstream among others, and
other geographical conditions, by the united influence of
which I believed the principal revolutions in the meteorolo-
gical state of the atmosphere at different geological periods
to have been brought about. The reader will find in the
eleventh edition of the Principles of Geology† a full discussion of the possible effects which the different distribution of
light and heat in different seasons and in the opposite
hemispheres may cause in the course of past astronomical
periods, owing chiefly to the varying eccentricity of the earth's
orbit, and in a smaller degree to changes in the obliquity
of the ecliptic. After giving due weight to all these cosmical
changes, it still appears to me that it is only by admitting the
frequent and paramount influence of geographical changes
on the earth itself that we can hope to explain those great revolutions of climate which wrapt large parts of the northern
hemisphere in a winding-sheet of continental ice.

But here another question arises, whether the eras at
which the maximum of cold was attained on the opposite
sides of the Atlantic were really contemporaneous?   We
have now discovered, not only that the glacial period was of
vast duration, but that it passed through various phases and

* 1st edit. ch. vii. ; 9th edit. ib.   † Chapters xii. and xiii. 11th edit.

oscillations of temperature; so that, although the chief
polishing and furrowing of the rocks and transportation of
erratics in Europe and North America may have taken place
contemporaneously, according to the ordinary language of
geology, or when the same testacea and the same pleistocene
assemblage of mammalia flourished, yet the extreme development of cold on the opposite sides of the ocean may not have
been strictly simultaneous, but, on the contrary, the one
may have preceded or followed the other by a thousand, or
more than a thousand, centuries.

It is probable that the greatest refrigeration of Norway,
Sweden, Scotland, Wales, the Vosges, and the Alps coincided very nearly in time; but when the Scandinavian and
Scotch mountains were encrusted with a general covering of
ice, similar to that now enveloping Greenland, this last
country may not have been in nearly so glacial a condition
as now, just as we find that the old icy crust and great
glaciers, which have left their mark on the mountains of
Norway and Sweden, have now disappeared, precisely at a
time when the accumulation of ice in Greenland is so
excessive. In other words, we see that in the present state
of the northern hemisphere, at the distance of about fifteen
hundred miles, two meridional zones, enjoying very different
conditions of temperature, may coexist, and we are, therefore, at liberty to imagine some former alterations of colder
and milder climates on the opposite sides of the ocean
throughout the pleistocene era of a compensating kind, the
cold on the one side corresponding to the milder temperature on the other.   By assuming such a succession of events
we can more easily explain why there has not been a greater
extermination of species, both terrestrial and aquatic, in
polar and temperate regions, during the glacial epoch, and
why so many species are common to pre-glacial and post-
glacial times.

The scope and limits of this volume forbid my pursuing these speculations and reasonings farther; but I trust I have said enough to show that the monuments of the glacial period, when more thoroughly investigated, will do much towards expanding our views as to the antiquity of the fauna and flora now contemporary with Man, and will therefore enable us the better to determine the time at which Man began to form part of the fauna of the northern hemisphere.

CHAPTER XIX.

RECAPITULATION OF GEOLOGICAL PROOFS OF MAN'S ANTIQUITY.

RECAPITULATION OF RESULTS ARRIVED AT IN THE EARLIER CHAPTERS —AGES OF STONE AND BRONZE—DANISH PEAT AND KITCHEN-MIDDENS —SWISS LAKE-DWELLINGS—LOCAL CHANGES IN VEGETATION AND IN THE WILD AND DOMESTICATED ANIMALS AND IN PHYSICAL GEOGRAPHY COEVAL WITH THE AGE OF BRONZE AND THE LATER STONE PERIOD— ESTIMATES OF THE POSITIVE DATE OF SOME DEPOSITS OF THE LATER STONE PERIOD—STONE AGE OF ST. ACHEUL AND AURIGNAC— MIGRATIONS OF MAN IN THAT PERIOD FROM THE CONTINENT TO ENGLAND IN POST-GLACIAL TIMES—SLOW RATE OF PROGRESS IN BARBAROUS AGES—DOCTRINE OF THE SUPERIOR INTELLIGENCE AND ENDOWMENTS OF THE ORIGINAL STOCK OF MANKIND CONSIDERED— OPINIONS OF THE GREEKS AND ROMANS, AND THEIR COINCIDENCE WITH THOSE OF THE MODERN PROGRESSIONIST—EARLY ORIENTAL AND EGYPTIAN CIVILISATION AND ITS DATE IN COMPARISON WITH THAT OF THE FIRST AND SECOND STONE PERIODS.

THE ages of stone and bronze, so called by archæologists, were spoken of in the earlier chapters of this work. That of bronze has been traced back to times anterior to the Roman occupation of Helvetia, Gaul, and other countries north of the Alps. When weapons of that mixed metal were in use, a somewhat uniform civilisation seems to have prevailed over a wide extent of central and northern Europe, and the long duration of such a state of things in Denmark and Switzerland is shown by the gradual improvement which took place in the useful and ornamental arts. Such progress is attested by the increasing variety of the forms, and the more perfect finish and tasteful decoration of the tools and utensils obtained from the more modern deposits of the bronze age (see fig 1, p. 10), those from the upper layers of peat, for

example, as compared to those found in the lower ones.
The greater number also of the Swiss lake-dwellings of the
bronze age (about seventy villages having been already
discovered), and the large population which some of them
were capable of containing, afford indication of a considerable
lapse of time, as does the thickness of the stratum of mud
in which, in some of the lakes, the works of art are entombed.
The unequal antiquity, also, of the settlements, is occasionally attested by the different degrees of decay which the
wooden stakes or piles have undergone, some of them projecting more above the mud than others, while all the piles
of the antecedent age of stone have rotted away quite down
to the level of the mud, such part of them only as was
originally driven into the bed of the lake having escaped
decomposition.*

Among the monuments of the stone period, which immediately preceded that of bronze, the polished hatchets
called celts are abundant, and were in very general use in
Europe before metallic tools were introduced. We learn,
from the Danish peat and shell-mounds, and from the older
Swiss lake-settlements, that the first inhabitants were hunters,
who fed almost entirely on game, but their food in after
ages consisted more and more of tamed animals, and, still
later, a more complete change to a pastoral state took place,
accompanied, as population increased, by the cultivation of
some cereals (p. 22).

Both the shells and quadrupeds, belonging to the later stone
period and to the age of bronze, consist exclusively of species
now living in Europe, the fauna being the same as that which
flourished in Gaul at the time when it was conquered by Julius
Cæsar, even the *Bos primigenius*, the only animal of which
the wild type is lost, being still represented, according to

---

* Troyon. Habitations lacustres. Lausanne, 1860.

Cuvier, Bell, and Rütimeyer, by one of the domesticated races of cattle now in Europe. (See p. 27.)

These monuments, therefore, whether of stone or bronze, belong to what I have termed geologically the Recent Period, the definition of which some may think rather too dependent on negative evidence, or on the non-discovery hitherto of extinct mammalia, such as the mammoth, which may one day turn up in a fossil state in some of the oldest peaty deposits, as, indeed, it has already in Torbay and at Holyhead (see p. 398). No doubt more of these exceptional cases will be met with in the course of future investigations, for we are still imperfectly acquainted with the entire fauna of the age of stone in Denmark, as we may infer from an opinion expressed by Steenstrup, that some of the instruments exhumed by antiquaries from the Danish peat are made of the bones and horns of the elk and reindeer. Yet no skeleton or uncut bone of either of those species has hitherto been observed in the same peat.

Nevertheless, the examination made by naturalists of the various Danish and Swiss deposits of the Recent Period has been so searching, that the finding in them of a stray elephant or rhinoceros, should it ever occur, would prove little more than that some few individuals lingered on, when the species was on the verge of extinction, and such rare exceptions would not render the classification above proposed inappropriate.

At the time when many wild quadrupeds and birds were growing scarce, and some of them becoming locally extirpated in Denmark, great changes were taking place in the vegetation. The pine, or Scotch fir, buried in the oldest peat, gave place at length to the oak; and the oak, after flourishing for ages, yielded, in its turn, to the beech; the periods when these three forest trees predominated in succession tallying pretty nearly with the ages of stone, bronze, and iron in Denmark (p. 17). In the same country, also, during the stone period, various fluctuations, as we have seen, occurred in physical geography. Thus, on the ocean side of certain islands, the old refuse-heaps, or 'kitchen-middens,' were destroyed by the waves, the cliffs having wasted away, while on the side of the Baltic, where the sea was making no encroachment, or where the land was sometimes gaining on the sea, such mounds remained uninjured. It was also shown, that the oyster, which supplied food to the primitive people, attained its full size in parts of the Baltic where it cannot now exist, owing to a want of saltness in the water, and that certain marine univalves and bivalves, such as the common periwinkle, mussel, and cockle, of which the castaway shells are found in the mounds, attained in the olden time their full dimensions, like the oysters, whereas the same species, though they still live on the coast of the inland sea adjoining the mounds, are dwarfed, and never half their natural size, the water being rendered too fresh for them by the influx of so many rivers.

Some archæologists and geologists of merit have endeavoured to arrive at positive dates, or an exact estimate of the minimum of time assignable to the later age of stone. These computations have been sometimes founded on changes in the level of the land, or on the increase of peat, as in the Danish bogs, or of the conversion of water into land by alluvial deposits, since certain lake-settlements in Switzerland were abandoned. Alterations also in the geographical distribution or preponderance of certain living species of animals and plants, have been taken into account in corroboration, as have the signs of progress in human civilisation, as serving to mark the lapse of time during the stone and bronze epochs.

M. Morlot has estimated with care the probable antiquity of three superimposed vegetable soils cut open at different

depths in the delta of the Tinière, each containing human
bones or works of art, belonging successively to the Roman,
Bronze, and later Stone periods. According to his estimate,
an antiquity of 7,000 years at least must be assigned to the
oldest of these remains, though believed to be long posterior
in date to the time when the mammoth and other extinct
mammalia flourished together with Man in Europe. (See
above p. 29 *et seq.*). Such computations of past time must
be regarded as tentative in the present state of our knowledge, and much collateral evidence will be required to
confirm them ; yet the results appear to me already to afford
a rough approximation to the truth.

Between the newer or recent division of the stone period
called Neolithic and the older or Pleistocene division, called
Palæolithic, there was evidently a vast interval of time—a
gap in the history of the past, into which many monuments
of intermediate date will one day have to be intercalated.
Of this kind are those caves in the south of France, in which
M. Lartet has lately found bones of the reindeer, associated
with works of art somewhat more advanced in style than
those of St. Acheul (p. 158). In the valley of the Somme,
we have seen that peat of great thickness exists in basins or
depressions conforming to the present contour and drainage
levels of the country.   This peat, which contains in its
upper layers Roman and Celtic memorials, has been of slow
growth and is long posterior in date to older gravels,
containing bones of the mammoth and a large number of
flint implements of a very rude and antique type. Some
of those gravels were accumulated in the channels of rivers
which flowed at higher levels, by a hundred feet, than the
present streams, and before the valley had attained its present
depth and form. No intermixture has been observed in
those ancient river beds of any of the polished weapons,
called Celts, or other relics of the more recent or Neolithic

stone period, nor any interstratified peat ; and the climate
of those Pleistocene ages, when Man coexisted in the north-
west of France and southern and central England with
many animals now extinct, appears to have been much
more severe in winter than it is now in the same region,
though far less cold than in the glacial period which immediately preceded.

We may presume that the time demanded for the gradual
dying out or extirpation of a large number of wild beasts
which figure in the Pleistocene strata, and are missing in
the Recent fauna, was of protracted duration, for we know
how tedious a task it is in our own times, even with the aid
of fire-arms, to exterminate a noxious quadruped, a wolf, for
example, in any region comprising within it an extensive
forest or a mountain chain. In many villages in the north
of Bengal, the tiger still occasionally carries off its human
victims, and the abandonment of late years by the natives of
a part of the Sunderbunds or lower delta of the Ganges,
which they once peopled, is attributed chiefly to the ravages
of the tiger. It is probable that causes more general and
powerful than the agency of Man, alterations in climate,
variations in the range of many species of animals, vertebrate
and invertebrate, and of plants, geographical changes in the
height, depth, and extent of land and sea, some or all of
these combined, have given rise, in a vast series of years, to
the annihilation, not only of many large mammalia, but to
the disappearance of the *Cyrena fluminalis*, once common
in the rivers of Europe, and to the different range or relative
abundance of other shells which we find in the European
drifts.

That the growing power of Man may have lent its aid
as the destroying cause of many Pleistocene species, must,
however, be granted ; yet, before the introduction of fire-
arms, or even the use of improved weapons of stone, it seems

more wonderful that the aborigines were able to hold their own against the cave-lion, hyæna, and wild bull, and to cope with such enemies, than that they failed to bring about their speedy extinction.

It is already clear that Man was contemporary in Europe with two species of elephant, now extinct, *E. primigenius* and *E. antiquus*, two, also, of rhinoceros, *R. tichorhinus* and *R. hemitœcus* (Falc.), *R. megurhinus*, at least one species of hippopotamus, the cave-bear, cave-lion, and cave-hyæna, various bovine, equine, and cervine animals now extinct, and many smaller carnivora, rodentia, and insectivora. While these were slowly passing away, the musk buffalo, reindeer, and other arctic species, which have survived to our times, were retreating northwards, from the valleys of the Thames and Seine, to their present more arctic haunts.

The human skeletons of the Belgian caverns of times co-eval with the mammoth and other extinct mammalia, do not betray any signs of a marked departure in their structure, whether of skull or limb, from the modern standard of certain living races of the human family. As to the remarkable Neanderthal skeleton (Ch. V. p. 83), it is at present too isolated and exceptional, and its age too uncertain, to warrant us in relying on its abnormal and ape-like characters as bearing on the question whether the further back we trace Man into the past, the more we shall find him approach in bodily conformation to those species of the anthropoid quadrumana which are most akin to him in structure.

In the descriptions already given of the geographical changes which the British Isles have undergone since the commencement of the glacial period (as illustrated by several maps, pp. 325-328), it has been shown that there must have been a free communication by land between the Continent and these islands, and between the several islands themselves, within the Pleistocene epoch, in order to account for the

Germanic fauna and flora having migrated into every part of
the area, as well as for the Scandinavian plants and animals
having retreated into the higher mountains. During some
part of the Pleistocene ages, the large pachyderms and accompanying beasts of prey, now extinct, wandered from the
Continent to England; and it is highly probable that France
was united with some part of the British Isles as late as
the period of the gravels of St. Acheul, and the era of those
engulfed rivers which, in the basin of the Meuse, near Liége,
swept into many a rent and cavern the bones of Man and
of the mammoth and cave-bear. There have been vast geographical revolutions in the times alluded to, and oscillations of land, during which the English Channel, which can
be shown by the Pagham erratics, and the old Brighton
beach (p. 330), to be of very ancient origin, may have been
more than once laid dry and again submerged. During some
one of these phases, Man may have crossed over, whether by
land or in canoes, or even on the ice of a frozen sea (as
Mr. Prestwich has hinted), for the winters of the period
of the higher level gravels of the valley of the Somme were
intensely cold.

The primitive people, who coexisted with the elephant and
rhinoceros in the valley of the Ouse at Bedford, and who
made use of flint tools of the Amiens type, certainly inhabited part of England, which had already emerged from
the waters of the glacial sea; and the fabricators of the flint
tools of Hoxne, in Suffolk, were also, as we have seen, post-
glacial. We may likewise presume, that the people of pleistocene date, who have left their memorials in the valley of
the Thames, were of corresponding antiquity, posterior to the
boulder clay, but anterior to the time when the rivers of that
region had settled into their present channels.

The vast distance of time which separated the origin of
the higher and lower level gravels of the valley of the Somme,

both of them rich in flint implements of similar shape (although those of oval form predominate in the newer gravels), leads to the conclusion that the state of the arts in those early times remained stationary for almost indefinite periods. There may, however, have been different degrees of civilisation, and in the art of fabricating flint tools, of which we cannot easily detect the signs in the first age of stone, and some contemporary tribes may have been considerably in advance of others. Those hunters, for example, who carved the representation of the mammoth on ivory found in the cavern of La Madelaine in Perigord may have been far less barbarous than the savages of St. Acheul. To a European who looks down from a great eminence on the products of the humble arts of the aborigines of all times and countries, the stone knives and arrows of the Red Indian of North America, the hatchets of the native Australian, the tools found in the ancient Swiss lake-dwellings, or those of the Danish kitchen-middens and of St. Acheul, may seem nearly all alike in rudeness, and very uniform in general character. But when we reflect how savage races continue for ages without making the slightest perceptible advance, we are prepared to recognise the signs of long lapses of time even in the difference of finish between very rude implements. The slowness of the progress of the arts of savage life is manifested by the fact, that the earlier instruments of bronze were modelled on the exact plan of the stone tools of the preceding age, notwithstanding the waste of precious metal caused by the unnecessary bulk, although such shapes would never have been chosen had metals been known from the first. The reluctance or incapacity of savage tribes to adopt new inventions, has been shown in the East, by their continuing to this day to use the same stone implements as their ancestors, after that mighty empires, where the use of metals in the arts was well known, had flourished for three thousand years in their neighbourhood.

We see in our own times, that the rate of progress in the arts and sciences proceeds in a geometrical ratio as knowledge increases, and so, when we carry back our retrospect into the past, we must be prepared to find the signs of retardation augmenting in a like geometrical ratio; so that the progress of a thousand years at a remote period, may correspond to that of a century in modern times, and in ages still more remote Man would more and more resemble the brutes in that attribute which causes one generation exactly to imitate in all its ways the generation which preceded it.

The extent to which even a considerably advanced state of civilisation may become fixed and stereotyped for ages, is the wonder of Europeans who travel in the East. One of my friends declared to me, that whenever the natives expressed to him a wish 'that he might live a thousand years,' the idea struck him as by no means extravagant, seeing that if he were doomed to sojourn for ever among them, he could only hope to exchange in ten centuries as many ideas, and to witness as much progress, as he could do at home in half a century.

It has sometimes happened that one nation has been conquered by another less civilised though more warlike, or that, during social and political revolutions, people have retrograded in knowledge. In such cases, the traditions of earlier ages, or of some higher and more educated caste which has been destroyed, may give rise to the notion of degeneracy from a primæval state of superior intelligence, or of science supernaturally communicated. But had the original stock of mankind been really endowed with such superior intellectual powers, and with inspired knowledge, and possessed the same improvable nature as their posterity, the point of advancement which they would have reached ere this would have been immeasurably higher. We cannot ascertain at

present the limits, whether of the beginning or the end, of
the first stone period, when Man coexisted with the extinct
mammalia, but that it was of great duration we cannot
doubt. During those ages there would have been time for
progress of which we can scarcely form a conception, and
very different would have been the character of the works of
art which we should now be endeavouring to interpret,—those
relics which we are now disinterring from the old gravel-pits
of St. Acheul, or from the Liége caves. In them, or in the
upraised bed of the Mediterranean, on the south coast of
Sardinia, instead of the rudest pottery or flint tools, so irregular in form as to cause the unpractised eye to doubt
whether they afford unmistakable evidence of design, we
should now be finding sculptured forms, surpassing in beauty
the master-pieces of Phidias or Praxiteles; lines of buried
railways or electric telegraphs, from which the best engineers
of our day might gain invaluable hints ; astronomical instruments and microscopes of more advanced construction than
any known in Europe, and other indications of perfection in
the arts and sciences, such as the nineteenth century has not
yet witnessed. Still farther would the triumphs of inventive
genius be found to have been carried, when the later deposits,
now assigned to the ages of bronze and iron, were formed.
Vainly should we be straining our imaginations to guess
the possible uses and meaning of such relics—machines, perhaps, for navigating the air or exploring the depths of the
ocean, or for calculating arithmetical problems, beyond the
wants or even the conception of living mathematicians.

The opinion entertained generally by the classical writers
of Greece and Rome, that Man in the first stage of his existence was but just removed from the brutes, is faithfully
expressed by Horace in his celebrated lines, which begin—

Quum prorepserunt primis animalia terris.—Sat., lib. i. 3, 99.

The picture of transmutation given in these verses, however

severe and contemptuous the strictures lavishly bestowed on
it by Christian commentators, accords singularly with the
train of thought which the modern doctrine of progressive
development has encouraged.

'When animals,' he says, 'first crept forth from the newly
formed earth, a dumb and filthy herd, they fought for acorns
and lurking-places with their nails and fists, then with clubs,
and at last with arms, which, taught by experience, they had
forged. They then invented names for things, and words
to express their thoughts, after which they began to desist
from war, to fortify cities, and enact laws.' They who in
later times have embraced a similar theory, have been led
to it by no deference to the opinions of their pagan predecessors, but rather in spite of very strong prepossessions in
favour of an opposite hypothesis, namely, that of the superiority of their original progenitors, of whom they believe
themselves to be the corrupt and degenerate descendants.

So far as they are guided by palæontology, they arrive
at this result by an independent course of reasoning; but they
have been conducted partly to the same goal as the ancients,
by ethnological considerations common to both, or by reflecting in what darkness the infancy of every nation is
enveloped, and that true history and chronology are the
creation, as it were, of yesterday. Thus the first Olympiad
is generally regarded as the earliest date on which we can
rely, in the past annals of mankind, only 776 years before
the Christian era.

When we turn from historical records to ancient monuments and inscriptions, the highest antiquity is claimed by
those of India and Egypt. Passing over the legendary and
mythical accounts of the Hindoos we find that they have lists
of kings which are calculated to go back between four and
five thousand years, and the Sanscrit language, in which their
Sacred Books are written, is of such remote antiquity that no

tradition remains of any people by whom it was originally spoken. In the same manner the temples, obelisks, cities, tombs, and pyramids of Egypt are many of them of such remote origin that their exact date, after they have been studied with so much patience and sagacity for centuries, remains uncertain and obscure. Nevertheless, by showing the advanced point which the civilisation of mankind had reached in the valley of the Nile, in times which were regarded by the Greeks, more than two thousand years ago, as lost in the night of ages, we may form some estimate of the minimum of time which a people such as the Egyptians must have required to emerge slowly from primæval barbarism, and reach, long before the first Olympiad, so high a degree of power and civilisation.

Sir George Cornewall Lewis, in his ' Historical Survey of the Astronomy of the Ancients,'* reminds us that Homer, in the Iliad, speaks of ' Egyptian Thebes, with its hundred gates, through each of which two hundred chariots went forth to battle,' and we may form an idea of the size which the great poet intended to ascribe to Thebes in Egypt, from the fact that Thebes in Bœotia, one of the first cities of Greece, was supposed to have only seven gates. Homer is believed to have flourished about eight centuries before the Christian era. At so early a period, therefore, the magnificence of Thebes had attracted the attention of the Greeks. But in the opinion of Egyptologists, there were great cities of still older date than Thebes; as, for example, Memphis, which, from the names of the kings on the oldest monuments now extant there, as compared with those in Thebes, is inferred to go back to remoter times. As to the speculations of Aristotle, in his ' Meteorics' (1, 14), that Memphis was probably the less ancient of the two, because the ground on which it stood was nearer the Mediterranean, and would therefore, at

* London, 1862.

a later period, be first redeemed from a watery and marshy
state, this argument, if it were available, would give an
extremely high antiquity to both cities, seeing the small
progress which the delta and alluvial deposits of the Nile
have made in the last two or three thousand years. It is only
in bays like that of Menzaleh, that any great amount of new
land has been gained, the general advance of the delta being
checked by a strong current of the Mediterranean, which,
running from the west, sweeps eastward the sediment brought
down by the great river, and prevents the land from en-
croaching farther on the sea. The slow subsidence also of
the land may be another powerful cause checking the advance
of the delta, and the desiccation of the inland country.

Aristotle remarks, that as Homer does not mention Memphis, the city either had no existence in the time of the poet
or was less considerable than Thebes. This observation is no
doubt just, so far as regards the comparative splendour of
the two cities, the one the metropolis of Upper and the
other of Lower Egypt in former times. But it has no
bearing whatever on the question of the existence of Memphis,
for Thebes is only alluded to incidentally as the grandest city
known to Homer. Achilles, when refusing to fight in consequence of the affront offered to him by Agamemnon
who had deprived him of his beautiful captive Briseis, is
made to exclaim, 'Not though you were to offer me the
wealth of Egyptian Thebes, with its hundred gates,' &c. &c.,
'would I stir ;' * and the allusion to Thebes in the Odyssey is
equally a passing one.† If a work like Strabo's 'Geography,'
compiled in the days of Homer, had come down to us, and
Thebes had been fully described without any mention being
made of Memphis, we might then have inferred the non-
existence of the latter city at that period.

But there are other monuments which carry us back far

* Iliad, ix. 381.   † Odyssey, iv. 127.

beyond the traditionary building of Thebes. We cannot
contemplate the average size and number of the pyramids now
extant (upwards of forty large and small), to say nothing of
the inscriptions upon them, without supposing them to have
been the work of a long succession of generations. Yet the
best authorities believe that these pyramids were built more
than 3,000 years before the birth of Christ,* and we find
evidence that the Egyptians had then attained a settled
government, an advanced state of the arts, and extended
commercial intercourse. Herodotus, who visited Egypt more
than 400 years before our era, was told by the priests that
Menes, the first Egyptian king who reigned before the building of the pyramids, and whose date Bunsen gives as 3620 B.C.
turned the Nile from its course under the Lybian mountains
into a new bed in the middle of the valley, and made the
dykes by which the land was reclaimed on which Memphis
afterwards stood.†

Works like these imply an indigenous civilisation which
must have been slowly matured long before the time when
Thebes attained wealth and consequence. ‘The empire of
Menes,’ says Bunsen, ‘with which regular chronology com-
mences, is based on two necessary and demonstrable strata
of early facts. The first is that which was requisite in order
to the establishment of a double kingdom of the Upper and
Lower Country. There existed registers of the princes of
both the one and the other consequently, prior to Menes;
Thinite princes in Upper Egypt, and Memphite princes, or
those of the Lower Country, who were immediately succeeded
by the imperial dynasties of Memphis. In order to be on the
safe side we will consider these two kingdoms as contemporaneous, not insisting on a few centuries one way or the other.
But even under these conditions we reach a period of 3,500

* Bunsen's ‘Egypt's place in the   † Martineau's ‘Eastern Life,’ vol. i.
World's History,’ vol. iv. p. 490.   p. 152, who quotes Herod. II. 99.

years, one which is not by any means too long for the number
and importance of the combinations which must have taken
place before a kingdom like that of Menes, with an established
language and religion—indeed with a regular hieroglyphical
and phonetic system of written characters—could have been
constituted.*

There has been much detraction both as to the antiquity
and character of the Egyptian civilisation and religious ideas.
Their claims to monotheism, or the worship of a Supreme God,
or God of gods, has been by some lost sight of in their
supposed adoration of the various inferior animals, such as
bulls and crocodiles.   But it can scarcely be doubted that
these were symbolical, and it is more than probable that there
was originally at least a deep meaning in such religious rites.

It is a valuable suggestion of a traveller in Egypt,† that
brute worship may have been a deification of instinct, or the
apparent working of the mind of the Creator through an unconscious medium.   Such an explanation is highly probable and
not unworthy of a people possessed of sufficient scientific knowledge to turn a river from its course and to prevent it by
embankments from inundating the sites of towns and cities,
who had already attained in those remote ages a peculiar style
of sculpture and architecture, hieroglyphics, and the practice
of embalming their dead, and who produced magnificent
temples and monuments, still unrivalled in size and grandeur.
In these temples are found pictorial representations of
battles and sieges, processions in which trophies are
carried and prisoners led captive; and if it be true, as
generally admitted, that throughout the historical period the
Egyptians were a peaceful and never a conquering people,
the wars to which these monuments would then refer must be
indefinitely remote.

* 'Egypt's place in the World's   † Miss Martineau's 'Eastern Life
History,' vol. iv. p. 15.   vol. ii. p. 57.

Nevertheless, geologically speaking, and in reference to the date of the first age of stone, these records of the valley of the Nile may be called extremely modern. Wherever excavations have been made into the Nile mud underlying the foundations of Egyptian cities, as, for example, sixty feet below the peristyle of the obelisk of Heliopolis, and generally in the alluvial plains of the Nile, the bones met with belong to living species of quadrupeds, such as the camel, dromedary, dog, ox, and pig, without, as yet, the association in any single instance of the teeth or bone of a lost species.

In like manner in all the countries bordering the Mediterranean, whether in Algeria, Spain, the south of France, Italy, Greece, Asia Minor, Sicily, or the islands of the Mediterranean generally, wherever the bones of extinct mammalia, such as the elephant, rhinoceros, and hippopotamus, have been found, it is not in the modern deltas of rivers or in the alluvial plains, now overflowed when the waters are high, that such fossil remains present themselves, but in situations corresponding to the ancient gravels of the valley of the Somme, in which the bones of the mammoth and the oldest type of flint implements occur.

If, therefore, the Egyptian monarch Necho, who sent an expedition to circumnavigate Africa, or some earlier king than he, had commanded his admiral to sail past the Pillars of Hercules, and then northwards as far as he could penetrate, leaving, before he set out on his return, some monument to · commemorate to after ages the Ultima Thule of his expedition at the most northern point reached by him, and if we had now discovered an obelisk of granite left by him at that era on the platform of St. Acheul, near Amiens, its foundations might well have occupied the precise position which the Gallo-Roman tombs now hold, as shown in *fig.* 28 (p. 183). If they had dug deep enough to exhume some teeth of the elephant, they might easily have seen that they differed from the teeth of their African species, and were distinct, like many other accompanying bones, from the animals then inhabiting the valley of the Somme, or that of the Nile.   The Paleolithic flint implements would then have lain buried in the old gravel as now, and the only geological distinction between those times and ours would be a diminished thickness of peat bordering the Somme, the upper layers of which would not contain, as now, Roman antiquities, and some beds below, in which hatchets called Celts now occur, might perhaps have been wanting; but, with this slight exception, the valley would have worn the same aspect as at the era when the Romans subdued Gaul.

PART III.

THE ORIGIN OF SPECIES AS BEARING UPON MAN'S PLACE IN NATURE.

—•—

CHAPTER XX.

THEORIES OF PROGRESSION AND TRANSMUTATION.

ANTIQUITY AND PERSISTENCY IN CHARACTER OF THE EXISTING RACES OF MANKIND—THEORY OF THEIR UNITY OF ORIGIN CONSIDERED—BEARING OF THE DIVERSITY OF RACES ON THE DOCTRINE OF TRANSMUTATION—DIFFICULTY OF DEFINING THE TERMS 'SPECIES' AND 'RACE' —LAMARCK'S INTRODUCTION OF THE ELEMENT OF TIME INTO THE DEFINITION OF A SPECIES—HIS THEORY OF VARIATION AND PROGRESSION—OBJECTIONS TO HIS THEORY, HOW FAR ANSWERED—ARGUMENTS OF MODERN WRITERS IN FAVOUR OF PROGRESSION IN THE ANIMAL AND VEGETABLE WORLD—THE OLD LANDMARKS SUPPOSED TO INDICATE THE FIRST APPEARANCE OF MAN, AND OF DIFFERENT CLASSES OF ANIMALS, FOUND TO BE ERRONEOUS—YET THE THEORY OF AN ADVANCING SERIES OF ORGANIC BEINGS NOT INCONSISTENT WITH FACTS—EARLIEST KNOWN FOSSIL MAMMALIA OF LOW GRADE—NO VERTEBRATA AS YET DISCOVERED IN THE OLDEST FOSSILIFEROUS ROCKS —OBJECTIONS TO THE THEORY OF PROGRESSION CONSIDERED—CAUSES OF THE POPULARITY OF THE DOCTRINE OF PROGRESSION AS COMPARED WITH THAT OF TRANSMUTATION.

WHEN speaking in a former work of the distinct races of mankind,[*] I remarked that, 'if all the leading varieties of the human family sprang originally from a single pair,' (a doctrine, to which then, as now, I could see no valid objection,) 'a much greater lapse of time was required for the

* 'Principles of Geology,' 7th ed., p. 637, 1847; see also 9th ed., p. 660. This passage, though altered in form, has been retained in substance in the 11th edition.

slow and gradual formation of such races as the Caucasian,
Mongolian, and Negro, than was embraced in any of the
popular systems of chronology.'

In confirmation of the high antiquity of two of these, I
referred to pictures on the walls of ancient temples in Egypt,
in which, a thousand years or more before the Christian era,
'the Negro and Caucasian physiognomies were portrayed
as faithfully, and in as strong contrast, as if the likenesses of
these races had been taken yesterday.' In relation to the
same subject, I dwelt on the slight modification which the
Negro has undergone, after having been transported from
the tropics, and settled for more than two centuries in the
temperate climate of Virginia. I therefore concluded that,
' if the various races were all descended from a single pair, we
must allow for a vast series of antecedent ages, in the course
of which the long-continued influence of external circumstances gave rise to peculiarities increased in many successive
generations, and at length fixed by hereditary transmission.'

So long as physiologists continued to believe that Man had
not existed on the earth above six thousand years, they
might, with good reason, withhold their assent from the
doctrine of a unity of origin of so many distinct races; but
the difficulty becomes less and less, exactly in proportion as
we enlarge our ideas of the lapse of time during which different communities may have spread slowly, and become
isolated, each exposed for ages to a peculiar set of conditions,
whether of temperature, or food, or danger, or ways of living.
The law of the geometrical rate of the increase of population
which causes it always to press hard on the means of subsistence, would ensure the migration, in various directions, of off-
shoots from the society first formed abandoning the area where
they had multiplied. But when they had gradually penetrated
to remote regions by land or water,—drifted sometimes by
storms and currents in canoes to an unknown shore,—barriers

of mountains, deserts, or seas, which oppose no obstacle to mutual intercourse between civilised nations, would ensure the complete isolation for tens or thousands of centuries of tribes in a primitive state of barbarism.

Some modern ethnologists, in accordance with the philosophers of antiquity, have assumed that men at first fed on the fruits of the earth, before even a stone implement or the simplest form of canoe had been invented. They may, it is said, have begun their career in some fertile island in the tropics, where the warmth of the air was such that no clothing was needed, and where there were no wild beasts to endanger their safety. But as soon as their numbers increased, they would be forced to migrate into regions less secure and blest with a less genial climate. Contests would soon arise for the possession of the most fertile lands, where game or pasture abounded, and their energies and inventive powers would be called forth, so that, at length, they would make progress in the arts.

But as ethnologists have failed, as yet, to trace back the history of any one race to the area where it originated, some zoologists of eminence have declared their belief, that the different races, whether they be three, five, twenty, or a much greater number, (for on this point there is an endless diversity of opinion,*) have all been primordial creations, having from the first been stamped with the characteristic features, mental and bodily, by which they are now distinguished, except where intermarriage has given rise to mixed or hybrid races. Were we to admit, say they, a unity of origin of such strongly marked varieties as the Negro and European, differing as they do in colour and bodily constitution, each fitted for distinct climates, and exhibiting some marked peculiarities in their osteological, and even in some details

* See Transactions of Ethnological Society, vol. i. 1861.

I F

of cranial and cerebral conformation, as well as in their
average intellectual endowments (see above, p. 96),—if, in
spite of the fact that all these attributes have been faithfully
handed down unaltered for hundreds of generations, we
are to believe that, in the course of time, they have all
diverged from one common stock, how shall we resist the
arguments of the transmutationist, who contends that all
closely allied species of animals and plants have in like
manner sprung from a common parentage, albeit that for
the last three or four thousand years they may have been
persistent in character? Where are we to stop, unless we
make our stand at once on the independent creation of those
distinct human races, the history of which is better known
to us than that of any of the inferior animals?

So long as Geology had not lifted up a part of the veil
which formerly concealed from the naturalist the history of
the changes which the animate creation had undergone in
times immediately antecedent to the Recent period, it was
easy to treat these questions as too transcendental, or as
lying too far beyond the domain of positive science to
require serious discussion. But it is no longer possible to
restrain curiosity from attempting to pry into the relations
which connect the present state of the animal and vegetable
worlds, as well as of the various races of mankind, with the
state of the fauna and flora which immediately preceded.

In the very outset of the enquiry, we are met with the
difficulty of defining what we mean by the terms ' species ' and
' race ; ' and the surprise of the unlearned is usually great,
when they discover how wide is the difference of opinion now
prevailing as to the significance of words in such familiar
use. But, in truth, we can come to no agreement as to such
definitions, unless we have previously made up our minds on
some of the most momentous of all the enigmas with which
the human intellect ever attempted to grapple.

It is now thirty years since I gave an analysis in the first edition of my 'Principles of Geology' (vol. ii. 1832) of the views which had been put forth by Lamarck, in the beginning of the century, on this subject. In that interval the progress made in zoology and botany, both in augmenting the number of known animals and plants, and in studying their physiology and geographical distribution, and, above all, in examining and describing fossil species, is so vast, that the additions made to our knowledge probably exceed all that was previously known; and what Lamarck then foretold has come to pass: the more new forms have been multiplied, the less are we able to decide what we mean by a variety, and what by a species. In fact, zoologists and botanists are not only more at a loss than ever how to define a species, but even to determine whether it has any real existence in nature, or is a mere abstraction of the human intellect, some contending that it is constant within certain narrow and impassable limits of variability, others that it is capable of indefinite and endless modification.

Before I attempt to explain a great step, which has recently been made by Mr. Darwin and his fellow-labourers in this field of enquiry, I think it useful to recapitulate here some of the leading features of Lamarck's system, without attempting to adjust the claims of some of his contemporaries (Geoffroy St. Hilaire in particular) to share in the credit of some of his original speculations.

From the time of Linnæus to the commencement of the present century, it seemed a sufficient definition of the term species to say, that 'a species consisted of individuals all resembling each other, and reproducing their like by generation.' But Lamarck, after having first studied botany with success, had then turned his attention to conchology, and soon became aware that in the newer (or tertiary) strata of the earth's crust there were a multitude of fossil species of shells,

some of them identical with living ones, others simply
varieties of the living, and which, as such, were entitled to
be designated, according to the ordinary rules of classification, by the same names. He also observed that other shells
were so nearly allied to living forms, that it was difficult not
to suspect that they had been connected by a common bond
of descent. He therefore proposed that the element of
time should enter into the definition of a species, and that it
should run thus: ' A species consists of individuals all resembling each other, and reproducing their like by generation, *so long as the surrounding conditions do not undergo
changes sufficient to cause their habits, characters, and forms
to vary.*' He came at last to the conclusion, that none of the
animals and plants now existing were primordial creations,
but were all derived from pre-existing forms, which, after
they may have gone on for indefinite ages reproducing their
like, had, at length, by the influence of alterations in climate
and in the animate world, been made to vary gradually,
and adapt themselves to new circumstances, some of them
deviating, in the course of ages, so far from their original
type as to have claims to be regarded as new species.

In support of these views, he referred to wild and cultivated plants, and to wild and domesticated animals, pointing
out how their colour, form, structure, physiological attributes, and even instincts, were gradually modified by exposure to new soils and climates, new enemies, modes of
subsistence, and kinds of food.

Nor did he omit to notice that the newly acquired peculiarities may be inherited by the offspring for an indefinite series
of generations, whether they be brought about naturally,—as
when a species, on the extreme verge of its geographical range,
comes into competition with new antagonists, and is subjected
to new physical conditions; or artificially,—as when, by the

act of the breeder or horticulturist, peculiar varieties of form or disposition are selected.

But Lamarck taught not only that species had been constantly undergoing changes from one geological period to another, but that there also had been a progressive advance of the organic world from the earliest to the latest times, from beings of the simplest to those of more and more complex structure, and from the lowest instincts up to the highest, and, finally, from brute intelligence to the reasoning powers of Man. The improvement in the grade of being had been slow and continuous, and the human race itself was at length evolved out of the most highly organised and endowed of the inferior mammalia.

In order to explain how, after an indefinite lapse of ages, so many of the lowest grades, of animal or plant, still abounded, he imagined that the germs or rudiments of living things, which he called monads, were continually coming into the world, and that there were different kinds of these monads for each primary division of the animal and vegetable kingdoms. This last hypothesis does not seem essentially different from the old doctrine of equivocal or spontaneous generation ; which, after all the able and conscientious experiments of Mr Bastian,* is still as much a matter of controversy as it was in the days of Lamarck.

Some of the laws which govern the appearance of new varieties were clearly pointed out by Lamarck. He remarked, for example, that as the muscles of the arm become strengthened by exercise or enfeebled by disuse, some organs may in this way, in the course of time, become entirely obsolete, and others previously weak become strong and play a new or more leading part in the organisation of a species. And so with instincts, where animals experience new

---

* Bastian. Beginnings of Life. 1872.

dangers they become more cautious and cunning, and trans-
mit these acquired faculties to their posterity. But not
satisfied with such legitimate speculations, the French
philosopher conceived that by repeated acts of volition
animals might acquire new organs and attributes, and that
in plants, which could not exert a will of their own, certain
subtle fluids or organising forces might operate so as to
work out analogous effects.

After commenting on these purely imaginary causes, I
pointed out in 1832, as the two great flaws in Lamarck's
attempt to explain the origin of species, first, that he had
failed to adduce a single instance of the initiation of a new
organ in any species of animal or plant; and secondly, that
variation, whether taking place in the course of nature or
assisted artificially by the breeder and horticulturist, had
never yet gone so far as to produce two races sufficiently
remote from each other in physiological constitution as to be
sterile when intermarried, or, if fertile, only capable of producing sterile hybrids, &c.*

To this objection Lamarck would, no doubt, have answered
that there had not been time for bringing about so great an
amount of variation ; for when Cuvier and some other of his
contemporaries appealed to the embalmed animals and plants
taken from Egyptian tombs, some of them 3,000 years old,
which had not experienced in that long period the slightest
modification in their specific characters, he replied that the
climate and soil of the valley of the Nile had not varied in the
interval, and that there was therefore no reason for expecting
that we should be able to detect any change in the fauna and
flora. ' But if,' he went on to say, ' the physical geography,
temperature, and other conditions of life, had been altered
in Egypt as much as we know from geology has happened in

* Principles of Geology, 1st ed., vol. ii. ch. ii.

other regions, some of the same animals and plants would have deviated so 'far from their pristine types as to be thought entitled to take rank as new and distinct species.'

Although I cited this answer of Lamarck, in my account of his theory,* I did not, at the time, fully appreciate the deep conviction which it displays of the slow manner in which geological changes have taken place, and the insignificance of thirty or forty centuries in the history of a species, and that, too, at a period when very narrow views were entertained of the extent of past time by most of the ablest geologists, and when great revolutions of the earth's crust, and its inhabitants, were generally attributed to sudden and violent catastrophes.

While, in 1832, I argued against Lamarck's doctrine of the gradual transmutation of one species into another, I agreed with him in believing that the system of changes now in progress in the organic world would afford, when fully understood, a complete key to the interpretation of all the vicissitudes of the living creation in past ages. I contended against the doctrine, then very popular, of the sudden destruction of vast multitudes of species, and the abrupt ushering into the world of new batches of plants and animals.

I endeavoured to sketch out (and it was, I believe, the first systematic attempt to accomplish such a task) the laws which govern the extinction of species, with a view of showing that the slow, but ceaseless, variations now in progress in physical geography, together with the migration of plants and animals into new regions, must, in the course of ages, give rise to the occasional loss of some of them, and eventually cause an entire fauna and flora to die out; also, that we must infer from geological data, that the places thus left vacant from time to time are filled up without delay by new forms,

* Principles of Geology, 1st ed., vol. ii. p. 567, 1832.

adapted to new conditions, sometimes by immigration from
adjoining provinces, sometimes by new creations. Among
the many causes of extinction enumerated by me, were the
power of hostile species, diminution of food, mutations in
climate, the conversion of land into sea, and of sea into land,
&c. I firmly opposed Brocchi's hypothesis, of a decline in
the vital energy of each species ; * maintaining that there
was every reason to believe that the reproductive powers of
the last surviving representatives of a species were as
vigorous as those of their predecessors, and that they were
as capable, under favourable circumstances, of repeopling the
earth with their kind. The manner in which some species
are now becoming scarce and dying out, one after the other,
appeared to me to favour the doctrine of the fixity of the
specific character, showing a want of pliancy and capability
of varying, which ensured their annihilation whenever changes
sufficiently adverse to their well-being occurred ; time not
being allowed for such a transformation as might be conceived capable of adapting them to the new circumstances,
and of converting them into what naturalists would call, new
species.†

But while rejecting transmutation, I was equally opposed
to the popular theory that the creative power had diminished
in energy, or that it had been in abeyance ever since Man had
entered upon the scene. That a renovating force, which had
been in full operation for millions of years, should cease to
act while the causes of extinction were still in full activity, or
even intensified by the accession of Man's destroying power,
seemed to me in the highest degree improbable. The only
point on which I doubted was, whether the force might not
be intermittent instead of being, as Lamarck supposed, in

* Principles of Geology, 1st ed.   † Laws of Extinction, Principles of
vol. ii, ch. viii. ; and 11th ed. vol. ii.   Geology, 1st ed. 1832, vol. ii. chap. v.
p. 270.   to xi. inclusive ; and 11th ed. p. 437.

ceaseless operation. Might not the births of new species, like
the deaths of old ones, be sudden? Might they not still escape our observation? If the coming in of one new species,
and the loss of one other which had endured for ages, should
take place annually, still, assuming that there are a million
of animals and plants living on the globe, it would require,
I observed, a million of years to bring about a complete
revolution in the fauna and flora. In that case, I imagined
that, although the first appearance of a new form might be as
abrupt as the disappearance of an old one, yet naturalists
might never yet have witnessed the first entrance on the stage
of a large and conspicuous animal or plant, and as to the
smaller kinds, many of them might be conceived to have
stolen in unseen, and to have spread gradually over a wide
area, like species migrating into new provinces.*

It may now be useful to offer some remarks on the very
different reception which the twin branches of Lamarck's
development theory, namely, progression and transmutation,
have met with, and to enquire into the causes of the popularity of the one, and the great unpopularity of the other.
We usually test the value of a scientific hypothesis by the
number and variety of the phenomena of which it offers a
fair or plausible explanation. If transmutation, when thus
tested, has decidedly the advantage over progression, and yet
is comparatively in disfavour, we may reasonably suspect that
its reception is retarded, not so much by its own inherent de-
merits, as by some apprehended consequences which it is
supposed to involve, and which run counter to our preconceived opinions.

Theory of Progression.

In treating of this question, I shall begin with the doctrine
of progression, a concise statement of which, so far as it relates

* Principles of Geology, 1st ed. 1832, vol. ii. ch. xi.

to the animal kingdom, was thus given forty-two years ago by Professor Sedgwick, in his anniversary address to the Geological Society of London in 1831.

'In the repeated and almost entire changes of organic types,' he says, 'in the successive strata of the earth; in the absence of mammalia in the older, and their very rare appearance in the newer secondary groups; in the diffusion of warm-blooded quadrupeds (frequently of unknown genera) in the older tertiary system, and in their great abundance (and frequently of known genera) in the upper portions of the same series; and lastly, in the recent appearance of Man on the surface of the earth' . . . we have a series of proofs the most emphatic and convincing that the approach to the present system of things has been gradual, and that there has been a progressive development of organic structure subservient to the purposes of life.

'This historical development,' writes the same author, in 1850, 'of the forms and functions of organic life during successive epochs, seems to mark a gradual evolution of creative power, manifested by a gradual ascent towards a higher type of being.' 'But the elevation of the fauna of successive periods was not made by transmutation, but by creative additions; and it is by watching these additions that we get some insight into Nature's true historical progress, and learn that there was a time when Cephalopoda were the highest types of animal life, the primates of this world; that Fishes next took the lead, then Reptiles; and that during the secondary period they were anatomically raised far above any forms of the reptile class now living in the world. Mammals were added next, until Nature became what she now is, by the addition of Man.'[*]

It cannot be too much insisted upon, when considering such generalizations as this of the late Woodwardian Professor,

* Professor Sedgwick's Discourse   Cambridge, Preface to 5th ed. pp. xliv,
on the Studies of the University of   cliv. ccxvi. 1850.

that although palæontological research ever since the days of
Lamarck has been carrying back to earlier and earlier dates
the appearance of the successive animal types upon the earth,
and leading us to assign a higher antiquity both to Man and
the older fossil mammalia, fish, and reptiles, yet the chronological succession as here laid down holds good in all essential
particulars.

The progressive theory was propounded in the following
terms by the late Hugh Miller in his 'Footprints of the
Creator.'

'It is of itself an extraordinary fact without reference to
other considerations, that the order adopted by Cuvier in his
" Animal Kingdom," as that in which the four great classes of
vertebrate animals, when marshalled according to their rank
and standing, naturally range, should be also that in which
they occur in order of time. The brain which bears an
average proportion to the spinal cord of not more than two
to one, comes first,—it is the brain of the fish; that which
bears to the spinal cord an average proportion of two and a
half to one succeeded it,—it is the brain of the reptile; then
came the brain averaging as three to one,—it is that of the
bird. Next in succession came the brain that averages as
four to one,—it is that of the mammal; and last of all there
appeared a brain that averages as twenty-three to one,—
reasoning, calculating Man had come upon the scene.'*

Without insisting on the perfect accuracy of these
cerebral proportions, which may be only approximations to
the truth, the development in chronological order may, I
believe, be accepted as here stated as a fair argument in
support of the theory of progression.

M. Agassiz, in his Essay on Classification, has devoted a
chapter to the ' Parallelism between the Geological Succession

* Footprints of the Creator, p. 283. Edinburgh, 1849.

of Animals and Plants and their present relative Standing ; '
in which he has expressed a decided opinion that, within the
limits of the orders of each great class, there is a coincidence
between their relative rank in organisation and the order
of succession of their representatives in time.[*]

Professor Owen, in his Palæontology, advanced similar
views, and remarked, in regard to the vertebrata, that there
is much positive as well as negative evidence in support of
the doctrine of an advance in the scale of being, from ancient
to more modern geological periods. We observe, for example,
in the triassic, oolitic, and cretaceous strata, not only an
absence of placental mammalia, but the presence of innumerable reptiles, some of large size, terrestrial and aquatic,
herbivorous and predaceous, fitted to perform the functions
now discharged by the mammalia.

The late Professor Bronn, of Heidelberg, after passing in
review more than 24,000 fossil animals and plants, which he
had classified and referred each to their geological position
in his ' Index Palæontologicus,' came to the conclusion that,
in the course of time, there had been introduced into the
earth more and more highly organised types of animal and
vegetable life ; the modern species being, on the whole, more
specialised, i.e. having separate organs, or parts of the body,
to perform different functions, which, in the earlier periods
and in beings of simpler structure, were discharged in common by a single part or organ.

Professor Adolphe Brongniart, in an essay published in
1849, on the botanical classification and geological distribution of the genera of fossil plants,[†] arrives at similar results
as to the progress of the vegetable world from the earliest
periods to the present.   He does not pretend to trace an

* Contributions to Natural History of United States, Part I.—Essay
on Classification, p. 108.

† Tableau des Genres de Végétaux
fossiles, &c. Dictionnaire Universel
d'Histoire Naturelle. Paris, 1849.

exact historical series from the sea-weed to the fern, or from the fern again to the conifers and cycads, and lastly, from those families to the palms and oaks, but he, nevertheless, points out that the cryptogamic forms, especially the acrogens, predominate among the fossils of the primary formations, the carboniferous especially, while the gymnosperms or coniferous and cycadeous plants abound in all the strata, from the Trias to the Wealden inclusive; and lastly, the more highly developed angiosperms, both monocotyledonous and dicotyledonous, do not become abundant until the tertiary period. It is a remarkable fact, as he justly observes, that the angiospermous exogens, which comprise four-fifths of living plants, —a division to which all our native European trees, except the Coniferæ, belong, and which embrace all the Compositæ, Leguminosæ, Umbelliferæ, Cruciferæ, Heaths, and so many other families,—are wholly unrepresented by any fossils hitherto discovered in the primary and secondary formations from the Silurian to the Oolitic inclusive. It is not till we arrive at the cretaceous period that they begin to appear, sparingly at first, and only playing a conspicuous part, together with the palms and other endogens, in the tertiary epoch.

When commenting on the eagerness with which the doctrine of progression was embraced from the close of the last century to the time when I first attempted, in 1830, to give some account of the prevailing theories in geology, I observed, that far too much reliance was commonly placed on the received dates of the first appearances of certain orders or classes of animals or plants, such dates being determined by the age of the stratum in which we then happened to have discovered the earliest memorials of such types. At that time (1830), it was taken for granted that Man had not coexisted with the mammoth and other extinct mammalia, yet now that we have traced back the signs of his existence to the Pleistocene era, and may even anticipate the finding of his remains on some future day in the Pliocene period, the theory of progression is not shaken.

In the beginning of this century it was one of the canons of the popular geological creed, that the first warm-blooded quadrupeds which had inhabited this planet were those derived from the Eocene gypsum of Montmartre in the suburbs of Paris, almost all of which Cuvier had shown to belong to extinct genera. This dogma continued in force for more than a quarter of a century, in spite of the discovery in 1818 of a marsupial quadruped in the Stonesfield strata, a member of the lower oolite, near Oxford. Some disputed the authority of Cuvier himself, as to the mammalian character of the fossil; others the accuracy of those who had assigned to it so ancient a place in the chronological series of rocks. Since that period great additions have been made to our knowledge of the existence of land quadrupeds in the olden times. We have ascertained that, in Eocene strata older than the gypsum of Paris, no less than four distinct sets of placental mammalia have flourished; namely, first, those of the Headon series in the Isle of Wight, from which fourteen species have been procured; secondly, those of the antecedent Bagshot and Bracklesham beds, which have yielded, together with the contemporaneous 'calcaire grossier' of Paris, twenty species; thirdly, the still older beds of Kyson, near Ipswich, and those of Herne Bay, at the mouth of the Thames, in which seven species have been found; and fourthly, the plastic clay or lignite formation, which has supplied ten species.[*]

We can scarcely doubt that we should already have traced back the evidence of this class of fossils much farther had not our enquiries been arrested, first, by the vast gap between the tertiary and secondary formations, and then by the marine nature of the cretaceous rocks.

* Lyell's Student's Elements, 1871, p. 306.

The mammalia next in antiquity, of which we have any
cognisance, are those of the upper oolite of Purbeck, discovered between the years 1854 and 1857, and comprising
no less than fourteen species, referable to eight or nine
genera; one of them, *Plagiaulax*, considered by Dr. Falconer
to have been a herbivorous marsupial. The whole assemblage appears, from the joint observations of Professor Owen
and Dr. Falconer, to indicate a low grade of quadruped, probably of the marsupial type. They were, for the most part,
diminutive, the two largest not much exceeding our common
hedgehog and polecat in size.

Next anterior in age are the mammalia of the Lower Oolite
of Stonesfield, of which four species are known, also very
small, and probably marsupial, with one exception, the
Stereognathus ooliticus, which, according to Professor Owen's
conjecture, may have been a hoofed quadruped and placental, though, as we have only half of the lower jaw with
teeth, and the molars are unlike any living type, such an
opinion is, of course, hazarded with due caution.

Still older than the above are some fossil quadrupeds of
small size, found in the Upper Trias of Stuttgart in Germany, and more lately by Mr. C. Moore in beds of corresponding age near Frome, which are also of a very low grade,
like the living myrmecobius of Australia. Beyond this limit
our knowledge of the highest class of vertebrata does not as yet
extend into the past, but the frequent shifting back of the old
land-marks, nearly all of them once supposed in their turn to
indicate the date of the first appearance of warm-blooded
quadrupeds on this planet, should serve as a warning to us
not to consider the goal at present reached by palæontology
as one beyond which they who come after us are never
destined to pass.

On the other hand, it may be truly said, in favour of progression, that, after all these discoveries, the doctrine is not

gainsaid, for the less advanced marsupials precede the more
perfect placental mammalia in the order of their appearance on the earth.

If the three localities where the most ancient mammalia
have been found,—Purbeck, Stonesfield and Stuttgart—had
belonged all of them to formations of the same age, we
might well have imagined so limited an area to have been
peopled exclusively with pouched quadrupeds, just as Australia now is, while other parts of the globe were inhabited
by placentals, for Australia now supports one hundred and
sixty species of marsupials, while the rest of the continents
and islands are tenanted by about seventeen hundred species
of mammalia, of which only forty-six are marsupial, namely,
the opossums of North and South America. But the great
difference of age of the strata in each of these three localities
seems to indicate the predominance throughout a vast lapse
of time, (from the era of the Upper Trias to that of the
Purbeck beds,) of a low grade of quadrupeds; and this persistency of similar generic and ordinal types in Europe while
the species were changing, and while the fish, reptiles, and
mollusca were undergoing vast modifications, raises a strong
presumption that there was also a vast extension in space of
the same marsupial forms during that portion of the secondary
epoch which has been termed ' the age of reptiles.'

As to the class Reptilia, some of the orders which prevailed when the secondary rocks were formed are confessedly
much higher in their organisation than any of the same
class now living. If the less perfect ophidians, or snakes,
which now abound on the earth had taken the lead in those
ancient days among the land reptiles, and the Deinosaurians
had been contemporary with Man, there can be no doubt
that the progressionist would have seized upon this fact with
unfeigned satisfaction as confirmatory of his views. Now
that the order of succession is precisely reversed, and that

the age of the Iguanodon was long anterior to that of the
Eocene Palæophis and living boa, while the crocodile is in
our own times the highest representative of its class, a retrograde movement in this important division of the vertebrata
must be admitted. It may perhaps be accounted for by the
power acquired by the placental mammalia, when they
became dominant, a power before which the class of vertebrata next below them, as coming most directly in competition with them, may, more than any other, have given
way.

For no less than thirty-four years it had been a received
axiom in palæontology, that reptiles had never existed before
the Permian or Magnesian limestone period, when at length,
in 1844, this supposed barrier was thrown down, and carboniferous reptiles, terrestrial and aquatic, of several genera,
were brought to light. Still, no traces of this class have
yet been detected in rocks as ancient as those in which the
oldest fish have been found.

As to fossil representatives of the ichthyic type, the most
ancient were not supposed, before 1838, to be of a date
anterior to the Coal, but they have since been traced
back, first to the Devonian, and then to the Upper Silurian
rocks. No remains, however, of them or of any vertebrate
animal have yet been discovered in the Lower Silurian strata,
rich as these are in invertebrate fossils, nor in the still older
primordial zone of Barrande, still less in the Cambrian or
Laurentian rocks; so that we seem authorised to conclude,
though not without considerable reserve, that the vertebrate type was extremely scarce, if not wholly wanting,
in those epochs often spoken of as 'primitive,' but which, if
the Development Theory be true, were probably the last
of a long series of antecedent ages in which living beings
flourished.

As to the Mollusca, which afford the most unbroken series of geological medals, the highest of that class, the cephalopoda, abounded in older Silurian times, comprising several hundred species of chambered univalves. Had there been strong prepossessions against the progressive theory, it would probably have been argued that when these cephalopods abounded, and the siphonated gasteropods were absent, a higher order of zoophagous mollusca discharged the functions afterwards performed by an inferior order in the secondary, tertiary, and post-tertiary seas. But I have never seen this view suggested as adverse to the doctrine of progress, although much stress has been laid on the fact, that the Silurian brachiopoda, creatures of a lower grade, formerly discharged the functions of the' existing lamellibranchiate bivalves, which are higher in the scale.

It is said truly that the ammonite, orthoceras, and nautilus of these ancient rocks were of the tetrabranchiate division, and none of them so highly organised as the belemnite and other dibranchiate cephalopods which afterwards appeared, and some of which now flourish in our seas. Therefore, we may infer that the simplest forms of the cephalopoda took precedence of the more complex in time. But if we embrace this view, we must not forget that there are living cephalopoda, such as the octopods, which are devoid of any hard parts, whether external or internal, and which could leave behind them no fossil memorials of their existence; so that we must make a somewhat arbitrary assumption, namely, that at a remote era, no such dibranchiata were in being, in order to avail ourselves of this argument in favour of progression. On the other hand, it is true that in the ' primordial zone ' of Barrande not even the shell-bearing tetrabranchiates whose hard parts might have been preserved have yet been discovered.

In regard to plants, although the generalisation, above cited, of M. Adolphe Brongniart (p. 444) is probably true, there has been a tendency in the advocates of progression to push the inferences deducible from known facts, in support of their favourite dogma, somewhat beyond the limits which the evidence justifies. Dr. Hooker observes, in his recent introductory essay on the flora of Australia, that it is impossible to establish a parallel between the successive appearances of vegetable forms in time, and their complexity of structure or specialisation of organs as represented by the successively higher groups in the natural method of classification. He also adds that the earliest recognisable cryptogams are not only the highest now existing, but have more highly differentiated vegetative organs than any subsequently appearing, and that the dicotyledonous embryo and perfect exogenous wood, with the highest specialised tissue known (the coniferous with glandular tissue), preceded the monocotyledonous embryo and endogenous wood in date of appearance on the globe—facts wholly opposed to the doctrine of progression, and which can only be set aside on the supposition that they are fragmentary evidence of a time farther removed from the origin of vegetation than from the present day.*

It would be an easy task to multiply objections to the theory now under consideration ; but from this I refrain, as I regard it not only as a useful, but rather, in the present state of science, as an indispensable hypothesis, and one which, though destined hereafter to undergo many and great modifications, will never be overthrown.

It may be thought almost paradoxical that writers who are most in favour of transmutation (Mr. Darwin and Dr. J. Hooker, for example) are nevertheless among those who are most cautious, and one would say timid, in their mode of espousing the doctrine of progression ; while, on the other hand, the most zealous advocates of progression are oftener than

* Flora of Australia, Introductory Essay, p. xxi. London, 1859. Published separately.

n u 2

not very vehement opponents of transmutation. We might
have anticipated a contrary leaning on the part of both, for
to what does the theory of progression point? It supposes
a gradual elevation in grade of the vertebrate type, in the
course of ages, from the most simple ichthyic form to that
of the placental mammalia and the coming upon the stage
last in the order of time of the most anthropomorphous
mammalia, followed by the human race—this last thus
appearing as an integral part of the same continuous series of
acts of development, one link in the same chain, the crowning
operation as it were of one and the same series of manifesta-
tions of creative power. If the dangers apprehended from
transmutation arise from the too intimate connection which
it tends to establish between the human and merely animal
natures, it might have been expected that the progressive
development of organisation, instinct, and intelligence might
have been unpopular, as likely to pioneer the way for the reception of the less favoured doctrine. But the true explanation of the seeming anomaly is this, that no one can believe
in transmutation who is not profoundly convinced that all
we know in palæontology is as nothing compared with what
we have yet to learn, and they who regard the record as so
fragmentary, and our acquaintance with the fragments which
are extant as so rudimentary, are apt to be astounded at
the confidence placed by the progressionists in data which
must be defective in the extreme. But exactly in proportion as the completeness of the record and our knowledge of
it are overrated, in that same degree are many progressionists
unconscious of the goal towards which they are drifting.
Their faith in the fullness of the annals leads them to
regard all breaks in the series of organic existence, or in
the sequence of the fossiliferous rocks, as proofs of original
chasms and leaps in the course of nature,—signs of the intermittent action of the creational force, or of catastrophes which

devastated the habitable surface.   They do not doubt that
there is a continuity of plan, but they believe that it exists
in the Divine mind alone, and they are therefore without
apprehension that any facts will be discovered which would
imply a material connection between the outgoing organisms
and the incoming ones.

CHAPTER XXI.

ON THE ORIGIN OF SPECIES BY VARIATION AND NATURAL SELECTION.

MR. DARWIN'S THEORY OF THE ORIGIN OF SPECIES BY NATURAL SELECTION—MEMOIR BY MR. WALLACE—MANNER IN WHICH FAVOURED RACES PREVAIL IN THE STRUGGLE FOR EXISTENCE—FORMATION OF NEW RACES BY BREEDING—HYPOTHESES OF DEFINITE AND INDEFINITE MODIFIABILITY EQUALLY ARBITRARY—COMPETITION AND EXTINCTION OF RACES — PROGRESSION NOT A NECESSARY ACCOMPANIMENT OF VARIATION — DISTINCT CLASSES OF PHENOMENA WHICH NATURAL SELECTION EXPLAINS — UNITY OF TYPE, RUDIMENTARY ORGANS, GEOGRAPHICAL DISTRIBUTION, RELATION OF THE EXTINCT TO THE LIVING FAUNA AND FLORA, AND MUTUAL RELATIONS OF SUCCESSIVE GROUPS OF FOSSIL FORMS—LIGHT THROWN ON EMBRYOLOGICAL DEVELOPMENT BY NATURAL SELECTION—WHY LARGE GENERA HAVE MORE VARIABLE SPECIES THAN SMALL ONES—DR. HOOKER ON THE EVIDENCE AFFORDED BY THE VEGETABLE KINGDOM IN FAVOUR OF CREATION BY VARIATION —STEENSTRUP ON ALTERNATE GENERATION—HOW FAR THE DOCTRINE OF INDEPENDENT CREATION IS OPPOSED TO THE LAWS NOW GOVERNING THE MIGRATION OF SPECIES.

FOR many years after the promulgation of Lamarck's doctrine of progressive development, geologists were much occupied with the question whether the past changes in the animate and inanimate world were brought about by sudden and paroxysmal action, or gradually and continuously, by causes differing neither in kind nor degree from those now in operation.

An anonymous author published in 1844 'The Vestiges of Creation,' a treatise, written in a clear and attractive style, which made the English public familiar with the leading views of Lamarck on transmutation and progression, but brought no new facts or original line of argument to support those views, or to combat the principal objections which the scientific world entertained against them.

No decided step in this direction was made until the publication in 1858 of two papers, one by Mr. Darwin and another by Mr. Wallace, followed in 1859 by Mr. Darwin's celebrated work on 'The Origin of Species by Means of Natural Selection; or, the Preservation of favoured Races in the Struggle for Life.' The author of this treatise had for twenty previous years strongly inclined to believe that variation and the ordinary laws of reproduction were among the secondary causes always employed by the Author of Nature, in the introduction from time to time of new species into the world, and he had devoted himself patiently to the collecting of facts, and making of experiments in zoology and botany, with a view of testing the soundness of the theory of transmutation. Part of the MS. of his projected work was read to Dr. Hooker as early as 1844, and some of the principal results were communicated to me on several occasions. Dr. Hooker and I had repeatedly urged him to publish without delay, but in vain, as he was always unwilling to interrupt the course of his investigations; until at length Mr. Alfred R. Wallace, who had been engaged for years in collecting and studying the animals of the East Indian archipelago, thought out, independently for himself, one of the most novel and important of Mr. Darwin's theories. This he embodied in an essay 'On the Tendency of Varieties to depart indefinitely from the original Type.' It was written at Ternate, in February 1858, and sent to Mr. Darwin, with a request that it might be shown to me if thought sufficiently novel and interesting. Dr. Hooker and I were of opinion that it should be immediately printed, and we succeeded in persuading Mr. Darwin to allow one of the MS. chapters of is 'Origin of Species,' entitled 'On the Tendency of Species to form Varieties, and on the Perpetuation of Species and

Varieties by natural Means of Selection,' to appear at the
same time.*

By reference to these memoirs it will be seen that both
writers begin by applying to the animal and vegetable worlds
the Malthusian doctrine of population, or its tendency to increase in a geometrical ratio, while food can only be made to
augment even locally in an arithmetical one. There being,
therefore, no room or means of subsistence for a large proportion of the plants and animals which are born into the
world, a great number must annually perish. Hence there
is a constant struggle for existence among the individuals
which represent each species, and the vast majority can
never reach the adult state, to say nothing of the multitudes
of ova and seeds, which are never hatched or allowed to
germinate. Of birds it is estimated that the number of
those which die every year equals the aggregate number by
which the species to which they respectively belong is on the
average permanently represented.

The trial of strength, which must decide what individuals
are to survive and what to succumb, occurs in the season
when the means of subsistence are fewest, or enemies most
numerous, or when the individuals are enfeebled by climate
or other causes; and it is then that those varieties which
have any, even the slightest, advantage over others come off
victorious. They may often owe their safety to what would
seem to a casual observer a trifling difference, such as a darker
or lighter shade of colour rendering them less visible to a
species which preys upon them, or sometimes to attributes
more obviously advantageous, such as greater cunning, or
superior powers of flight or swiftness of foot. These peculiar
qualities and faculties, bodily and instinctive, may enable
them to outlive their less favoured rivals, and being trans-

* See Proceedings of Linnæan Society, 1858.

mitted by the force of inheritance to their offspring, will
constitute new races, or what Mr. Darwin calls 'incipient
species.' If one variety, being in other respects just equal
to its competitors, happens to be more prolific, some of its
offspring will stand a greater chance of being among those
which will escape destruction, and their descendants, being
in like manner very fertile, will continue to multiply at the
expense of all less prolific varieties.

As breeders of domestic animals, when they choose certain
varieties in preference to others to breed from, speak techni-
cally of their method as that of 'selecting,' Mr. Darwin calls
the combination of natural causes, which may enable certain
varieties of wild animals or plants to prevail over others of
the same species, 'natural selection.'

A breeder finds that a new race of cattle with short horns
or without horns may be formed, in the course of several
generations, by choosing varieties having the most stunted
horns as his stock from which to breed ; so nature, by altering,
in the course of ages, the conditions of life, the geographical
features of a country, its climate, the associated plants and
animals, and, consequently, the food and enemies of a species
and its mode of life, may be said, by this means, to select
certain varieties best adapted for the new state of things.
Such new races may often supplant the original type from
which they have diverged, although that type may have been
perpetuated without modification for countless anterior ages
in the same region, so long as it was in harmony with the
surrounding conditions then prevailing.

Lamarck, when speculating on the origin of the long neck
of the giraffe, imagined that quadruped to have stretched
himself up in order to reach the boughs of lofty trees, until
by continued efforts, and longing to reach higher, he obtained
an elongated neck. Mr. Darwin and Mr. Wallace simply
suppose that, in a season of scarcity, the longer-necked indi-

viduals, having the advantage in this respect over the rest of
the herd, as being able to browse on foliage out of their
reach, survived them, and transmitted their peculiarity of
cervical conformation to their successors.

By the multiplying of slight modifications in the course
of thousands of generations, and by the handing down of
the newly-acquired peculiarities by inheritance, a greater and
greater divergence from the original standard is supposed to
be effected, until what may be called a new species, or, in a
greater lapse of time, a new genus, will be the result.

Every naturalist admits that there is a general tendency in
animals and plants to vary ; but it is usually taken for granted,
though we have no means of proving the assumption to be
true, that there are certain limits beyond which each species
cannot pass under any circumstances, or in any number of
generations. Mr. Darwin and Mr. Wallace say that the
opposite hypothesis, which assumes that every species is
capable of varying indefinitely from its original type, is not
a whit more arbitrary, and has this manifest claim to be pre-
ferred, that it will account for a multitude of phenomena
which the ordinary theory is incapable of explaining.

We have no right, they say, to assume, should we find that
a variable species can no longer be made to vary in a certain
direction, that it has reached the utmost limit to which it
might, under more favourable conditions, or if more time
were allowed, be made to diverge from the parent type, and
this view is supported by the fact, that our oldest domestic
animals and cultivated plants, those which have varied most
widely from the original parent stock, still continue to produce new varieties and show no sign whatever of ceasing to
vary.

Hybridisation is not considered by Mr. Darwin as a cause
of new species, but rather as tending to keep variation within bounds. Varieties which are nearly allied cross readily

with each other, and with the parent stock, and such crossing tends to keep the species true to its type, while forms
which are less nearly related, although they may intermarry,
produce no mule offspring capable of perpetuating their kind.

The competition of races and species, observes Mr. Darwin,
is always most severe between those which are most closely
allied and which fill nearly the same place in the economy of
nature. Hence, when the conditions of existence are modified, the original stock runs great risk of being superseded
by some one of its modified offshoots. The new race or
species may not be absolutely superior in the sum of its
powers and endowments to the parent stock, and may even be
more simple in structure and of a lower grade of intelligence,
as well as of organisation, provided, on the whole, it happens
to have some slight advantage over its rivals. Progression,
therefore, is not a necessary accompaniment of variation and
natural selection, though, when a higher organisation happens to be coincident with superior fitness to new conditions,
the new species will have greater power and a greater chance
of permanently maintaining and extending its ground. One
of the principal claims of Mr. Darwin's theory to acceptance
is, that it enables us to dispense with a law of progression
as a necessary accompaniment of variation. It will account
equally well for what is called degradation, or a retrograde
movement towards a simpler structure, and does not require
Lamarck's continual creation of monads; for this was a
necessary part of his system, in order to explain how, after
the progressive power had been at work for myriads of ages,
there were as many beings of the simplest structure in existence as ever.

Mr. Darwin argues, and with no small success, that
all true classification in zoology and botany is, in fact,
genealogical, and that community of descent is the hidden
bond which naturalists have been unconsciously seeking,

while they often imagined that they were looking for some
unknown plan of creation.

As the 'Origin of Species'[*] is in itself a condensed
abstract of a much larger work not yet published, I could
not easily give an analysis of its contents within narrower
limits than those of the original, but it may be useful to
enumerate briefly some of the principal classes of phenomena
on which the theory of 'Natural Selection' would throw
light.

In the first place, it would explain, says Mr. Darwin, the
unity of type which runs through the whole organic world,
and why there is sometimes a fundamental agreement in
structure in the same class of beings which is quite independent of their habits of life, for such structure, derived by
inheritance from a remote progenitor, has been modified, in
the course of ages, in different ways, according to the conditions of existence. It would also explain why all living and
extinct beings are united, by complex radiating and circuitous
lines of affinity with one another, into one grand system;[†]
also, there having been a continued extinction of old races
and species in progress, and a formation of new ones by variation, why in some genera which are largely represented, or to
which a great many species belong, many of these are closely
but unequally related; also, why there are distinct geographical
provinces of species of animals and plants, for, after long
isolation by physical barriers, each fauna and flora, by varying
continually, must become distinct from its ancestral type,
and from the new forms assumed by other descendants which
have diverged from the same stock.

The theory of indefinite modification would also explain
why rudimentary organs are so useful in classification, being
the remnants preserved by inheritance of organs which the

* Origin of Species, 6th ed. Introduction, p. 1.
† Ibid. p. 417.

ancestors of the present species once used—as in the case of
the rudiments of eyes in insects and reptiles inhabiting dark
caverns, or of the wings of birds and beetles which have lost
all power of flight. In such cases the affinities of species are
often more readily discerned by reference to these imperfect
structures than by others of much more physiological importance to the individuals themselves.

The same hypothesis would explain why there are no mammalia in islands far from continents, except bats, which can
reach them by flying ; and also why the birds, insects, plants,
and other inhabitants of islands, even when specifically
unlike, usually agree generically with those of the nearest
continent, it being assumed that the original stock of such
species came by migration from the nearest land.

Variation and natural selection would also afford a key to a
multitude of geological facts otherwise wholly unaccounted
for, as, for example, why there is generally an intimate connection between the living animals and plants of each great
division of the globe and the extinct fauna and flora of the
post-tertiary or tertiary formations of the same region ; as, for
example, in North America, where we not only find among the
living mollusca peculiar forms foreign to Europe, such as Gna-
thodon and Fulgar (a subgenus of Fusus), but meet also with
extinct species of those same genera in the tertiary fauna of the
same part of the world. In like manner, among the mammalia
we find in Australia not only living kangaroos and wombats,
but fossil individuals of extinct species of the same genera.
So also there are recent and fossil sloths, armadilloes, and other
edentata in South America, and living and extinct species
of elephant, rhinoceros, tiger, and bear in the great Europeo-
Asiatic continent. The theory of the origin of new species
by variation will also explain why a species which has once
died out never reappears, and why the fossil fauna and flora
recede farther and farther from the living type in propor-

tion as we trace it back to remoter ages. It would also
account for the fact, that when we have to intercalate a new
set of fossiliferous strata between two groups previously
known, the newly-discovered fossils serve to fill up gaps
between specific or generic types previously familiar to us,
supplying often the missing links of the chain, which, if
transmutation is accepted, must once have been continuous.

One of the most original speculations in Mr. Darwin's
work is derived from the fact that, in the breeding of
animals, it is often observed that at whatever age any variation first appears in the parent, it tends to reappear at a
corresponding age in the offspring. Hence the young individuals of two races which have sprung from the same
parent stock are usually more like each other than the
adults. Thus the puppies of the greyhound and bull-dog
are much more nearly alike in their proportions than the
grown-up dogs, and in like manner the foals of the cart and
racehorse than the adult individuals. For the same reason
we may understand why the species of the same genus,
or genera of the same family, resemble each other more
nearly in their embryonic than in their more fully developed
state, or how it is that in the eyes of most naturalists the
structure of the embryo is even more important in classification than that of the adult, 'for the embryo is the animal in
its less modified state, and in so far it reveals the structure
of its progenitor. In two groups of animals, however much
they may at present differ from each other in structure and
habits, if they pass through the same or similar embryonic
stages, we may feel assured that they have both descended
from the same or nearly similar parents, and are therefore in
that degree closely related. Thus community in embryonic
structure reveals community of descent, however much the
structure of the adult may have been modified.'*

* Darwin, Origin, &c. 8th ed. pp. 388-398.

If then there had been a system of progressive development, the successive changes through which the embryo of a species of a high class, a mammifer, for example, now passes, may be expected to present us with a picture of the stages through which, in the course of ages, that class of animals has successively passed in advancing from a lower to a higher grade. Hence the embryonic states exhibited one after the other by the human individual bear a certain amount of resemblance to those of the fish, reptile, and bird before assuming those of the highest division of the vertebrata.

Mr. Darwin, after making a laborious analysis of many floras, found that those genera which are represented by a large number of species contain a greater number of variable species, relatively speaking, than the smaller genera, or those less numerously represented. This fact he adduces in support of his opinion that varieties are incipient species, for he observes that the existence of the larger genera implies, in the period immediately preceding our own, that the manufacturing of species has been active, in which case we ought generally to find the same forces still in full activity, more especially as we have every reason to believe the process by which new species are produced is a slow one.[*]

Dr. Hooker tells us that he was long disposed to doubt this result, as he was acquainted with so many variable small genera, but after examining Mr. Darwin's data, he was compelled to acquiesce in his generalisation.[†]

It is one of those conclusions, to verify which requires the investigation of many thousands of species, and to which exceptions may easily be adduced, both in the animal and vegetable kingdoms, so that it will be long before we can expect it to be thoroughly tested, and, if true, fairly appreciated. Among the most striking exceptions will be some

[*] Origin of Species, 6th ed. ch. ii.
[†] Introductory Essay on Flora of Australia, p. vi.

genera still large, but which are beginning to decrease, the
conditions which were favourable to their former predominance having already begun to change.   To many, this doctrine of Natural Selection, or 'the preservation of favoured
races in the struggle for life,' seems so simple, when once
clearly stated, and so consonant with known facts and
received principles, that they have difficulty in conceiving
how it can constitute a great step in the progress of science.
Such is often the case with important discoveries, but in
order to assure ourselves that the doctrine was by no means
obvious, we have only to refer back to the writings of skilful
naturalists who attempted, in the earlier part of the nineteenth century, to theorise on this subject, before the invention of this new method of explaining how certain forms
are supplanted by new ones, and in what manner these
last are selected out of innumerable varieties, and rendered
permanent.

*Dr. Hooker, on the Theory of ' Creation by Variation' as
applied to the Vegetable Kingdom.*

Of Dr. Hooker, whom I have often cited in this chapter,
Mr. Darwin has spoken in the Introduction to his 'Origin of
Species,' as one ' who had, for fifteen years, aided him in every
possible way, by his large stores of knowledge, and his excellent judgment.'   This distinguished botanist published his
' Introductory Essay to the Flora of Australia ' * in December
1859, the year after the memoir on ' Natural Selection ' was
communicated to the Linnæan Society, and a month after
the appearance of the ' Origin of Species.'

Having, in the course of his extensive travels, studied the
botany of arctic, temperate, and tropical regions, and written on the flora of India, which he had examined at all

* Introductory Essay, &c.   Lovell Reeve, London, 1859.

heights above the sea, from the plains of Bengal to the limits
of perpetual snow in the Himalaya, and having especially
devoted his attention to 'geographical varieties,' or those
changes of character which plants exhibit, when traced over
wide areas and seen under new conditions; being also prac-
tically versed in the description and classification of new
plants, from various parts of the world, and having been
called upon carefully to consider the claims of thousands of
varieties to rank as species, no one was better qualified by
observation and reflection to give an authoritative opinion on
the question, whether the present vegetation of the globe is
or is not in accordance with the theory which Mr. Darwin
has proposed. We cannot but feel, therefore, deeply inter-
ested when we find him making the following declaration :

'The mutual relations of the plants of each great botanical
province, and, in fact, of the world generally, are just such as
would have resulted if variation had gone on operating
throughout indefinite periods, in the same manner as we see
it act in a limited number of centuries, so as gradually to give
rise in the course of time, to the most widely divergent forms.'

In the same Essay, this author remarks, 'The element of
mutability pervades the whole Vegetable Kingdom ; no class,
nor order, nor genus of more than a few species claims absolute exemption from it, whilst the grand total of unstable
forms, generally assumed to be species, probably exceeds
that of the stable.' Yet he contends that species are neither
visionary, nor even arbitrary creations of the naturalist, but
realities, though they may not remain true for ever (p. 11).
The majority of them, he remarks, are so far constant,
'within the range of our experience,' and their forms and
characters so faithfully handed down, through thousands of
generations, that they admit of being treated as if they were
permanent and immutable. But the range of our experience is so limited, that it will 'not account for a single fact

in the present geographical distribution, or the origin of any one species of plant, nor for the amount of variation it has undergone, nor will it indicate the time when it first appeared, nor the form it had when created.' *

To what an extent the limits of species are indefinable, is evinced, he says, by the singular fact that, among those botanists who believe them to be immutable, the number of flowering plants is by some assumed to be 80,000, and by others over 150,000. The general limitation of species to certain areas, suggests the idea that each of them, with all their varieties, have sprung from a common parent, and have spread in various directions from a common centre. The frequency also of the grouping of genera within certain geographical limits, is in favour of the same law, although the migration of species may sometimes cause apparent exceptions to the rule, and make the same types appear to have originated independently at different spots.†

Certain genera of plants, which, like the brambles, roses, and willows in Europe, consist of a continuous series of varieties between the terms of which no intermediate forms can be intercalated, may be supposed to be newer types and on the increase, and therefore undergoing much variation; whereas genera which present no such perplexing gradations may be of older date, and may have been losing species and varieties by extinction. In this case, the annihilation of intermediate forms which once existed, makes it an easy task to distinguish those which remain.

It had usually been supposed by the advocates of the immutability of species, that domesticated races, if allowed to run wild, always revert to their parent type. Mr. Wallace had said in reply, that a domesticated species, if it loses the protection of Man, can only stand its ground in a wild state

---

* Hooker, Introductory Essay, Flora of Australia.   † Ibid. p. 13.

by resuming those habits, and recovering those attributes, which it may have lost when under domestication. If these faculties are so much enfeebled as to be irrecoverable, it will perish ; if not, and if it can adapt itself to the surrounding conditions, it will revert to the state in which Man first found it : for in one, two, or three thousand years, which may have elapsed since it was originally tamed, there will not have been time for such geographical, climatal, and organic changes, as would only be suited to a new race, or a new and allied species.

But in regard to plants, Dr. Hooker questions the fact of reversion. According to him, species in general do not readily vary ; but when they once begin to do so, the new varieties, as every horticulturist knows, show a great inclination to go on departing more and more from the old stock. As the best marked varieties of a wild species occur on the confines of the area which it inhabits, so the best marked varieties of a cultivated plant, are those last produced by the gardener. Cabbages, for example, wall fruit, and cerealia, show no disposition, when neglected, to assume the characters of the wild states of these plants. Hence the difficulty of determining what are the true parent species of most of our cultivated plants. Thus the finer kinds of apples, if grown from seed, degenerate and become crabs, but in so doing they do not revert to the original wild crab-apple, but become crab states of the varieties to which they belong.*

It would lead me into too long a digression, were I to attempt to give a fuller analysis of this admirable essay ; but I may add, that none of the observations are more in point, as bearing on the doctrine of what Dr. Hooker terms ' creation by variation,' than the great extent to which the internal characters and properties of plants, or their physiological

* Introductory Essay, Flora of Australia, p. ix.

H H 2

constitution, are capable of being modified, while they exhibit
externally no visible departure from the normal form. Thus,
in one region a species may possess peculiar medicinal qualities which it wants in another, or it may be hardier and better
able to resist cold. The average range in altitude, says Dr.
Hooker, of each species of flowering plant in the Himalayan
Mountains, whether in the tropical, temperate, or Alpine
region, is 4,000 feet, which is equivalent to twelve degrees
of isothermals of latitude. If an individual of any of these
species be taken from the upper limits of its range and
carried to England, it is found to be better able to stand our
climate than those from the lower or warmer stations.
When several of these internal or physiological modifications
are accompanied by variation in size, habits of growth, colour
of the flowers, and other external characters, and these are
found to be constant in successive generations, botanists may
well begin to differ in opinion as to whether they ought to
regard them as distinct species or not.

Alternate Generation.

Hitherto, no rival hypothesis has been proposed as a sub-
stitute for the doctrine of transmutation; for what we
term 'independent creation,' or the direct intervention of the
Supreme Cause, must simply be considered as an avowal that
we deem the question to lie beyond the domain of science.

The discovery by Steenstrup of alternate generation
enlarges our views of the range of metamorphosis through
which a species may pass, so that some of its stages (as when
a Sertularia and a Medusa interchange) deviate so far from
others as to have been referred by able zoologists to distinct
genera, or even families. But in all these cases the organism,
after running through a certain cycle of change, returns to
the exact point from which it set out, and no new form or

species is thereby introduced into the world. The only secondary cause, therefore, which has, as yet, been even conjecturally brought forward, to explain how, in the ordinary course of nature, a new specific form may be generated is, as Lamarck declared, 'variation,' and this has been rendered a far more probable hypothesis by the way in which Natural Selection is shown to give intensity and permanency to certain varieties.

Independent Creation.

When I formerly advocated the doctrine that species were primordial creations, and not derivative, I endeavoured to explain the manner of their geographical distribution, and the affinity of living forms to the fossil types nearest akin to them in the tertiary strata of the same part of the globe, by supposing that the creative power, which originally adapts certain types to aquatic and others to terrestrial conditions, has, at successive geological epochs, introduced new forms best suited to each area and climate, so as to fill the places of those which may have died out.

In that case, although the new species would differ from the old (for these would not be revived, having been already proved, by the fact of their extinction, to be incapable of holding their ground), still, they would resemble their predecessors generically. For, as Mr. Darwin states in regard to new races, those of a dominant type inherit the advantages which made their parent species flourish in the same country, and they likewise partake in those general advantages which made the genus to which the parent species belonged, a large genus in its own country.

We might, therefore, by parity of reasoning, have anticipated that the creative power, adapting the new types to the new combination of organic and inorganic conditions of a given region, such as its soil, climate, and inhabitants, would

introduce new modifications of the old types,—marsupials,
for example, in Australia, new sloths and armadilloes in South
America, new heaths at the Cape, new roses in the northern,
and new calceolarias in the southern hemisphere.   But to
this line of argument Mr. Darwin and Dr. Hooker reply, that
when animals or plants migrate into new countries, whether
assisted by man, or without his aid, the most successful
colonisers appertain by no means to those types which are
most allied to the old indigenous species.   On the contrary,
it more frequently happens that members of genera, orders,
or even classes, distinct and foreign to the invaded country,
make their way most rapidly, and become dominant at the
expense of the endemic species.   Such is the case with the
placental quadrupeds in Australia, and with horses and many
foreign plants in the pampas of South America, and number-
less instances in the United States and elsewhere, which
might easily be enumerated.   Hence, the transmutationists
infer that, the reason why these foreign types, so peculiarly
fitted for these regions, have never before been developed
there, is simply that they were excluded by natural barriers.
But these barriers of sea, or desert, or mountain, could never
have been of the least avail, had the creative force acted
independently of material laws, or had it not pleased the
Author of Nature that the origin of new species should be
governed by some secondary causes analogous to those which
we see preside over the appearance of new varieties, which
never appear except as the offspring of a parent stock very
closely resembling them.

CHAPTER XXII.

OBJECTIONS TO THE HYPOTHESIS OF TRANSMUTATION CONSIDERED.

Theory of Transmutation—Absence of Intermediate Links.

THE most obvious and popular of the objections urged against the theory of transmutation may be thus expressed : If the extinct species of plants and animals of the later geological periods were the progenitors of the living species, and gave origin to them by variation and natural selection, where are all the intermediate forms, fossil and living, through which the lost types must have passed during their conversion into the living ones?   And why do we not find almost everywhere passages between the nearest allied

species and genera, instead of such strong lines of demarcation, and often wide intervening gaps?

We may consider this objection under two heads:—

First, To what extent are the gradational links really
wanting in the living creation or in the fossil world, and how
far we may expect to discover such as are missing by future
research?

Secondly, Are the gaps more numerous than we ought
to anticipate, allowing for the original defective state of the
geological records, their subsequent dilapidation, and our
slight acquaintance with such parts of them as are extant,
and allowing also for the rate of extinction of races and
species now going on, and which has been going on since the
commencement of the tertiary period?

First, As to the alleged absence of intermediate varieties
connecting one species with another; every zoologist and
botanist who has engaged in the task of classification has
been occasionally thrown into this dilemma,—if I make
more than one species in this group, I must, to be consistent,
make a great many. Even in a limited region like the British
Isles, this embarrassment is continually felt.

Scarcely any two botanists, for example, can agree as to
the number of roses, still less as to how many species of
bramble we possess. Of the latter genus, *Rubus*, there is
one set of forms, respecting which it is still a question
whether it ought to be regarded as constituting three species
or thirty-seven. Mr. Bentham adopts the first alternative,
and Mr. Babington the second, in their well-known treatises
on British plants.

We learn from Dr. Hooker that at the antipodes, both in
New Zealand and Australia, this same genus *Rubus* is represented by several species rich in individuals and remarkable
for their variability. When we consider how, as we extend
our knowledge of the same plant over a wider area, new

geographical varieties commonly present themselves, and
then endeavour to imagine the number of forms of the genus
Rubus which may now exist, or probably have existed, in
Europe and in regions intervening between Europe and
Australia, comprehending all which may have flourished in
tertiary and post-tertiary periods, we shall perceive how little
stress should be laid on arguments founded on the assumed
absence of missing links in the flora as it now exists.

If in the battle of life the competition is keenest between
closely allied varieties and species, as Mr. Darwin contends,
many forms can never be of long duration, nor have a wide
range, and these must often pass away without leaving behind
them any fossil memorials. In this manner we may account
for many breaks in the series which no future researches will
ever fill up.

Davidson on Fossil Brachiopoda.

It is from fossil conchology more than from any other
department of the organic world that we may hope to derive
traces of a transition from certain types to others, and fossil
memorials of all the intermediate shades of form. We may
especially hope to gain this information from the study of
some of the lower groups, such as the *Brachiopoda,* which are
persistent in type, so that the thread of our enquiry is
less likely to be interrupted by breaks in the sequence
of the fossiliferous rocks. The splendid monograph by
Mr. Davidson, on the British Brachiopoda, illustrates, in
the first place, the tendency of certain generic forms in this
division of the mollusca to be persistent throughout the
whole range of geological time yet known to us ; for the four
genera *Rhynchonella, Crania, Discina,* and *Lingula* have
been traced through the Silurian, Devonian, Carboniferous,
Permian, Jurassic, Cretaceous, Tertiary, and Recent periods,

and still retain in the existing seas the identical shape and
character which they exhibited in the earliest formations.
On the other hand, other brachiopoda have gone through in
shorter periods a vast series of transformations, so that
distinct specific, and even generic names have been given to
the same varying form, according to the different aspects
and characters it has put on in successive sets of strata.

In proportion as materials for comparison have accumulated, the necessity of uniting species, previously regarded as distinct, under one denomination has become
more and more apparent. Mr. Davidson, accordingly, after
studying not less than 260 reputed species from the British
carboniferous rocks, has been obliged to reduce that number to 100, to which he has added 20 species either entirely
new, or new to the British strata; but he declares his conviction that, when our knowledge of these 120 brachiopoda
is more complete, a further reduction of species will take
place.

Speaking of one of these forms, which he calls *Spirifera
trigonalis*, he says that it is so dissimilar to another extreme
of the series, S. *crassa*, that in the first part of his memoir
(published some ten years ago) he described them as distinct,
and the idea of confounding them together must, he admits,
appear absurd to those who have never seen the intermediate
links, such as are presented by S. *bisulcata*, and at least four
others with their varieties, most of them shells formerly
recognised as distinct by the most eminent palæontologists,
but respecting which these same authorities now agree with
Mr. Davidson in uniting them into one species.*

The same species has sometimes continued to exist under
slightly modified forms throughout the whole of the Lower
and Upper Silurian as well as the entire Devonian and Car-

* Monograph on British Brachiopoda, Palæontographical Society. p. 222.

boniferous periods, as in the case of the shell generally known
as *Leptæna depressa*, which we must now call, in obedience
to the law of priority of nomenclature, *Strophomena
rhomboidalis* Wahlenberg. No less than fifteen commonly received species are demonstrated by Mr. Davidson, by
the aid of a long series of transitional forms, to appertain to
this one type; and it is acknowledged by some of the best
writers that they were induced, on purely theoretical grounds,
to give distinct names to some of the varieties now suppressed,
merely because they found them in rocks so widely remote
in time, that they deemed it contrary to analogy to suppose
that the same species could have endured so long: a fallacious
mode of reasoning, analogous to that which leads some zoologists and botanists to distinguish by specific names slight
varieties of living plants and animals met with in very remote
countries, as in Europe and Australia, for example; it being
assumed that each species has had a single birth-place or
area of creation, and that they could not by migration have
gone from the northern to the southern hemisphere across
the intervening tropics.

Examples are also given by Mr. Davidson of species which
pass from the Devonian into the Carboniferous, and from that
again into the Permian rocks. The vast longevity of such
specific forms has not been generally recognised in consequence of the change of names, which they have undergone
when derived from such distant formations, as when *Atrypa
unguicularis* assumes, when derived from a carboniferous
rock, the name of *Spirifera Urii*, besides several other synonyms, and then when it reaches the Permian period, takes the
name of *Spirifera Clannyana* King; all of which forms the
author of the monograph, now under consideration, asserts to
be one and the same.

No geologist will deny that the distance of time which
separates some of the eras above alluded to, or the dates of

the earliest and latest appearances of some of the fossils
above mentioned, must be reckoned by millions of years.
According to Mr. Darwin's views, it is only by having at our
command the records of such enormous periods, that we can
expect to be able to point out the gradations which unite
very distinct specific forms. But the advocate of transmutation must not be disappointed if, when he has succeeded in
obtaining some of the proofs which he was challenged to produce, they make no impression on the mind of his opponent.
All that will be conceded is that specific variation in the
Brachiopoda, at least, has a wider range than was formerly
suspected. So long as several allied species were brought
nearer and nearer to each other, considerable uneasiness might
have been felt as to the reality of species in general; but when
fifteen or more are once fairly merged in one group, constituting in the aggregate a single species, one and indivisible,
and capable of being readily distinguished from every other
group at present known, all misgivings are at an end. Implicit
trust in the immutability of species is then restored, and the
more insensible the shades from one extreme to the other—in
a word, the more complete the evidence of transition—the
more nugatory does the argument derived from it appear.
It then simply resolves itself into one of those exceptional
instances of what is called a protean form.

Thirty years ago a great London dealer in shells, himself
an able naturalist, told me that there was nothing he had so
much reason to dread, as tending to depreciate his stock in
trade, as the appearance of a good monograph on some large
genus of mollusca; for, in proportion as the work was executed
in a philosophical spirit, it was sure to injure him, every
reputed species pronounced to be a mere variety becoming
from that time unsaleable. Fortunately, so much progress
has since been made in England in estimating the true ends
and aims of science, that specimens indicating a passage

between forms usually separated by wide gaps, whether in the recent or fossil fauna, are now eagerly sought for, and often more prized than the mere normal or typical forms.

It is clear, that the more ancient the existing mollusca, or the farther back into the past we can trace the remains of shells still living, the more easy it becomes to reconcile with the doctrine of transmutation the distinctness in character of the majority of living species. For, what we want is time, first, for the gradual formation, and then for the extinction of races and allied species, occasioning gaps between the survivors.

In the year 1830, I announced, on the authority of M. Deshayes, that about one fifth of the mollusca of the Falunian or Upper Miocene strata of Europe, belonged to living species. Although the soundness of that conclusion was afterwards called in question by two or three eminent conchologists (and by the late M. Alcide d'Orbigny among others), it has been since confirmed by the majority of living naturalists, and is well borne out by the copious evidence on the subject laid before the public in the magnificent work edited by M. Hörnes, and published under the auspices of the Austrian Government, 'On the Fossil Shells of the Vienna Basin.'

The collection of tertiary shells from which those descriptions and beautiful figures were taken is almost unexampled for the fine state of preservation of the specimens, and the care with which all the varieties have been compared. It is now admitted that about one third of these Miocene forms, univalves and bivalves included, agree specifically with living mollusca, so that much more than the enormous interval which divides the Miocene from the Recent period must be taken into our account when we speculate on the origin by transmutation of the shells now living, and the disappearance by extinction of intermediate varieties and species.

Miocene Plants and Insects related to recent Species.

Geologists were acquainted with about three hundred species of marine shells from the ' Falunian' strata on the banks of the Loire, before they knew anything of the contemporary insects and plants. At length, as if to warn us against inferring from negative evidence the poverty of any ancient set of strata in organic remains proper to the land, a rich flora and entomological fauna were suddenly revealed to us characteristic of Central Europe during the Upper Miocene period. This result followed the determination of the true position of the Oeninghen beds in Switzerland, and of certain formations of ' Brown Coal ' in Germany.

Professor Heer, who has described nearly five hundred species of fossil plants from Oeninghen, besides many more from other Miocene localities in Switzerland,[*] estimates the phenogamous species, which must have flourished in Central Europe at that time, at 3,000, and the insects as having been more numerous in the same proportion as they now exceed the plants in all latitudes. This European Miocene flora was remarkable for the preponderance of arborescent and shrubby evergreens, and comprised many generic types no longer associated together in any existing flora or geographical province. Some genera, for example, which are at present restricted to America, coexisted in Switzerland with forms now peculiar to Asia, and with others at present confined to Australia.

Professor Heer has not ventured to identify any of this vast assemblage of Miocene plants and insects with living species, so far at least as to assign to them the same specific names, but he presents us with a list of what he terms

* Heer, Flora Tertiaria Helvetiae, 1859; and Gaudin's French translation, with additions, 1861.

homologous forms, which are so like the living ones, that he
supposes the one to have been derived genealogically from
the others. He hesitates indeed as to the manner of the
transformation, or the precise nature of the relationship,
'whether the changes were brought about by some influence
exerted continually for ages, or whether at some given
moment the old types were struck with a new image.'

Among the homologous plants alluded to are forty species,
of which both the leaves and fruits are preserved, and 'thirty
others, known at present by their leaves only. In the first
list we find many American types, such as the tulip tree
(*Liriodendron*), the deciduous cypress (*Taxodium*), the red
maple, and others, together with Japanese forms, such as a
cinnamon, which is very abundant. And what is worthy of
notice, some of these fossils so closely allied to living plants
occur not only in the Upper, but even some few of them as
far back in time as the Lower Miocene formations of Switzerland and Germany, which are probably as distant from the
Upper Miocene or Oeninghen beds as are the latter from our
own era.

Some of the fossil plants to which Professor Heer has
given new names have been regarded as recent species by
other eminent naturalists. Thus, one of the trees allied to
the elm had been called by Unger *Planera Richardi*, a species
which now flourishes in the Caucasus and Crete. Professor
Heer had attempted to distinguish it from the living tree by
the greater size of its fruit, but this character he confessed
did not hold good, when he had an opportunity (1861) of
comparing all the varieties of the living *Planera Richardi*
which Dr. Hooker laid before him in the rich herbarium
of Kew.

As to the 'homologous insects' of the Upper Miocene
period in Switzerland, we find among them, mingled with
genera now wholly foreign to Europe, some very fami-

liar forms, such as the common glowworm, *Lampyris nocti-
luca*, Linn., the dung-beetle, *Geotrupes stercorarius*, Linn.,
the ladybird, *Coccinella septempunctata*, Linn., the ear-
wig, *Forficula auricularia*, Linn., some of our common
dragon-flies, as *Libellula depressa*, Linn., the honey-bee,
Apis mellifera, Linn., the cuckoo spittle insect, *Aphrophora
spumaria*, Linn., and a long catalogue of others, to all of
which Professor Heer has given new names, but which some
entomologists may regard as mere varieties until some
stronger reasons are adduced for coming to a contrary
opinion.

Several of the insects above enumerated, like the common ladybird, are well known at present to have a very wide
range, over nearly the whole of the Old World for example,
without varying, and might, therefore, be expected to have
been persistent throughout many successive changes of the
earth's surface and climate. Yet we may fairly anticipate
that even the most constant types will have undergone some
modifications in passing from the Miocene to the Recent
epoch, since in the former period the geography and climate
of Europe, the heights of the Alps, and the general fauna and
flora were so different from what they now are. But the
deviation may not exceed that which would generally be
expressed by what is called a well-marked variety.

Before I pass on to another topic, it may be well to answer
a question which may have occurred to the reader ; how it
happens that we remained so long ignorant of the vegetation
and insects of the Upper Miocene period in Europe? The
answer may be instructive to those who are in the habit of un-
derrating the former richness of the organic world wherever
they happen to have no evidence of its condition. A large part
of the Upper Miocene insects and plants alluded to have been
met with at Ooninghen, near the Lake of Constance, in two or
three spots embedded in thinly laminated marls, the entire

thickness of which scarcely exceeds three or four feet, and in two quarries of very limited dimensions. The rare combination of causes which have led to the faithful preservation of so many treasures of a perishable nature in so small an area, appear to have been the following: first, a river flowing into a lake; secondly, storms of wind, by which 'leaves, and sometimes the boughs of trees, were torn off, and floated by the stream into the lake; and thirdly, a constant supply of carbonate of lime in solution from mineral springs, the calcareous matter, when precipitated to the bottom, mingling with fine mud, and thus forming the fossiliferous marls.

In his work entitled 'The Naturalist of the Amazons,' Mr. Bates mentions having observed on the Tapajos river in Brazil the dead or half-dead bodies of ants heaped up in a line an inch or two in height and breadth for miles along the beach.[*] He also informs me that on the sandy shores of Lake Ega, on the Upper Amazons, he saw on several occasions sloping ridges of dead insects of all orders piled up on the margin of the lake. This sudden destruction of whole shoals of insects is caused, he says, by a sudden chill and squall occurring in the night over a wide expanse of water, after a hot evening. The insects are tempted to fly by the sultry weather, the chill and storm overtake them, and they are cast into the water, the waves of which wash their bodies on to the lee-shore. Sand is also often thrown up at the same time, and some of the insects are thus buried a little above the water-line.

Species of Insects in Britain and North America, represented by distinct Varieties.

If we compare the living British insects with those of the American continent, we frequently find that even those species which are considered to be identical are, neverthe-

* Naturalist of the Amazons, 1863, vol. ii. p. 85.

I I

less, varieties of the European types. I have noticed this
fact when speaking of the common English butterfly, *Vanessa
atalanta*, or ' red admirable,' which I saw flying about the
woods of Alabama in mid-winter. I was unable to detect
any difference myself; but all the American specimens which
I took to the British Museum were observed by Mr. Double-
day to exhibit a slight peculiarity in the colouring of a
minute part of the anterior wing,* a character first detected
by Mr. T. F. Stephens, who has also discovered that similar
slight, but equally constant variations distinguish other Lepi-
doptera now inhabiting the opposite sides of the Atlantic,
insects which, nevertheless, he and Mr. Westwood and
the late Mr. Kirby have always agreed to regard as mere
varieties of the same species.

Mr. T. V. Wollaston, in treating of the variation of insects
in maritime situations and small islands, has shown how the
colour, growth of the wings, and many other characters,
undergo modification under the influence of local conditions,
continued for long periods of time; † and Mr. Edwin Brown
has lately called our attention to the fact, that the insects of
the Shetland Isles present slight deviations from the corresponding types occurring in Great Britain, but far less marked
than those which distinguish the American from the European
varieties.‡ In the case of Shetland, Mr. Brown remarks, a
land communication may well be supposed to have prevailed
with Scotland at a more modern era than that between
Europe and America. In fact, we have seen that Shetland
can hardly fail to have been united with Scotland after the
commencement of the Glacial period (see map, p. 328);
whereas a communication between the north of Europe by
Iceland and Greenland (which, as before stated, once enjoyed

---

* Lyell's Second Visit to the   Species, &c. London, Van Voorst, 1856.
United States, vol. ii. p. 293.   ‡ Transactions of Northern Ento-
† Wollaston, on the Variation of   mological Society, 1862.

a genial climate), must have been anterior to the Glacial
epoch. A much larger isolation, and the impossibility of
varieties formed in the two separated areas crossing with each
other, would account, according to Mr. Darwin's theory, for
the much wider divergence observed in the specific types of
the two regions.

The reader will remember that at the commencement of
the Glacial period there was scarcely any appreciable difference between the molluscous fauna and that now living.
When therefore the events of the Glacial period, as described
in the earlier part of this volume, are duly pondered on, and
when we reflect that in the Upper Miocene formations the
living species of Mollusca constitute only one third of the
whole fauna, we see clearly by how high a figure we must
multiply the time in order to express the distance between
the Miocene period and our own days.

Forms intermediate between Reptiles and Birds.

It was formerly assumed, upon negative evidence, that
birds did not exist at the time of the formation of the
secondary rocks, until, in the year 1858, Mr. Lucas Barrett
found in the upper greensand of the cretaceous series, near
Cambridge, the femur, tibia, and some other bones of a
swimming bird rather larger than a pigeon, supposed by him
to be of the gull tribe, in which opinion he was afterwards
confirmed by Professor Owen.

Four years later (in 1862) a skeleton of a bird almost
entire was found in the great quarries of lithographic limestone at Pappenheim, near Solenhofen, in Bavaria, the rock
being a member of the Upper Oolite. This valuable
specimen is now in the British Museum, and has been named
by Professor Owen *Archæopteryx macrura.* It was at first
conjectured in Germany, before any experienced osteologist

had had an opportunity of inspecting it, that this fossil might be a feathered pterodactyl, flying reptiles having been often met with in the same stratum. But Professor Owen has shown* that it is a true bird, although in the length of the bones of the tail, and some minor points of its anatomy, it approaches more nearly to reptiles than any known living representative of the class Aves.

In all living birds the tail-feathers are arranged in fan-shaped order and attached to a coccygian bone, consisting of several vertebræ united together, although in the embryo state these same vertebræ are distinct. The greatest number is seen in the ostrich, which has eighteen caudal vertebræ in the fœtal state, which are reduced to nine in the adult bird, many of them having been anchylosed together. But in the Archæopteryx the tail, which is eleven and a half inches long and three and a half broad, is composed of twenty vertebræ, each of which supports a pair of quill feathers. Professor Owen therefore considers the tail of the Archæopteryx as exemplifying the persistency of what is now an embryonic character. The Archæopteryx also differs from all known birds in having two, if not three digits in the wing; but there is no trace of the fifth digit of the winged reptile. The head was at first supposed to be wanting; but Mr. Evans detected on the slab what seems to be the impression of the cranium and beak, much resembling in size and shape that of the jay or woodcock.

Another and more decided link between the classes Aves and Reptilia is presented by a reptile about two feet long, called *Compsognathus longipes*, also found in the Stonesfield slate. This small extinct reptile had, according to Mr. Huxley, a slight bird-like head, provided with numerous teeth; the hind limbs were large and disposed as in birds, and

* Proc. Roy. Soc., vol. xvi. p. 243.

the femur was shorter than the tibia, a circumstance in which Compsognathus is more ornithic than even the ordinary Dinosauria (Iguanodon, &c.), which nevertheless approach birds in their general structure. It is impossible, says Professor Huxley, 'to look at the conformation of this strange reptile, and to doubt that it hopped or walked, in an erect or semi-erect position, after the manner of a bird, to which its long neck, slight head, and small anterior limbs, must have given it an extraordinary resemblance.' *

A third example of a form intermediate between birds and reptiles has lately (1872) been described by Professor O. C. Marsh, of Yale College, Connecticut, from the Upper Cretaceous shale of Kansas, under the name of *Ichthyornis dispar*. The fossil is that of an adult individual, about the size of a pigeon; the scapular arch and the bones of the wings and legs all conform closely to the true ornithic type, and the whole of the skeleton is truly bird-like, with the exception that the vertebrae are all biconcave, and that there are well-developed teeth in both jaws. These teeth are implanted in distinct sockets, and those in the lower jaws number about twenty in each ramus. So completely did the portions of the lower jaws first discovered resemble those of reptiles, that Professor Marsh described them as reptilian in a previous paper, before the remaining parts of the jaw and other portions of the skeleton were laid bare by removing the surrounding shale.†

Thus, in the space of thirteen years, two specimens of reptile-birds and one of a bird-reptile have been brought to light; and when we remember that these forms do not merely present us with transitions between species or genera, or even orders, but that they link together two distinct classes of the

* Animals Intermediate between Birds and Reptiles, Royal Inst. Lecture, Feb. 7, 1868.   † American Journ. of Science, vol. v. Feb. 1873.

Animal Kingdom, it cannot be denied that we have already
made no small progress in lessening the force of that part of the
argument against transmutation which consists in the absence
of intermediate types. On the other hand, the exceptional
occurrence of these single specimens proves how easily the
study of different formations might be carried on for years
without one intermediate link being discovered, and warns
us of the danger of arguing upon negative evidence without
taking fully into account the extreme imperfection of the geological record presently to be insisted upon. It is not simply
that new formations are brought to light from year to year,
reminding us of the elementary state of our knowledge of
palæontology, but new types also of structure are discovered
in rocks the fossil contents of which were supposed to be
peculiarly well known.

The theory of transmutation, as first enunciated by
Lamarck, was impugned on the ground that no adequate causes
were adduced which could bring about the necessary modifications. Mr. Darwin, by the theory of Variation and Natural
Selection, supplied these causes; but still the opponents
of transmutation urged that no proofs were to be obtained in
the fossil world of those transitions which are assumed to have
taken place. These proofs we now see are gradually presenting themselves, few and far between, as might be expected
when the numerous causes of destruction at work in successive
geological ages are taken into account; although, as good
observers multiply in distant lands, we may expect that the
discovery of missing links will be more frequent.

Species of Mammalia Recent and Fossil.

But it may perhaps be said that the Mammalia afford more
conspicuous examples than do the plants, mollusca, insects,
or even reptiles and birds, of the wide gaps which separate

species and genera, and that if in this higher class such a
multitude of transitional forms had ever existed as would be
required to unite the Tertiary and recent species into one
series or network of allied or transitional forms, they could
not so entirely have escaped observation, whether in the fossil
or living fauna. A zoologist who entertains such an opinion
would do well to devote himself to the study of some one
genus of Mammalia, such as the elephant, rhinoceros, hippopotamus, bear, horse, ox, or deer; and after collecting all
the materials he can get together respecting the extinct and
recent species, decide for himself whether the present state
of science justifies his assuming that the chain could never
have been continuous, the number of the missing links being
so great.

Among the extinct species formerly contemporary with
man, no fossil quadruped has so often been alluded to in this
work as the mammoth, *Elephas primigenius*. From a monograph on the proboscidians by Dr. Falconer, it appears that this
species represents one extreme of a type of which the Pliocene
Mastodon Borsoni represents the other. Between these
extremes there are already enumerated by Dr. Falconer no
less than twenty-six species, some of them ranging as far
back in time as the Miocene period, others still living, like
the Indian and African forms. Two of these species, however, he has always considered as doubtful, *Stegodon Ganesa*,
probably a mere variety of one of the others, and *Elephas
priscus* of Goldfuss, founded partly on specimens of the
African elephant assumed by mistake to be fossil, and partly
on some aberrant forms of *E. antiquus*.

The first effect of the intercalation of so many intermediate forms between the two most divergent types has been
to break down almost entirely the generic distinction between Mastodon and Elephant. Dr. Falconer, indeed, observes that Stegodon (one of several subgenera which he

has founded) constitutes an intermediate group, from which
the other species diverge through their dental characters,
on the one side into the Mastodons, and on the other
into the Elephants.* The next result is to diminish the
distance between the several members of each of these
groups.

Dr. Falconer discovered that no less than four species
of elephant were formerly confounded together under the
title of *Elephas primigenius*, whence its supposed ubiquity
in post-Pliocene times, or its wide range over half the
habitable globe. But even when this form has been thus
restricted in its specific characters, it has still its geographical
varieties; for the mammoth's teeth brought from America
may in most instances, according to Dr. Falconer, be distinguished from those proper to Europe. On this American
variety Dr. Leidy has conferred the name of *E. Americanus*.
Another form allied to the mammoth (*E. meridionalis*)
existed, as we have seen, before the Glacial period, or
at the time when the buried forest of Cromer and the
Norfolk cliffs (see above, p. 258) were deposited; and the
Swiss geologists have lately found remains of the mammoth
in their country, both in pre-glacial and post-glacial formations.

Since the publication of Dr. Falconer's monograph, two
other species of elephant, *E. mirificus*, Leidy, and *E. im-
perator*, have been obtained from the Pliocene formations of
the Niobrara Valley in Nebraska, one of which, however, may
possibly be found hereafter to be the same as *E. Columbi*,
Fale.

The fossil remains of no less than three distinct species
of elephant have been discovered in the island of Malta, in
caverns, by Admiral Spratt and Dr. Leith Adams. Two of
these species are of comparatively small size, though one is

* Geological Quarterly Journal, vol. xiii. p. 314, 1857.

rather larger than the other ; the third must have equalled
the African elephant in stature. The two dwarf species were
included by Dr. Falconer under the common name of
Elephas Melitensis, but they have since been shown by Mr.
Busk* to be distinct, the smaller being named by him
E. Falconeri. It would appear that while *E. Melitensis* in
its teeth and certain parts of the skeleton approaches the
Indian type, *E. Falconeri,* as pointed out by Dr. Falconer,
exhibits decidedly African affinities. The third or large
species, though distinct from *E. antiquus,* presents characters resembling it.

How much the difficulty of discriminating between the
fossil representatives of this genus may hereafter augment,
when all the species with their respective geographical
varieties are known, may be inferred from the following
fact :—Professor H. Schlegel, in a recently published memoir,
endeavours to show that the living elephant of Sumatra
agrees with that of Ceylon, but is a distinct species from that
of Continental India, being distinguishable by the number
of its dorsal vertebræ and ribs, the form of its teeth, and
other characteristics.† Dr. Falconer, on the other hand,
considers these two living species as mere geographical
varieties, the characters referred to not being constant, as
he has ascertained on comparing different individuals of
E. Indicus in different parts of Bengal, in which the ribs
vary from nineteen to twenty, and different varieties of
E. Africanus, in which they vary from twenty to twenty-
one.

An enquiry into the various species of the genus
Rhinoceros, recent and fossil, has led Dr. Falconer to
analogous results, as might be inferred from what was
said in p. 111, and as will be more fully seen in the

various papers on Rhinoceros in his Palæontological
Memoirs, vol ii.

Among the fossils brought in 1858 by Mr. Hayden from
the Niobrara Valley, Dr. Leidy describes a rhinoceros so like
the Asiatic species, *R. Indicus*, that he at first referred it to
the same; and, what is most singular, he remarks generally of
the Pliocene fauna of that part of North America, that it is
far more related in character to the Pleistocene and recent
fauna of Europe than to that now inhabiting the American
continent.   Another instance of a similar kind is suggested
to me by Mr. Busk, namely that of the *Ursus priscus*, whose
remains are found in caverns in this country and on the continent, and in gravel in Ireland. This bear is in all probability
identical with *Ursus ferox*, a species at present limited to the
western side of North America, and perhaps the nearest point
of Eastern Asia, but which would thus seem to have been at
one time contemporary with the Wapiti deer, musk-sheep,
and reindeer in the fauna of Western Europe.

It seems indeed more and more evident that when we
speculate in future on the pedigree of any extinct quadruped
which abounds in the drift or caverns of Europe, we shall
have to look to America as one of the principal sources
of information.   Forty years ago, if we had been searching for fossil types which might fill up a gap between two
species or genera of the horse tribe (or great family of the
Solipedes), we might have thought it sufficient to have got
together as ample materials as we could obtain from the
continents of Europe, Africa, and Asia. We might have presumed that as no living representative of the equine family,
whether horse, ass, zebra, or quagga, existed in North
or South America when those regions were first explored
by Europeans, a search in the transatlantic world for fossil
species might be dispensed with.   But how different is the
prospect now opening before us!   Mr. Darwin first detected

the remains of a fossil horse during his visit to South
America, since which two other fossil species have been met
with on the same continent; while in North America, in the
valley of the Nebraska alone, Mr. Hayden, besides a fossil
species not distinguishable from the domestic horse, has obtained, according to Dr. Leidy, representatives of five other
fossil genera of Solipedes. These Dr. Leidy names Hipparion,
Protohippus, Merychippus, Hypohippus, and Parahippus.
On the whole, no less than twelve equine species, belonging
to seven genera (including the Miocene *Anchitherium* of
Nebraska), have been already detected in the Tertiary and
post-Tertiary formations of the United States.*

Two of these, Hipparion and Anchitherium, were already
known in European Miocene formations. It is remarkable that
these fossil species afford us examples which prove that the
existing forms of the genus Equus have resulted from the
gradual modification of very different ancestral types. 'The
skeleton of the older Pliocene and newer Miocene Hipparion,'
says Mr. Huxley, 'very closely resembles that of an ass or a
moderate-sized horse.' But, among other anatomical differences, he points out that 'each limb possesses three complete
toes—one strong, median, and provided with a large hoof,
while the two lateral toes are so small that they do not
extend beyond the fetlock joint. In the fore limb, rudiments
of the first and fifth toes have been found. . . . In the genus
Anchitherium, all the remains of which are of older Miocene
(and perhaps newer Eocene) age, the skeleton in general is
still extraordinarily like that of a horse. . . But not only
are there three toes in each foot as in Hipparion, but the inner
and the outer toes are so large that they must have rested
on the ground. Thus, so far as the limbs are concerned,
the Anchitherium is just such a step beyond the Hipparion
as the Hipparion is beyond the horse, in the direction of a

* Proceedings of Academy of Natural Science, Philadelphia, for 1858, p. 59.

less specialized quadruped. . . In all those respects in which
Anchitherium departs from the modern Equine type, it
approaches that of the extinct Palæotheria ; and this is so
much the case, that Cuvier considered the remains of the
Anchitherium with which he was acquainted to be those of
a species of Palæotherium.' * It cannot be doubted, therefore, that we have here, so far as regards the genus Equus,
traces of those intermediate types the absence of which in a
fossil state is so strongly relied on by the opponents of the
theory of transmutation.

Professor Unger† and Heer‡ have advocated, on botanical
grounds, the former existence of an Atlantic continent during
some part of the Tertiary period, as affording the only plausible
explanation that can be imagined of the analogy between
the Miocene flora of Central Europe and the existing flora of
Eastern America.   Professor Oliver, on the other hand, after
showing how many of the American types found fossil in
Europe are common to Japan, inclines to the theory, first
advanced by Dr. Asa Gray, that the migration of species, to
which the community of types in the Eastern States of North
America and the Miocene flora of Europe is due, took place
when there was an overland communication from America to
Eastern Asia between the fiftieth and sixtieth parallels of
latitude, or south of Behring's Straits, following the direction
of the Aleutian Islands.§   By this course they may have
made their way, at any epoch, Miocene, Pliocene, or Pleistocene, antecedently to the Glacial epoch, to Amoorland, on
the east coast of Northern Asia.

We have already seen (p. 205) that a large proportion of
the living quadrupeds of Amoorland (34 out of 48) are specifically identical with those at present inhabiting the continent
of Western Europe and the British Isles.

* Huxley's Anatomy of Vertebrated
Animals, 1871, pp. 358–360.
† Die versunkene Insel Atlantis.
‡ Flora Tertiaria Helvetiæ.
§ Oliver, Lecture at the Royal In-
stitution, March 7, 1862.

A monograph on the hippopotamus, bear, ox, stag, or any other genus of Mammalia common in the European drift or caverns, might equally well illustrate the defective state of the materials at present at our command. We are rarely in possession of one perfect skeleton of any extinct species, still less of skeletons of both sexes, and of different ages. We usually know nothing of the geographical varieties of the Pleistocene and Pliocene species, least of all those successive changes of form which they must have undergone in the pre-Glacial epoch between the Upper Miocene and Pleistocene eras. Such being the poverty of our palæontological data, we cannot wonder that osteologists are at variance as to whether certain remains found in caverns are of the same species as those now living : whether, for example, the *Talpa fossilis* is really the common mole, the *Meles Morreni* the common badger, *Lutra antiqua* the otter of Europe, *Sciurus priscus* the squirrel, *Arctomys primigenia* the marmot, *Myoxus fossilis* the dormouse, Schmerling's *Felis Engihoulensis* the European lynx ; or whether *Ursus priscus* is not identical with the grizzly bear of North America, *Ursus ferox* ; and *Ursus spelæus* an extinct race of the living brown bear (*Ursus arctos*).

If at some future period all the above-mentioned species should be united with their allied congeners, it cannot fail to enlarge our conceptions of the modifications which a species is capable of undergoing in the course of time, although the same form may appear absolutely immutable within the narrow range of our experience.

In the 'Principles of Geology,' in 1833,* I stated that the longevity of species in the class Mollusca exceeded that in the Mammalia. It has been since found that this generalisation can be carried much farther, and that, in fact, the law

* 1st edit., vol. iii. pp. 46 and 140.

which governs the changes in organic beings is such, that the
lower their place in a graduated scale, or the simpler their
structure, the more persistent are they in form and organisation. I soon became aware of the force of this rule in
the class Mollusca, when I first attempted to calculate the
numerical proportion of recent species in the newer Pliocene
formations as compared to the older Pliocene, and of them
again as contrasted with the Miocene; for it appeared invariably that a greater number of the lamellibranchiate
bivalves could be identified with living species than of the
gasteropods, and of these last a greater number in the
lower division, that of entire-mouthed univalves, than in
that of the siphonated. In whatever manner the changes
have been brought about, whether by variation and natural
selection, or by any other causes, the rate of change has been
greater where the grade of organisation is higher.

It is only, therefore, where there is a full representation
of all the principal orders of Mollusca, or when we compare
those of corresponding grade, that we can fully rely on the percentage test, or on the proportion of recent to extinct species
as indicating the degree of relationship of two geographical
formations to the existing fauna.

The foraminifera, which exemplify the lowest stage of
animal existence, being akin to the sponges, exhibit, as we
learn from the researches of Dr. Carpenter and of Messrs.
Rupert Jones and Parker, extreme variability in their specific
forms, and yet these same forms are persistent throughout
vast periods of time, exceeding, in that respect, even the
brachiopodous Mollusca before mentioned (p. 473).

Dr. Hooker observes, in regard to plants of complex floral
structure, that they manifest their physical superiority in a
greater extent of variation, and in thus better securing a
succession of race—an attribute which in some senses he

regards as of a higher order than that indicated by mere complexity or specialisation of organs.*

As one of the consequences of this law, he says that species, genera, and orders are, on the whole, best limited in plants of higher grade, the dicotyledons better than the monocotyledons, and the dicblamydew better than the achlamydeæ.

Mr. Darwin remarks, ' We can, perhaps, understand the apparently quicker rate of change in terrestrial, and in more highly organised productions, compared with marine and lower productions, by the more complex relations of the higher beings to their organic and inorganic conditions of life.'†

If we suppose the Mammalia to be more sensitive than are the inferior classes of the Vertebrata to every fluctuation in the surrounding conditions, whether of the animate or inanimate world, it would follow that they would oftener be called upon to adapt themselves, by variation, to new conditions, or if unable to do so, to give place to other types. This would give rise to more frequent extinction of varieties, species, and genera, whereby the surviving types would be better limited, and the average duration of the same unaltered specific types would be lessened.

Absence of Mammalia in Islands considered in reference to Transmutation.

But if Mammalia vary, upon the whole, at a more rapid rate than animals lower in the scale of being, it must not be supposed that they can alter their habits and structures readily, or that they are convertible in short periods into new species. The extreme slowness with which such changes of habits and organisation take place, when new conditions

* Introductory Essay, &c., p. vii.   † Origin of Species, 6th ed. p. 291.

arise, appears to be well exemplified by the absence even of
small warm-blooded quadrupeds in islands far from continents,
however well such islands may be fitted by their dimensions
to support them.

Mr. Darwin has pointed to this absence of Mammalia as
favouring his views, observing that bats, which are the only
exceptions to the rule, might have made their way to distant
islands by flight, for they are often met with on the wing far
out at sea. Unquestionably the total exclusion of quadrupeds in general, which could only reach such isolated habitations by swimming, seems to imply that nature does not
dispense with the ordinary laws of reproduction when she
peoples the earth with new forms; for if what has been
called special creation were alone at work, we might
naturally look for squirrels, rabbits, polecats, and other
small vegetable feeders and the beasts which prey on them,
as often as for bats, in the spots alluded to.

On the other hand, I have found it difficult to reconcile
the antiquity of certain islands, such as those of the Madeiran
Archipelago, and those of still larger size in the Canaries,
with the total absence of small indigenous quadrupeds; for,
judging by ancient deposits of littoral shells, now raised high
above the level of the sea, several of these volcanic islands
(Porto Santo and the Grand Canary among others) must
have existed ever since the Upper Miocene period. But,
waiving all such claims to antiquity, it is at least certain
that since the close of the Newer Pliocene period, Madeira
and Porto Santo have constituted two separate islands, each
in sight of the other, and each inhabited by an assemblage of
land shells (*Helix, Pupa, Clausilia,* &c.), for the most part
different or proper to each island. About thirty-six fossil
species have been obtained in Madeira, and thirty-five in
Porto Santo, only eight of the whole being common to both
islands, and five of these eight are represented by distinct

varieties in each island.[*] In each the living land-shells are
equally distinct, and correspond, for the most part, with the
species found fossil in each island respectively.

Among the fossil species, one or two appear to be entirely extinct, and a larger number have disappeared from
the fauna of the Madeiran Archipelago, though still extant
in Africa and Europe. Many which were amongst the most
common in the Newer Pliocene period, have now become
the scarcest, and others formerly scarce are now most numerously represented. The variety-making force has been at
work with such energy,—perhaps we ought to say, has had
so much time for its development,—that almost every isolated rock within gun-shot of the shores has its peculiar
living forms, or those very marked races to which Mr. Lowe,
in his excellent description of the fauna, has given the name
of 'sub-species.'

Since the fossil shells were embedded in sand near the
coast, these volcanic islands have undergone considerable
alterations in size and shape by the wasting action of the
waves of the Atlantic beating incessantly against the cliffs, so
that the evidence of a vast lapse of time is derivable from
inorganic as well as from organic phenomena.

During this period no Mammalia, not even of small species,
excepting bats, have made their appearance, whether in
Madeira and Porto-Santo or in the larger and more numerous
islands of the Canarian group. It might have been expected,
from some expressions met with here and there in the 'Origin
of Species,' though not perhaps from a fair interpretation of
the whole tenor of the author's reasoning, that this dearth of
the highest class of Vertebrata is inconsistent with the powers
of Mammalia to accommodate their habits and structures to
new conditions. Why did not some of the bats, for example,

* See 'Principles of Geology,' 11th ed. vol. II. p. 428.

K K

after they had greatly multiplied, and were hard pressed by a
scarcity of insects on the wing, betake themselves to the
ground in search of prey, and, gradually losing their wings,
become transformed into non-volant insectivora? Mr. Darwin
tells me that he has learnt that there is a bat in India which
has been known occasionally to devour frogs. One might also
be tempted to ask, how it has happened that the seals which
swarmed on the shores of Madeira and the Canaries, before the
European colonists arrived there, were never induced, when
food was scarce in the sea, to venture inland from the shores,
and begin in Teneriffe, and the Grand Canary especially,
and other large islands, to acquire terrestrial habits, venturing
first a few yards inland, and then further and farther until
they began to occupy some of the ' places left vacant in the
economy of nature.' During these excursions, we might suppose some varieties, which had the skin of the webbed intervals of their toes less developed, to succeed best in walking
on the land, and in the course of several generations they
might exchange their present gait or manner of shuffling
along and jumping by aid of the tail and their fin-like extremities, for feet better adapted for running.

It is said that one of the bats in the island of Palma (one
of the Canaries) is of a peculiar species, and that some of the
Cheiroptera of the Pacific islands (or Oceanica) are even of
peculiar genera. If so, we seem, on organic as well as on
geological grounds, to be precluded from arguing that there
has not been time for great divergence of character. We
seem also entitled to ask why the bats and rodents of
Australia, which are spread so widely among the marsupials
over that continent, have never, under the influence of
the principle of progression, been developed into higher
placental types, since we have now ascertained that that
continent was by no means unfitted to sustain such Mammalia,
for these, when once introduced by Man, have run wild, and

become naturalised in many parts. The following answers may perhaps be offered to the above criticisms of some of Mr. Darwin's theoretical views.

First, as to the bats and seals: they are what zoologists call aberrant and highly specialised types, and therefore precisely those which might be expected to display a fixity and want of pliancy in their organisation, or the smallest possible aptitude for deviating in new directions towards new structures, and the acquisition of such altered habits as a change from aquatic to terrestrial or from volant to non-volant modes of living would imply.

Secondly, the same powers of flight which enabled the first bats to reach Madeira or the Canaries, would bring others from time to time from the African continent, which, mixing with the first emigrants and crossing with them, would check the formation of new races, or keep them true to the old types, as is found to be actually the case with the birds of Madeira and the Bermudas.

This would happen the more surely, if, as Mr. Darwin has endeavoured to prove, the offspring of races slightly varying are usually more vigorous than the progeny of parents of the same race, and would be more prolific, therefore, than the insular stock which had been for a long time breeding in and in.

The same cause would tend in a still more decided manner to prevent the seals from diverging into new races or 'incipient species,' because they range freely over the wide ocean, and, may therefore have continual intercourse with all other individuals of their species.

Thirdly, as to peculiar species, and even genera of bats in islands, we are perhaps too little acquainted at present with all the species and genera of the neighbouring continents to be able to affirm, with any degree of confidence, that the forms supposed to be peculiar do not exist elsewhere: those

M M 2

of the Canaries in Africa, for example. But what is still more important, we must bear in mind how many species and genera of Pleistocene Mammalia have everywhere become extinct by causes independent of Man. It is always possible, therefore, that some types of Cheiroptera, originally derived from the main land, have survived in islands, although they have gradually died out on the continents from whence they came ; so that it would be rash to infer that there has been time for the creation, whether by variation or other agency, of new species or genera in the islands in question.

As to the rodents and Cheiroptera of Australia, we are as yet too ignorant of the Pleistocene and Newer Pliocene fauna of that part of the world, to be able to decide whether the introduction of such forms dates from a remote geological time. We know, however, that, before the recent period, that continent was peopled with large kangaroos, and other herbivorous and carnivorous marsupials, of species long since extinct, their remains having been discovered in ossiferous caverns. The preoccupancy of the country by such indigenous tribes may have checked the development of the placental rodents and Cheiroptera, even were we to concede the possibility of such forms being convertible by variation and progressive development into higher grades of Mammalia.*

Imperfection of the Geological Record.

When treating (p. 190) of the dearth of human bones in alluvium containing flint implements in abundance, I pointed out that it is not part of the plan of Nature to write everywhere, and at all times, her autobiographical memoirs. On the contrary, her annals are local and exceptional from

* The subject of Oceanic Islands and the peculiar character of their natural productions is discussed at some length in the 'Principles of Geology,' 11th ed. vol. ii. chap. xli.

the first, and portions of them are afterwards ground into mud, sand, and pebbles, to furnish materials for new strata. Even of those ancient monuments now forming the crust of the earth, which have not been destroyed by rivers and the waves of the sea, or which have escaped being melted by volcanic heat, three-fourths lie submerged beneath the ocean, and are inaccessible to Man; while of those which form the dry land, a great part are hidden for ever from our observation by mountain masses, thousands of feet thick, piled over them.

Mr. Darwin has truly said that the fossiliferous rocks known to geologists consist, for the most part, of such as were formed when the bottom of the sea was subsiding. This downward movement protects the new deposits from denudation, and allows them to accumulate to a great thickness; whereas sedimentary matter, thrown down where the sea-bottom is rising, must almost invariably be swept away by the waves as fast as the land emerges.

When we reflect, therefore, on the fractional state of the annals which are handed down to us, and how little even these have as yet been studied, we may wonder that so many geologists should attribute every break in the series of strata, and every gap in the past history of the organic world, to catastrophes and convulsions of the earth's crust, or to leaps made by the creational force from species to species, or from class to class. For it is clear that, even had the series of monuments been perfect and continuous at first (an hypothesis quite opposed to the analogy of the working of causes now in action), it could not fail to present itself to our eyes in a broken and disconnected state.

Those geologists who have watched the progress of discovery during the last half century, can best appreciate the extent to which we may still hope by future exertion to fill up some of the wider chasms which now interrupt the

regular sequence of fossiliferous rocks. The determination, for
example, of late years of the true place of the Hallstadt and
St. Cassian beds on the north and south flanks of the Austrian
Alps, has revealed to us, for the first time, the marine fauna
of a period (that of the Upper Trias) of which, until lately,
but little was known. In this case, the palæontologist is called
upon suddenly to intercalate about 800 species of Mollusca
and Radiata, between the fauna of the Lower Lias and that of
the Middle Trias. The period in question was previously
believed, even by many a philosophical geologist, to have
been comparatively barren of organic types. In England,
France, and Northern Germany, the only known strata of
Upper Triassic date had consisted almost entirely of fresh
or brackish-water beds, in which the bones of terrestrial and
amphibious reptiles were the most characteristic fossils. The
new fauna was, as might have been expected, in part peculiar, not a few of the species of Mollusca being referable
to new genera: while some species were common to the
older, and some to the newer rocks. On the whole, the new
forms have helped greatly to lessen the discordance, not only
between the Lias and Trias, but also generally between palæozoic and neozoic formations. Thus the genus Orthoceras has
been for the first time recognised in a neozoic deposit, and
with it we find associated, for the first time, large ammonites
with foliated lobes, a form never seen before below the Lias;
also the Ceratite, a family of cephalopods never before met
with in the Upper Trias, and never before in the same stratum
with such lobed ammonites.

We can now no longer doubt, that should we hereafter have
an opportunity of studying an equally rich marine fauna of
the age of the Lower Trias (or Bunter Sandstein), the marked
hiatus which still separates the Triassic and Permian eras
would to a great extent disappear.

CHAPTER XXIII.

ORIGIN AND DEVELOPMENT OF LANGUAGES AND SPECIES COMPARED.

ARYAN HYPOTHESIS AND CONTROVERSY — THE RACES OF MANKIND CHANGE MORE SLOWLY THAN THEIR LANGUAGES — THEORY OF THE GRADUAL ORIGIN OF LANGUAGES — DIFFICULTY OF DEFINING WHAT IS MEANT BY A LANGUAGE AS DISTINCT FROM A DIALECT — GREAT NUMBER OF EXTINCT AND LIVING TONGUES — NO EUROPEAN LANGUAGE A THOUSAND YEARS OLD — GAPS BETWEEN LANGUAGES, HOW CAUSED — IMPERFECTION OF THE RECORD — CHANGES ALWAYS IN PROGRESS — STRUGGLE FOR EXISTENCE BETWEEN RIVAL TERMS AND DIALECTS — CAUSES OF SELECTION — EACH LANGUAGE FORMED SLOWLY IN A SINGLE GEOGRAPHICAL AREA — MAY DIE OUT GRADUALLY OR SUDDENLY — ONCE LOST CAN NEVER BE REVIVED — MODE OF ORIGIN OF LANGUAGES AND SPECIES A MYSTERY — SPECULATIONS AS TO THE NUMBER OF ORIGINAL LANGUAGES OR SPECIES UNPROFITABLE.

THE supposed existence, at a remote and unknown period, of a language conventionally called the Aryan, has of late years been a favourite subject of speculation among German philologists, and Professor Max Müller has given us lately the most improved version of this theory, and has set forth the various facts and arguments by which it may be defended, with his usual perspicuity and eloquence. He observes that if we knew nothing of the existence of Latin, — if all historical documents previous to the fifteenth century had been lost, — if tradition even was silent as to the former existence of a Roman empire, a mere comparison of the Italian, Spanish, Portuguese, French, Wallachian, and Rhætian dialects would enable us to say that at some time there must have been a language, from which these six modern dialects derive their origin in common. Without this supposition it would be impossible to account for their structure and composition, as, for example, for the forms of the auxiliary verb 'to be,' all evidently varieties of one common type, while it is equally clear that no one of the six affords the original form from which the others could have been borrowed. So also in none of the six languages do we find the elements of which these verbal and other forms could have been composed; they must have been handed down as relics from a former period, they must have existed in some antecedent language, which we know to have been the Latin.

But, in like manner, he goes on to show, that Latin itself, as well as Greek, Sanscrit, Zend (or Bactrian), Lithuanian, old Sclavonic, Gothic, and Armenian are also eight varieties of one common and more ancient type, and no one of them could have been the original from which the others were borrowed. They have all such an amount of mutual resemblance, as to point to a more ancient language, the Aryan, which was to them what Latin was to the six Romance languages. The people who spoke this. unknown parent speech, of which so many other ancient tongues were off-shoots, must have migrated at a remote era to widely separated regions of the Old World, such as Northern Asia, Europe, and India south of the Himalaya.*

The soundness of some parts of this Aryan hypothesis has lately been called in question by Mr. Crawfurd, on the ground that the Hindoos, Persians, Turks, Scandinavians, and other people referred to as having derived not only words but grammatical forms from an Aryan source, belong each of them to a distinct race, and all these races have, it is said, preserved their peculiar characters unaltered from the earliest dawn of history and tradition. If, therefore, no

* Max Müller, Comparative Mythology.   Oxford Essays, 1856.

appreciable change has occurred in three or four thousand
years, we should be obliged to assume a far more remote
date for the first branching off of such races from a common
stock than the supposed period of the Aryan migrations, and
the dispersion of that language over many and distant
countries.

But Mr. Crawfurd has, I think, himself helped us to
remove this stumbling-block, by admitting that a nation
speaking a language allied to the Sanscrit (the oldest of the
eight tongues alluded to), once probably inhabited that
region situated to the north-west of India, which within the
period of authentic history has poured out its conquering
hordes over a great extent of Western Asia and Eastern
Europe. The same people, he says, may have acted the
same part in the long, dark night which preceded the dawn
of tradition.* These conquerors may have been few in
number when compared to the populations which they
subdued. In such cases the new settlers, although reckoned
by tens of thousands, might merge in a few centuries into the
millions of subjects which they ruled. It is an acknowledged
fact, that the colour and features of the Negro or European
are entirely lost in the fourth generation, provided that no
fresh infusion of one or other of the two races takes place.
The distinctive physical features, therefore, of the Aryan
conquerors might soon wear out and be lost in those of the
nations they overran ; yet many of the words, and, what is
more in point, some of the grammatical forms of their language, might be retained by the masses which they had
governed for centuries, these masses continuing to preserve
the same features of race which had distinguished them long
before the Aryan invasions.

There can be no question that if we could trace back any

* Crawfurd, Transactions of the Ethnological Society, vol. i. 1861.

set of cognate languages now existing to some common point
of departure, they would converge and meet sooner in some
era of the past than would the existing races of mankind;
in other words, races change much more slowly than languages. But, according to the doctrine of transmutation, to
form a new species would take an incomparably longer period
than to form a new race. No language seems ever to last
for a thousand years, whereas many a species seems to have
endured for hundreds of thousands. A philologist, therefore,
who is contending that all living languages are derivative
and not primordial, has a great advantage over a naturalist
who is endeavouring to inculcate a similar theory in regard
to species.

It may not be uninstructive, in order fairly to appreciate
the vast difficulty of the task of those who advocate transmutation in natural history, to consider how hard it would
be even for a philologist to succeed, if he should try to
convince an assemblage of intelligent but illiterate persons
that the language spoken by them, and all those talked by
contemporary nations, were modern inventions, moreover
that these same forms of speech were still constantly undergoing change, and none of them destined to last for ever.

We will suppose him to begin by stating his conviction,
that the living languages have been gradually derived from
others now extinct, and spoken by nations which had immediately preceded them in the order of time, and that those
again had used forms of speech derived from still older ones.
They might naturally exclaim, 'How strange it is that you
should find records of a multitude of dead languages, that a
part of the human economy which in our own time is so
remarkable for its stability, should have been so inconstant in
bygone ages! We all speak as our parents and grandparents
spoke before us, and so, we are told, do the Germans and
French. What evidence is there of such incessant variation

in remoter times? and, if it be true, why not imagine that
when one form of speech was lost, another was suddenly and
supernaturally created by a gift of tongues or confusion of
languages, as at the building of the Tower of Babel? Where
are the memorials of all the intermediate dialects which
must have existed, if this doctrine of perpetual fluctuation
be true? And how comes it that the tongues now spoken
do not pass by insensible gradations the one into the other,
and into the dead languages of dates immediately antecedent?

'Lastly, if this theory of indefinite modifiability be sound,
what meaning can be attached to the term language, and
what definition can be given of it so as to distinguish a
language from a dialect?'

In reply to this last question, the philologist might confess
that the learned are not agreed as to what constitutes a language as distinct from a dialect. Some believe that there
are 4,000 living languages, others that there are 6,000, so
that the mode of defining them is clearly a mere matter of
opinion. Some contend, for example, that the Danish,
Norwegian, and Swedish form one Scandinavian tongue,
others that they constitute three different languages, others
that the Danish and Norwegian are one,—mere dialects of the
same language, but that Swedish is distinct.

The philologist, however, might fairly argue that this very
ambiguity was greatly in favour of his doctrine, since if languages had all been constantly undergoing transmutation,
there ought often to be a want of real lines of demarcation
between them. He might, however, propose that he and his
pupils should come to a understanding that two languages
should be regarded as distinct whenever the speakers of them
are unable to converse together, or freely to exchange ideas,
whether by word or writing. Scientifically speaking, such a
test might be vague and unsatisfactory, like the test of species
by their capability of producing fertile hybrids; but if the

pupil is persuaded that there are such things in nature as distinct languages, whatever may have been their origin, the definition above suggested might be of practical use, and enable the teacher to proceed with his argument.

He might begin by undertaking to prove that none of the widely spoken languages of modern Europe were a thousand years old. No English scholar, he might say, who has not specially given himself up to the study of Anglo-Saxon can interpret the documents in which the chronicles and laws of England were written in the days of King Alfred, so that we may be sure that none of the English of the nineteenth century could converse with the subjects of that monarch if these last could now be restored to life. The difficulties encountered would not arise merely from the intrusion of French terms, in consequence of the Norman conquest, because that large portion of our language (including the articles, pronouns, &c.), which is Saxon has also undergone great transformations by abbreviation, new modes of pronunciation, spelling, and various corruptions, so as to be unlike both ancient and modern German. They who now speak German, if brought into contact with their Teutonic ancestors of the ninth century, would be quite unable to converse with them, and, in like manner, the subjects of Charlemagne could not have exchanged ideas with the Goths of Alaric's army, or with the soldiers of Arminius in the days of Augustus Cæsar. So rapid indeed has been the change in Germany, that the epic poem called the Nibelungen Lied, once so popular, and only seven centuries old, cannot now be enjoyed, except by the erudite.

If we then turn to France, we meet again with similar evidence of ceaseless change. There is a treaty of peace still extant a thousand years old, between Charles the Bald and King Louis of Germany (dated A.D. 841), in which the German king takes an oath in what was the French tongue of that day, while the French king swears in the

German of the same era, and neither of these on the would now convey a distinct meaning to any but the learned in these two countries. So also in Italy, the modern Italian cannot be traced back much beyond the time of Dante, or some six centuries before our time. Even in Rome, where there had been no permanent intrusion of foreigners, such as the Lombard settlers of German origin in the plains of the Po, the common people of the year 1000 spoke quite a distinct language from that of their Roman ancestors or their Italian descendants, as is shown by the celebrated Chronicle of the monk Benedict, of the convent of St. Andrea on Mount Soracte, written in such barbarous Latin, and with such strange grammatical forms, that it requires a profoundly skilled linguist to decipher it.*

Having thus established the preliminary fact, that the greater number of the tongues now spoken were not in existence ten centuries ago, and that the ancient languages have passed through many a transitional dialect before they settled into the forms now in use, the philologist might bring forward proofs of the great numbers both of lost and living forms of speech.

Strabo tells us that in his time, in the Caucasus alone (a chain of mountains not longer than the Alps, and much narrower), there were spoken at least seventy languages. At the present period the number, it is said, would be still greater, if all the distinct dialects of those mountains were reckoned. Several of these Caucasian tongues admit of no comparison with any known living or dead Asiatic or European language. Others which are not peculiar are obsolete forms of known languages, such as the Georgian, Mongolian, Persian, Arabic, and Tartarian. It seems that as often as conquering hordes swept over that part of Asia, always coming

* See G. Pertz, Monumenta Germanica, vol. iii.

from the north and east, they drove before them the inhabitants of the plains, who took refuge in some of the retired valleys and high mountain fastnesses, where they maintained their independence, as do the Circassians in our time in spite of the power of Russia.

In the Himalayan Mountains, from Assam to its extreme north-western limit, and generally in the more hilly parts of British India, the diversity of languages is surprisingly great, impeding the advance of civilisation and the labours of the missionary. In South America and Mexico, Alexander Humboldt reckoned the distinct tongues by hundreds, and those of Africa are said to be equally numerous. Even in China, some eighteen provincial dialects prevail, almost all deviating so much from others that the speakers are not mutually intelligible, and besides these there are other distinct forms of speech in the mountains of the same empire.

The philologist might next proceed to point out that the geographical relations of living and dead languages favour the hypothesis of the living ones having been derived from the extinct, in spite of our inability, in most instances, to adduce documentary evidence of the fact or to discover monuments of all the intermediate and transitional dialects which must have existed. Thus he would observe that the modern Romance languages are spoken exactly where the ancient Romans once lived or ruled, and the Greek of our days where the older classical Greek was formerly spoken. This latter language may no doubt be taken as a case of survival, being considered by philologists to be rather a corruption, than an evolution from the ancient language, differing in this from French, Italian, Portuguese, and Spanish, which are distinct from the old Latin tongue, on which they are all based. Exceptions to this rule of the geographical constancy of ancient languages and the modern forms derived from them might be detected, but they would be explicable by reference to colonisation and conquest.

As to the many and wide gaps sometimes encountered between the dead and living languages, we must remember that it is not part of the plan of any people to preserve memorials of their forms of speech expressly for the edification of posterity. Their MSS. and inscriptions serve some present purpose, are occasional and imperfect from the first, and are rendered more fragmentary in the course of time, some being intentionally destroyed, others lost by the decay of the perishable materials on which they are written ; so that to question the theory of all known languages being derivative on the ground that we can rarely trace a passage from the ancient to the modern through all the dialects which must have flourished one after the other in the intermediate ages, implies a want of reflection on the laws which govern the recording as well as the obliterating processes.

But another important question still remains to be considered, namely, whether the trifling changes which can alone be witnessed by a single generation, can possibly represent the working of that machinery which, in the course of many centuries, has given rise to such mighty revolutions in the forms of speech throughout the world. Every one may have noticed in his own lifetime the stealing in of some slight alterations of accent, pronunciation or spelling, or the introduction of some words borrowed from a foreign language to express ideas of which no native term precisely conveyed the import. He may also remember hearing for the first time some cant terms or slang phrases, which have since forced their way into common use, in spite of the efforts of the purist. But he may still contend that, 'within the range of his experience,' his language has continued unchanged, and he may believe in its immutability in spite of minor variations. The real question, however, at issue is, whether there are

any limits to this variability. He will find on farther
investigation, that new technical terms are coined almost
daily in various arts, sciences, professions, and trades, that
new names must be found for new inventions, that many of
these acquire a metaphorical sense, and then make their
way into general circulation, as 'stereotyped,' for instance,
which would have been as meaningless to the men of the
seventeenth century as would the new terms and images
derived from steamboat and railway travelling to the men of
the eighteenth.

If the numerous words, idioms, and phrases, many of
them of ephemeral duration, which are thus invented by the
young and old in various classes of society, in the nursery,
the school, the camp, the fleet, the courts of law and the
study of the man of science or literature, could all be
collected together and put on record, their number in one
or two centuries might compare with the entire permanent
vocabulary of the language. It becomes, therefore, a curious
subject of enquiry, what are the laws which govern not only
the invention, but also the 'selection' of some of these
words or idioms, giving them currency in preference to
others?—for as the powers of the human memory are
limited, a check must be found to the endless increase and
multiplication of terms, and old words must be dropped
nearly as fast as new ones are put into circulation.
Sometimes the new word or phrase, or a modification of the
old ones, will entirely supplant the more ancient expressions,
or, instead of the latter being discarded, both may flourish
together, the older one having a more restricted use.

Although the speakers may be unconscious that any great
fluctuation is going on in their language,—although when we
observe the manner in which new words and phrases are
thrown out, as if at random or in sport, while others get into
vogue, we may think the process of change to be the result

of mere chance,—there are, nevertheless, fixed laws in action, by which, in the general struggle for existence, some terms and dialects gain the victory over others. The slightest advantage attached to some new mode of pronouncing or spelling, from considerations of brevity or euphony, may turn the scale, or more powerful causes of selection may decide which of two or more rivals shall triumph and which succumb. Among these are fashion, or the influence of an aristocracy, whether of birth or education, popular writers, orators, preachers,—a centralised government organising its schools expressly to promote uniformity of diction, and to get the better of provincialisms and local dialects. Between these dialects, which may be regarded as so many 'incipient languages,' the competition is always keenest when they are most nearly allied, and the extinction of any one of them destroys some of the links by which a dominant tongue may have been previously connected with some other widely distinct one. It is by the perpetual loss of such intermediate forms of speech that the great dissimilarity of the languages which survive is brought about. Thus, if Dutch should become a dead language, English and German would be separated by a wider gap.

Some languages which are spoken by millions, and spread over a wide area, will endure much longer than others which have never had a wide range, especially if the tendency to incessant change in one of these dominant tongues is arrested for a time by a standard literature. But even this source of stability is insecure, for popular writers themselves are great innovators, sometimes coining new words, and still oftener new expressions and idioms, to embody their own original conceptions and sentiments, or some peculiar modes of thought and feeling characteristic of their age. Even when a language is regarded with superstitious veneration as the vehicle of divine truths and religious precepts, and which has

prevailed for many generations, it will be incapable of permanently maintaining its ground. Hebrew had ceased to be
a living language before the Christian era. Sanscrit, the
sacred language of the Hindoos, shared the same fate, in
spite of the veneration in which the Vedas are still held, and
in spite of many a Sanscrit poem once popular and national.

The Christians of Constantinople and the Morea still hear
the New Testament and their liturgy read in ancient Greek,
while they speak a dialect in which Paul might have preached
in vain at Athens. So in the Roman Catholic Church, the
Italians pray in one tongue and talk another. Luther's translation of the Bible acted as a powerful cause of 'selection,'
giving at once to one of many competing dialects (that of
Saxony) a prominent and dominant position in Germany;
but the style of Luther has, like that of our English Bible,
already become somewhat antiquated.

If the doctrine of gradual transmutation be applicable to
languages, all those spoken in historical times must each of
them have had a closely allied prototype ; and accordingly,
whenever we can thoroughly investigate their history, we
find in them some internal evidence of successive additions
by the invention of new words or the modification of old
ones. Proofs also of borrowing are discernible, letters being
retained in the spelling of some words which have no longer
any meaning as they are now pronounced,—no connection
with any corresponding sounds. Such redundant or silent
letters, once useful in the parent speech, have been aptly compared by Mr. Darwin to rudimentary organs in living beings,
which, as he interprets them, have at some former period
been more fully developed, having had their proper functions
to perform in the organising of a remote progenitor.

If all known languages are derivative and not primordial
creations, they must each of them have been slowly elaborated
in a single geographical area. No one of them can have had

two birthplaces. If one were carried by a colony to a distant region, it would immediately begin to vary unless frequent intercourse was kept up with the mother country. The descendants of the same stock, if perfectly isolated, would in five or six centuries, perhaps sooner, be quite unable to converse with those who remained at home, or with those who may have migrated to some distant region, where they were shut out from all communication with others speaking the same tongue.

A Norwegian colony which settled in Iceland in the ninth century, maintained its independence for about 400 years, during which time the old Gothic which they at first spoke became corrupted and considerably modified. In the meantime the natives of Norway, who had enjoyed much commercial intercourse with the rest of Europe, acquired quite a new speech, and looked on the Icelandic as having been stationary, and as representing the pure Gothic original of which their own was an off-shoot.

A German colony in Pennsylvania was cut off from frequent communication from Europe for about a quarter of a century, during the wars of the French Revolution between 1792 and 1815. So marked had been the effect even of this brief and imperfect isolation, that when Prince Bernhard of Saxe Weimar travelled among them a few years after the peace, he found the peasants speaking as they had done in Germany in the preceding century,[*] and retaining a dialect which at home had already become obsolete.

Even after the renewal of the German emigration from Europe when I travelled in 1841 among the same people in the retired valleys of the Alleghanies, I found the newspapers full of terms half English and half German, and many an Anglo-Saxon word which had assumed a Teutonic dress, as

[*] Travels of Prince Bernhard of Saxe Weimar, in North America, in 1825 and 1826, p. 123.

'fencen,' to fence, instead of umzäunen, 'flauer' for flour, instead of mehl, and so on. What with the retention of terms no longer in use in the mother country, and the borrowing of new ones from neighbouring states, there might have arisen in Pennsylvania in five or six generations, but for the influx of new comers from Germany, a mongrel speech equally unintelligible to the Anglo-Saxon and to the inhabitants of the European fatherland.

If languages resemble species in having had each their 'specific centre' or single area of creation, in which they have been slowly formed, so each of them is alike liable to slow or sudden extinction. They may die out very gradually in consequence of transmutation, or abruptly by the extermination of the last surviving representatives of the unaltered type. We know in what century the last dodo perished, and we know that in the seventeenth century the language of the Red Indians of Massachusetts, into which Father Eliot had translated the Bible, and in which Christianity was preached for several generations, ceased to exist, the last individuals by whom it was spoken having at that period died without issue.[*] But if just before that event the white man had retreated from the continent, or had been swept off by an epidemic, those Indians might soon have repeopled the wilderness, and their copious vocabulary and peculiar forms of expression might have lasted without important modification to this day. The extinction, however, of languages in general is not abrupt, any more than that of species. It will also be evident from what has been said, that a language which has once died out can never be revived, since the same assemblage of conditions can never be restored even among the descendants of the same stock, much less simultaneously among all the surrounding nations with whom they may be in contact.

[*] Lyell, Travels in North America, vol. i. p. 260. 1845.

We may compare the persistency of languages, or the tendency of each generation to adopt without change the vocabulary of its predecessor, to the force of inheritance in the organic world, which causes the offspring to resemble its parents. The inventive power which coins new words or modifies old ones, and adapts them to new wants and conditions as often as these arise, answers to the variety-making power in the animate creation.

Progressive improvement in language is a necessary consequence of the progress of the human mind from one generation to another. As civilisation advances, a greater number of terms are required to express abstract ideas, and words previously used in a vague sense, so long as the state of society was rude and barbarous, gradually acquire more precise and definite meanings, in consequence of which several terms must be employed to express ideas and things. which a single word had before signified, though somewhat loosely and imperfectly.

The farther this subdivision of function is carried, the more complete and perfect the language becomes, just as species of higher grade have special organs, such as eyes, lungs, and stomach, for seeing, breathing, and digesting, which in simpler organisms are all performed by one and the same part of the body.*

When we have satisfied ourselves that all the existing languages, instead of being primordial creations, or the direct gifts of a supernatural power, have been slowly elaborated, partly by the modification of pre-existing dialects, partly by borrowing terms at successive periods from numerous foreign sources, and partly by new inventions made, some of them deliberately, and some casually and as it were fortuitously, —when we have discovered the principal causes of selection,

* See Herbert Spencer's Psychology and Scientific Essays.

which have guided the adoption or rejection of rival names
for the same things and ideas, rival modes of pronouncing
the same words and provincial dialects competing one with
another,—wo are still very far from comprehending all the
laws which have governed the formation of each language.

It was a profound saying of William Humboldt, that
'Man is Man only by means of speech, but in order to invent
speech he must be already Man.' Other animals may be
able to utter sounds more articulate and as varied as the
click of the Bushman, but voice alone can never enable
brute intelligence to acquire language.

When wo consider the complexity of every form of speech
spoken by a highly civilised nation, and discover that the
grammatical rules and the inflections which denote number,
time, and quality are usually the product of a rude state of
society—that the savage and the sage, the peasant and the
man of letters, the child and the philosopher, have worked
together, in the course of many generations, to build up a
fabric which has been truly described as a wonderful instrument of thought, a machine, the several parts of which are
so well adjusted to each other as to resemble the product of
one period and of a single mind,—we cannot but look upon
the result as a profound mystery, and one of which the
separate builders have been almost as unconscious as are the
bees in a hive of the architectural skill and mathematical
knowledge which is displayed in the construction of the
honeycomb.

In our attempts to account for the origin of species, wo
find ourselves still sooner brought face to face with the
working of a law of development of so high an order as to
stand nearly in the same relation as the Deity himself to
man's finite understanding, a law capable of adding new and
powerful forces, such as the moral and intellectual faculties
of the human race, to a system of nature which had gone on

for millions of years without the intervention of any cause
capable of producing analogous effects. If we confound
'Variation' or 'Natural Selection' with such creational laws,
we deify secondary causes or immeasurably exaggerate their
influence.

Yet we ought by no means to undervalue the importance
of the step which will have been made, should it hereafter
become the generally received opinion of men of science (as I
fully expect it will), that the past changes of the organic
world have been brought about by the subordinate agency of
such causes as 'Variation' and 'Natural Selection.' All
our advances in the knowledge of Nature have consisted of
such steps as these, and we must not be discouraged because
greater mysteries remain behind wholly inscrutable to us.

If the philologist is asked whether in the beginning of
things there was one or five, or a greater number of languages,
he may answer that, before he can reply to such a question,
it must be decided whether the origin of Man was single,
or whether there were many primordial races. But he may
also observe, that if mankind began their career in a rude
state of society, their whole vocabulary would be limited to
a few words, and that if they then separated into several
isolated communities, each of these would soon acquire an
entirely distinct language, some roots being lost and others
corrupted and transformed beyond the possibility of subsequent identification, so that it might be hopeless to expect
to trace back the living and dead languages to one starting
point, even if that point were of much more modern date
than we have now good reason to suppose. In like manner
it may be said of species, that if those first formed were of
very simple structure, and they began to vary and to lose
some organs by disuse and acquire new ones by development, they might soon differ as much as so many distinctly
created primordial types. It would therefore be a waste of

time to speculate on the number of original monads or germs from which all plants and animals were subsequently evolved, more especially as the oldest fossiliferous strata known to us may be the last of a long series of antecedent formations, which once contained organic remains. It was not till geologists ceased to discuss the condition of the original nucleus of the planet, whether it was solid or fluid, and whether it owed its fluidity to aqueous or igneous causes, that they began to achieve their great triumphs ; and the vast progress which has recently been made in showing how the living species may be connected with the extinct by a common bond of descent, has been due to a more careful study of the actual state of the living world, and to those monuments of the past in which the relics of the animate creation of former ages are best preserved and least mutilated by the hand of time.

CHAPTER XXIV.

BEARING OF THE DOCTRINE OF TRANSMUTATION ON THE ORIGIN
OF MAN, AND HIS PLACE IN THE CREATION.

WHETHER MAN CAN BE REGARDED AS AN EXCEPTION TO THE RULE
IF THE DOCTRINE OF TRANSMUTATION BE EMBRACED FOR THE REST
OF THE ANIMAL KINGDOM—ZOOLOGICAL RELATIONS OF MAN TO OTHER
MAMMALIA — SYSTEMS OF CLASSIFICATION — TERM QUADRUMANOUS,
WHY DECEPTIVE—WHETHER THE STRUCTURE OF THE HUMAN BRAIN
ENTITLES MAN TO FORM A DISTINCT SUB-CLASS OF THE MAMMALIA
—INTELLIGENCE OF THE LOWER ANIMALS COMPARED TO THE IN-
TELLECT AND REASON OF MAN—HELPLESSNESS OF THE INFANT ORANG-
UTAN—GROUNDS ON WHICH MAN HAS BEEN REFERRED TO A DISTINCT
KINGDOM OF NATURE—IMMATERIAL PRINCIPLE COMMON TO MAN AND
ANIMALS — NON-DISCOVERY OF INTERMEDIATE LINKS AMONG FOSSIL
ANTHROPOMORPHOUS SPECIES—PROBABLE CAUSES WHICH HAVE LED TO
THE WIDE CHASM WHICH SEPARATES MAN FROM THE BRUTES APES—
HALLAM ON THE COMPOUND NATURE OF MAN, AND HIS PLACE IN THE
CREATION—GREAT INEQUALITY OF MENTAL ENDOWMENT IN DIFFERENT
HUMAN RACES AND INDIVIDUALS DEVELOPED BY VARIATION AND ORDI-
NARY GENERATION—TRANSMUTATION AND THEOLOGY.

SOME of the opponents of transmutation who are well
versed in Natural History admit, that though they
consider the doctrine to be untenable, it is not without its
practical advantages as a 'useful working hypothesis,' suggesting good experiments and observations, and aiding us to
retain in the memory a multitude of facts respecting the
geographical distribution of genera and species, both of
animals and plants, and the succession in time of organic
remains, and many other phenomena which, but for such a
theory, would be wholly without a common bond of relationship.

It is in fact conceded by many eminent zoologists and

botanists, as before explained, that whatever may be the nature of the species-making power or law, its effects are of such a character as to imitate the results which variation, guided by natural selection, would produce, if only we could assume with certainty that there are no limits to the variability of species. But as the anti-transmutationists are persuaded that such limits do exist, they regard the hypothesis as simply a provisional one, and expect that it will one day be superseded by another cognate theory, which will not require us to assume the former continuity of the links which have connected the past and present states of the organic world, or the outgoing with the incoming species.

In like manner, many of those who hesitate to give in their full adhesion to the doctrine of progression, the other twin branch of the development theory, and who even object to it, as frequently tending to retard the reception of new facts supposed to militate against opinions solely founded on negative evidence, are, nevertheless, agreed that on the whole it is of great service in guiding our speculations. Indeed, it cannot be denied that a theory which establishes a connection between the absence of all relics of vertebrata in the oldest fossiliferous rocks, and the presence of man's remains in the newest, which affords a more than plausible explanation of the successive appearance in strata of intermediate age of the fish, reptile, bird, and mammifer, has no ordinary claims to our favour as comprehending the largest number of positive and negative facts gathered from all parts of the globe, and extending over countless ages, that science has perhaps ever attempted to embrace in one grand generalisation.

But will not transmutation, if adopted, require us to include the human race in the same continuous series of developments, so that we must hold that Man himself has been derived by an unbroken line of descent from some one

of the inferior animals? Mr. Darwin, in his late work on
the 'Descent of Man,' has said, in reply to this question,
that 'the similarity between man and the lower animals in
embryonic development as well as in innumerable points of
structure and constitution—the rudiments which he retains
and the abnormal reversions to which he is occasionally
liable—are facts which cannot be disputed. They have long
been known, but until recently they told us nothing with
respect to the origin of man. Now when viewed by the light
of our knowledge of the whole organic world their meaning is
unmistakeable. The great principle of evolution stands up
clear and firm, when these groups of facts are considered in
connection with others, such as the mutual affinities of the
members of the same group, their geographical distribution in past and present times, and their geological succession. . . We are forced to admit that the close resemblance
of the embryo of man to that, for instance, of a dog—the construction of his skull, limbs, and whole frame, independently
of the uses to which the parts may be put, on the same plan
with that of other mammals—the occasional reappearance
of various structures, for instance of several distinct muscles,
which man does not normally possess, but which are common
to the quadrumana, and a crowd of analogous facts—all
point in the plainest manner to the conclusion that man is
the co-descendant with other mammals of a common progenitor.' *

We certainly cannot escape from such a conclusion
without abandoning many of the weightiest arguments
which have been urged in support of variation and natural
selection, considered as the subordinate causes by which
new types have been gradually introduced into the earth.
Many of the gaps which separate the most nearly allied

* Descent of Man, vol. ii. p. 385. 1871.

genera and orders of mammalia are, in a physical point of
view, as wide as those which divide Man from the mammalia
most nearly akin to him, and the extent of his isolation,
whether we regard his whole nature or simply his corporeal
attributes, must be considered before we can discuss the
bearing of transmutation upon his origin and place in the
creation.

Systems of Classification.

In order to qualify ourselves to judge of the degree of
affinity in physical organisation between Man and the lower
animals, we cannot do better than study those systems of
classification which have been proposed by the most eminent
teachers of natural history. Of these an elaborate and
faithful summary has been drawn up by the late Isidore
Geoffroy St. Hilaire, which the reader will do well to
consult.[*]

He begins by passing in review numerous schemes of
classification, each of them having some merit, and most
of which have been invented with a view of assigning to
Man a separate place in the system of Nature, as, for
example, by dividing animals into rational and irrational, or
the whole organic world into three kingdoms, the human,
the animal, and the vegetable,—an arrangement defended
on the ground that Man is raised as much by his intelligence
above the animals as are these by their sensibility above
plants. Admitting that these schemes are not unphilo-
sophical, as duly recognising the double nature of Man (his
moral and intellectual, as well as his physical attributes),
Isidore G. St. Hilaire observes that little knowledge has been
imported by them. We have gained, he says, much more
from those masters of the science who have not attempted any

[*] Histoire Naturale Générale des Règnes organiques. Paris, vol. A. 1856.

compromise between two distinct orders of ideas, the physical
and psychological, and who have confined their attention
strictly to Man's physical relation to the lower animals.

Linnæus led the way in this field of enquiry by comparing
Man and the apes, in the same manner as he compared these
last with the carnivores, ruminants, rodents, or any other
division of warm-blooded quadrupeds. After several modifications of his original scheme, he ended by placing Man as
one of the many genera in his order Primates, which
embraced not only the apes and lemurs, but the bats also,
as he found these last to be nearly allied to some of the
lowest forms of the monkeys. But all modern naturalists,
who retain the order Primates, agree to exclude from it the
bats or Cheiroptera; and most of them class Man as one of
several families of the order Primates. In this, as in most
systems of classification, the families of modern zoologists
and botanists correspond with the genera of Linnæus.

Blumenbach, in 1779, proposed to deviate from this
course, and to separate Man from the apes as an order apart,
under the name of Bimana, or two-handed. In making this
innovation he seems at first to have felt that it could not be
justified without calling in psychological considerations to
his aid, to strengthen those which were purely anatomical;
for, in the earliest edition of his 'Manual of Natural History,'
he defined Man to be 'animal rationale, loquens, erectum,
bimanum,' whereas in later editions he restricted himself
entirely to the two last characters, namely, the erect position
and the two hands, or 'animal erectum, bimanum.'

The terms 'bimanous' and 'quadrumanous' had been
already employed by Buffon, in 1769, but not applied in a
strict zoological classification till so used by Blumenbach.
Twelve years later, Cuvier adopted the same order Bimana
for the human family, while the apes, monkeys, and lemurs
constituted a separate order called Quadrumana.

Isidore G. St. Hilaire objects to this innovation that it
is a half measure which can satisfy neither party, and
proceeds to show how, in spite of the great authority of
Blumenbach and Cuvier, a large proportion of modern
zoologists of note have rejected the order Bimana, and have
regarded Man simply as a family of one and the same order,
Primates.

Professor Huxley maintains that the term 'Quadrumanous' may lead to erroneous conclusions, if it is held to
mean that the hind hand of a monkey is anatomically
homologous with the hand rather than with the foot of man.
He has shown that in all essential parts, in the number and
position of the bones and the arrangement and attachments
of the muscles, the hind foot of the Orang and Gorilla is
strictly comparable with the foot of Man, although the great
toe is so modified as to have the position and motions of a
human thumb. It becomes, therefore, functionally a perfect
hand, while it remains homologically a true foot.[*]

Professor Huxley therefore rejects the term 'Quadrumana,'
as leading to serious misconception, and regards Man as one
of the families of the Primates. This method of classification
he shows to be equally borne out by an appeal to another
character on which so much reliance has always been placed
in classification, as affording in the mammalia the most
trustworthy indications of affinity, namely, the dentition.

'The number of teeth in the Gorilla and all the Old
World monkeys except the lemurs is thirty-two, the same as
in Man, and the general pattern of their crowns the same.
But besides other distinctions, the canines in all but Man
project in the upper or lower jaws almost like tusks. But

---

[*] Professor Huxley's third lecture 'On the Motor organs of Man com-pared with those of other Animals,' de-livered in the Royal School of Mines. In Jermyn Street (March 1861), has been embodied with the rest of the course in his work entitled, 'Evidence as to Man's Place in Nature.' Williams and Norgate. London.

all the American apes have four more teeth in their permanent set, or thirty-six in all, so that they differ in this respect more from the Old World apes than do these last from Man.'

Whether the Structure of the Human Brain entitles Man to form a distinct Sub-class of the Mammalia.

In consequence of these and many other zoological considerations, the order Bimana had already been declared in 1856, by Isidore G. St. Hilaire, in his history of the science above quoted (p. 524), 'to have become obsolete,' even though sanctioned by the great names of Blumenbach and Cuvier. But in opposition to the new views Professor Owen announced, the year after the publication of G. St. Hilaire's work, that he had been led by purely anatomical considerations to separate Man from the other Primates and from the mammalia generally as a distinct *sub-class*, thus departing farther from the classification of Blumenbach and Cuvier than they had ventured to do from that of Linnæus.*

The proposed innovation was based on certain peculiarities in the form, size, and position of the lobes of the brain, which, it was alleged, distinguished man from all the other Primates. The facts, as stated by Professor Owen, were, however, disputed by many eminent anatomists, and a discussion on the subject was carried on for several years, which led to the examination of the brains of a large number of apes and monkeys, and added much to our knowledge of their structure. Ultimately, in 1862, Professor Owen modified his views, and virtually abandoned the attempt to place man in a distinct sub-class from the ape on account of radical differences in cerebral characters. A

---

* Owen, Proc. Linnæan Soc., London, 1857, vol. viii. p. 20.

difference in degree has alone been established, as, for
example, the vast increase of the brain in Man, as compared
with that of the highest ape, ' in absolute size, and the still
greater superiority in relative size to the bulk and weight of
the body.' *

If we ask why this character, though well known to
Cuvier and other great anatomists before our time, was not
considered by them to entitle Man, physically considered, to
claim a more distinct place in the group called Primates, than
that of a separate order, or, according to others, a separate
genus or family only, we shall find the answer thus concisely
stated by Professor Huxley in his new work, before cited:—

'So far as I am aware, no human cranium belonging to
an adult man has yet been observed with a less cubical
capacity than 62 cubic inches, the smallest cranium observed
in any race of men, by Morton, measuring 63 cubic inches;
while on the other hand, the most capacious gorilla skull
yet measured has a content of not more than 34½ cubic
inches. Let us assume, for simplicity's sake, that the
lowest man's skull has twice the capacity of the highest
gorilla's. No doubt this is a very striking difference, but
it loses much of its apparent systematic value when viewed
by the light of certain other equally indubitable facts respecting cranial capacities.

'The first of these is, that the difference in the volume of
the cranial cavity of different races of mankind is far greater,
absolutely, than that between the lowest man and the highest
ape, while relatively it is about the same; for the largest
human skull measured by Morton contained 114 cubic
inches, that is to say, had very nearly double the capacity of
the smallest, while its absolute preponderance of over 50
cubic inches is far greater than that by which the lowest

* Owen, Brit. Association Reports, Cambridge, 1862, and Medical Times,
Oct. 11, 1862, p. 373.

adult male human cranium surpasses the largest of the
gorillas ($62 - 34\frac{1}{2} = 27\frac{1}{2}$). Secondly, the adult crania of
gorillas which have as yet been measured, differ among
themselves by nearly one-third, the maximum capacity being
34·5 cubic inches, the minimum 24 cubic inches; and
thirdly, after making all due allowance for difference of size,
the cranial capacities of some of the lower apes fall nearly as
much, relatively, below those of the higher apes, as the
latter fall below Man.'*

Are we then to conclude, that differences in mental power
have no intimate connection with the comparative volume of
the brain? We cannot draw such an inference, because the
highest and most civilised races of Man exceed in the average
of their cranial capacity the lowest races, the European
brain, for example, being larger than that of the negro, and
somewhat more convoluted and less symmetrical; and those
apes, on the other hand, which approach nearest to Man in the
form and volume of their brain being more intelligent than
the Lemurs, or still lower divisions of the mammalia, such as
the Rodents and Marsupials, which have smaller brains. But
the extraordinary intelligence of the elephant and dog, so far
exceeding that of the larger part of the Quadrumana, although
their brains are of a type much more remote from the human,
may serve to convince us how far we are as yet from understanding the real nature of the dependence of intellectual
superiority on cerebral structure.

Professor Rolleston, in reference to this subject, remarks,
that ' even if it were to be proved that the differences between
Man's brain and that of the ape's are differences entirely of
quantity, there is no reason, in the nature of things, why so
many and such weighty differences in degree should not
amount to a difference in kind.'

* Huxley, ' Evidence as to Man's Place in Nature,' p. 76. London, 1863.

M M

'Differences of degree and differences of kind are, it is true, mutually exclusive terms in the language of the schools; but whether they are so also in the laboratory of Nature, we may very well doubt.' [*]

The same physiologist suggests, that as there is considerable plasticity in the human frame, not only in youth and during growth, but even in the adult, we ought not always to take for granted, as some advocates of the development theory seem to do, that each advance in physical power depends on an improvement in bodily structure, for why may not the soul, or the higher intellectual and moral faculties, play the first instead of the second part in a progressive scheme?

Intelligence of the lower Animals compared to that of Man.

Ever since the days of Leibnitz, metaphysicians who have attempted to draw a line of demarcation between the intelligence of the lower animals and that of Man, or between instinct and reason, have experienced difficulties analogous to those which the modern anatomist encounters when he tries to distinguish the brain of an ape from that of Man by some characters more marked than those of mere size and weight, which vary so much in individuals of the same species, whether simian or human.

Professor Agassiz, after declaring that as yet we scarcely possess the most elementary information requisite for a scientific comparison of the instincts and faculties of animals with those of Man, confesses that he cannot say in what the mental faculties of a child differ from those of a young chimpanzee. He also observes, that 'the range of the

[*] Report of a Lecture delivered at the Royal Institution, by Professor George Rolleston, On the Brain of Man and Animals. Medical Gazette, March 15, 1862, p. 262.

passions of animals is as extensive as that of the human
mind, and I am at a loss' he says 'to perceive a difference
of kind between them, however much they may differ in
degree, and in the manner in which they are expressed.
The gradations of the moral faculties among the higher
animals and Man are, moreover, so imperceptible, that to
deny to the first a certain sense of responsibility and con-
sciousness, would certainly be an exaggeration of the
difference between animals and Man. There exists, besides,
as much individuality among animals within their respective
capabilities as among Man; as every sportsman, or every
keeper of menageries, or every farmer and shepherd can
testify, who has had a large experience with wild, or tamed,
or domesticated animals. This argues strongly in favour of
the existence in every animal of an immaterial principle,
similar to that which, by its excellence and superior endowments, places Man so much above animals. Yet the
principle exists unquestionably, and whether it be called
soul, reason, or instinct, it presents, in the whole range of
organised beings, a series of phenomena closely linked together, and upon it are based not only the higher manifes-
tations of the mind, but the very permanence of the specific
differences which characterise every organ. Most of the
arguments of philosophy in favour of the immortality of
Man apply equally to the permanency of this principle in
other living beings.' * Without pretending to offer a
decided opinion on the transcendental questions raised in
the concluding part of this citation, they cannot be lost sight
of when we are comparing man with the inferior animals.

Professor Huxley, when commenting on a passage in
Professor Owen's memoir, above cited (p. 527), argues that
there is a unity in psychical as in physical plan among ani-

* Contributions to the Natural History of the United States of North
America, vol. i. part i. pp. 60, 61.

M M 2

mated beings, and adds, that although he cannot go so far as
to say that 'the determination of the difference between
Homo and Pithecus is the anatomist's difficulty,' yet no
impartial judge can doubt that the roots, as it were, of those
great faculties which confer on Man his immeasurable superiority above all other animate things are traceable far down
into the animate world. The dog, the cat, and the parrot,
return love for our love, and hatred for our hatred. They
are capable of shame and of sorrow, and, though they may
have no logic nor conscious ratiocination, no one who has
watched their ways can doubt that they possess that power
of rational cerebration which evolves reasonable acts from
the premises furnished by the senses—a process which takes
fully as large a share as conscious reason in human activity.*

Since the last edition of this work appeared Mr. Darwin
has published his 'Descent of Man,' in which he discusses in
great detail the relation of man to the lower animals.   Not
only does he show close resemblances between every part of
their structure, and in their whole process of development, but
gives instances of the possession, by animals, of the rudiments
of almost every human mental faculty. The sense of beauty, for
instance, is shown to be present in the Bower birds, who deco-
rate their playing-passages; as well as by the manner in which
male birds attract the females by displaying their ornamental
plumes.   The germs also of the sympathetic and moral
faculties are shown to exist in animals, by those rare cases in
which blind birds are fed by their companions,† and by the
instance of a small monkey, at the Zoological Gardens, coming
to the rescue of a keeper who was attacked by a baboon.

But Mr. Darwin admits that the difference between the
mind of the lowest man and of the highest ape is immense,
although he maintains that it is one of degree, not of kind.

* Natural History Review, No. 1, p. 68, January 1861.
† 'Descent of Man,' vol. I. p. 77.

'The belief in God,' he says ' has often been advanced as not only the greatest, but the most complete of all the distinctions between man and the lower animals. It is however impossible, as we have seen, to maintain that this belief is innate or instinctive in man. On the other hand, a belief in all-pervading spiritual agencies (many of them cruel and malignant) seems to be universal; and apparently follows from a considerable advance in the reasoning powers of man, and from a still greater advance in his faculties of imagination, curiosity and wonder. . . The idea of a universal and beneficent Creator of the Universe, does not seem to arise in the mind of man until he has been elevated by long-continued culture.' *

One of the features that have always been considered to be most characteristic of the human race, is the long period of almost complete helplessness we pass through in our infancy. In this particular it appears that there is a considerable difference between the lower and the higher apes, the latter, in this respect, as in their whole structure, more nearly approaching man. While residing in Borneo, Mr. Alfred R. Wallace had the opportunity of making the comparison, he having secured a young Orang-Utan or Mias a few months old, and also a young monkey of lower grade and corresponding age. He thus compares the habits and dispositions of the two animals: 'After I had had the little Mias about three weeks, I fortunately obtained a young hare-lip monkey (*Macacus cynomolgus*) which, though small, was very active, and could feed itself. I placed it in the same box with the Mias, and they immediately became excellent friends, neither exhibiting the least fear of the other. The little monkey would sit upon the other's stomach, or even on its face, without the least regard to its feelings.

* 'Descent of Man,' vol. ii. p. 395.

While I was feeding the Mias, the monkey would sit by,
picking up all that was spilt, and occasionally putting out its
hands to intercept the spoon ; and as soon as I had finished,
would pick off what was left sticking to the lips of the Mias,
and then pull open its mouth to see if any still remained
inside ; afterwards lying down on the poor creature's
stomach as on a comfortable cushion.   The little helpless
Mias would submit to all these insults with the most exem-
plary patience, only too glad to have something warm near it,
which it could clasp affectionately in its arms. . . It was
curious to observe the different actions of these two animals
which could not have differed much in age.   The Mias, like
a very young baby, lying on its back quite helpless, rolling
lazily from side to side, stretching out all four hands into
the air, wishing to grasp something, but hardly able to guide
its fingers to any definite object ; and when dissatisfied,
opening wide its almost toothless mouth, and expressing its
wants by a most infantine scream.   The little monkey, on the
other hand, in constant motion ; running and jumping about
wherever it pleased, examining everything around it, seizing
hold of the smallest objects with the great precision, balancing
itself on the edge of the box, or running up a post, and
helping itself to anything eatable that came in its way.
There could hardly be a greater contrast, and the baby Mias
looked more baby-like by the comparison.'[*]   When we
remember how far, in those faculties which resemble the
human, the adult Mias is destined to surpass the Macacus,
we seem to perceive in the anecdotes here given the dawn of
that contrast between the precocity of instinct which marks
the inferior animals, and the slowness of development which
characterises the human infant.

[*] Wallace. 'The Malay Archipelago,' 1st ed. 1869, vol. i. p. 69.

*Grounds for referring Man to a distinct Kingdom of
Nature.*

Few if any of the authors above cited, while they admit
so fully the analogy which exists between the faculties of
Man and the inferior animals, are disposed to underrate the
enormous gap which separates Man from the brutes, and if
they scarcely allow him to be referable to a distinct order,
and much less to a separate sub-class, on purely physical
grounds, it does not follow that they would object to the
reasoning of M. Quatrefages, who says, in his work on the
'Unity of the Human Species,' that Man must form a kingdom by himself if once we permit his moral and intellectual
endowments to have their due weight in classification.

As to his organisation, he observes, 'We find in the
mammalia nearly absolute identity of anatomical structure,
bone for bone, muscle for muscle, nerve for nerve—
similar organs performing like functions. It is not by a
vertical position on his feet, the *os sublime* of Ovid, which
he shares with the penguin, nor by his mental faculties,
which though more developed, are fundamentally the same
as those of animals, nor by his powers of perception, will,
memory, and a certain amount of reason, nor by articulate
speech, which he shares with birds and some mammalia, and
by which they express ideas comprehended not only by
individuals of their own species but often by Man, nor is it
by the faculties of the heart, such as love and hatred, which
are also shared by quadrupeds and birds, but it is by something completely foreign to the mere animal, and belonging
exclusively to Man, that we must establish a separate kingdom
for him (p. 21). These distinguishing characters,' he goes on
to say, 'are the abstract notion of good and evil, right and
wrong, virtue and vice, or the moral faculty, and a belief
in a world beyond ours, and in certain mysterious beings, or

a Being of a higher nature than ours, whom we ought to
fear or revere; in other words, the religious faculty.'—P. 23.

By these two attributes, the moral and the religious, not
common to man and the brutes, M. Quatrefages proposes to
distinguish the human from the animal kingdom.

But he omits to notice one essential character, which
Dr. Sumner, the late Archbishop of Canterbury, brought out
in strong relief fifty years ago, in his ' Records of Creation.'
' There are writers,' he observes, ' who have taken an extraordinary pleasure in levelling the broad distinction which
separates Man from the Brute Creation. Misled to a false
conclusion by the infinite variety of Nature's productions,
they have described a chain of existence connecting the
vegetable with the animal world, and the different orders of
animals one with another, so as to rise by an almost imperceptible gradation from the tribe of Simiæ to the lowest of
the human race, and from these upwards to the most refined.
But if a comparison were to be drawn, it should be taken,
not from the upright form, which is by no means confined to
mankind, nor even from the vague term reason, which cannot always be accurately separated from instinct, but from
that power of progressive and improvable reason, which is
Man's peculiar and exclusive endowment.

' It has been sometimes alleged, and may be founded on
fact, that there is less difference between the highest brute
animal and the lowest savage than between the savage and
the most improved Man. But, in order to warrant the pretended analogy, it ought to be also true that this lowest
savage is no more capable of improvement than the Chimpanzee or Orang-outang.

' Animals,' he adds, ' are born what they are intended to
remain. Nature has bestowed upon them a certain rank,
and limited the extent of their capacity by an impassable
decree. Man she has empowered and obliged to become the

artificer of his own rank in the scale of beings by the peculiar gift of improvable reason.'*

We have seen that Professor Agassiz, in his Essay on Classification, above cited (p. 531), speaks of the existence in every animal of 'an immaterial principle similar to that which, by its excellence and superior endowments, places man so much above animals;' and he remarks, 'that most of the arguments of philosophy in favour of the immortality of Man, apply equally to the permanency of this principle in other living beings.'

Although the author has no intention by this remark to impugn the truth of the great doctrine alluded to, it may be well to observe, that if some of the arguments in favour of a future state are applicable in common to Man and the lower animals, they are by no means those which are the weightiest and most relied on. It is no doubt true that, in both, the identity of the individual outlasts many changes of form and structure which take place during the passage from the infant to the adult state, and from that to old age; and the loss again and again of every particle of matter which had entered previously into the composition of the body during its growth, and the substitution of new elements in their place, while the individual remains always the same, carries the analogy a step farther. But beyond this we cannot push the comparison. We cannot imagine this world to be a place of trial and moral discipline for any of the inferior animals, nor can any of them derive comfort and happiness from faith in a hereafter. To Man alone is given this belief, so consonant to his reason, and so congenial to the religious sentiments implanted by nature in his soul, a doctrine which tends to raise him morally and intellectually in the scale of being, and the fruits of which are, therefore, most opposite in character to those which grow out of error and delusion.

* Records of Creation, vol. ii. chap. ii. 2nd ed. 1816.

Absence of intermediate fossil anthropomorphous species.

The opponents of the theory of transmutation sometimes
argue that, if there had been a passage by variation from the
lower Primates to Man, the geologist ought ere this to have
detected some fossil remains of the intermediate links of the
chain. But what we have said respecting the absence of
gradational forms between the recent and pliocene mammalia
(p. 493), may serve to show the weakness in the present state
of science of any argument based on such negative evidence,
especially in the case of Man, since we have not yet searched
those pages of the great book of nature, in which alone we
have any right to expect to find records of the missing links
alluded to. The countries of the anthropomorphous apes
are the tropical regions of Africa, and the islands of Borneo
and Sumatra, lands which may be said to be quite unknown
in reference to their pliocene and pleistocene mammalia.
Man is an old world type, and it is not in Brazil, the only
equatorial region where ossiferous caverns have yet been explored, that the discovery, in a fossil state, of extinct forms
allied to the human, could be looked for. Lund, a Danish
naturalist, found in Brazil, not only extinct sloths and armadilloes, but extinct genera of fossil monkeys; but all of the
American type, and therefore, widely departing in their dentition and some other characters from the Primates of the old
world.[*]

At some future day, when many hundred species of extinct
quadrumana may have been brought to light, the naturalist
may speculate with advantage on this subject ; at present we
must be content to wait patiently, and not to allow our judgment respecting transmutation to be influenced by the want
of evidence, which it would be contrary to analogy to look

[*] See above, p. 479.

CHAP. XLIV.   FOSSIL ANTHROPOMORPHOUS SPECIES.   A30

for in pleistocene deposits in any districts, which as yet we
have carefully examined.   For, as we meet with extinct
kangaroos and wombats in Australia, extinct llamas and
sloths in South America, so in equatorial Africa, and in
certain islands of the East Indian Archipelago, may we hope
to meet hereafter with lost types of the anthropoid Primates,
allied to the gorilla, chimpanzee, and orang-outang.

Europe, during the pliocene period, seems not to have
enjoyed a climate fitting it to be the habitation of the quadrumanous mammalia ; but we no sooner carry back our researches into miocene times, where plants and insects, like
those of Oeninghen, and shells, like those of the faluns of the
Loire, would imply a warmer temperature both of sea and
land, than we begin to discover fossil apes and monkeys north
of the Alps and Pyrenees.   Among the few species already
detected, two at least belong to the anthropomorphous class.
One of these, the Dryopithecus of Lartet, a gibbon or long-
armed ape, about equal to man in stature, was obtained in the
year 1856 in the upper miocene strata at Sansan, near the
foot of the Pyrenees in the South of France, and one bone
of the same ape has been since procured from a deposit of
corresponding age at Eppelsheim near Darmstadt,in a latitude
answering to that of the southern counties of England.*   M.
Rütimeyer,† an able osteologist, referred to in the earliest
chapters of this work, has also discovered in eocene strata,
in the Swiss Jura, the jawbone of a monkey) Cænopithecus
lemuroides) allied in some points to the Mycetes or howling
monkey of America and in others to the lemurs.   But according to the doctrine of progression it is not in these miocene
or eocene strata, but in those of pliocene and pleistocene
date, in more equatorial regions, that there will be the
greatest chance of discovering hereafter some species more
highly organised than the gorilla and chimpanzee.

* Owen, 'Geologist,' November
1862.
† Rütimeyer, 'Fauna Säugethiere,'
&c. Zurich, 1862.

Probable Causes which have led to the wide Chasm which separates Man from the higher Apes.

We have already obtained evidence of the existence of Man at so remote a period that there has since been time for many conspicuous mammalia once his contemporaries to die out, and this, even before the era of the earliest historical records. Yet, in spite of the long lapse of prehistoric ages, during which he must have flourished on the earth, there is no proof of any perceptible change in his bodily structure. If, therefore, he ever diverged from some unreasoning brute ancestor, we must suppose him to have existed at a far more distant epoch, possibly on some continents or islands now submerged beneath the ocean. The immensity of this period might enable us to understand how it is that man, notwithstanding his close resemblance in anatomical structure to the higher apes, yet rises so far above them in symmetry of form no less than in mental and moral faculties. Lamarck first pointed out that, as soon as some family of the higher apes had advanced in organization and intellect beyond its fellows, a severe competition would immediately arise between it and the tribes nearest to it in rank, which would lead to the repression and final extinction of the latter. According to Mr. Darwin, this depends upon the general law, that competition is always keenest between those varieties or species which are most nearly allied; and one of the consequences of this law is seen in the fact, that, as a rule, the most closely allied forms do not inhabit the same region, but are almost always restricted to adjacent regions, where they form what naturalists term 'representative species.' A more familiar illustration is to be found in the fact (long supposed to be inexplicable) that a lower race of mankind inevitably disappears when in contact with one higher in physical and mental capacity. The Red Indian and the Australian disappear before the Anglo-Saxon, just as the woolly-haired Negritos of the Philippine Islands and the Malay Peninsula disappear before the superior Malay race. Now if a similar struggle has been going on ever since the first advance occurred, which led to the development of Man, and each successive step in his onward march has in like manner led to the extinction of the inferior race above whom he rose, this will go far to explain why the various races of man, although they present considerable minor differences, yet all agree in those marked characteristics which distinguish mankind from the highest brutes.

'Man' says Mr Wallace, 'by the mere capacity of clothing himself, and making weapons and tools, has taken away from nature that power of slowly but permanently changing the external form and structure, in accordance with changes in the external world, which she exercises over all other animals. As the competing races by which they are surrounded, the climate, the vegetation, or the animals which serve them for food are slowly changing, they must undergo a corresponding change in their structure, habits, and constitution, to keep them in harmony with the new conditions—to enable them to live and maintain their numbers. But Man does this by means of his intellect alone, the variations of which enable him, with an unchanged body, still to keep in harmony with the changing universe.' * Thus, while the progressive change in a dominant species of the lower animals is always marked by divergence of character leading to a multiplicity of specific forms, in the case of Man it seems to have led to a continued development in one direction, and the production of a single specific form of so perfect and dominant a charac-

* 'Contributions to Natural Selection,' 1871, p. 314.

ter, that all diverging types were rapidly extinguished by
competition with it.

The superiority of Man to the higher apes is more
marked in the brain than in any other part of his structure,
and this leads to the conclusion, that for many long ages
the human frame has remained almost stationary, with the
exception of the increased capacity and dimensions of the
brain and a corresponding improvement in intellectual
power. If this is the case, we may expect to find traces of
Man's past developmental history by discovering crania
of a greatly diminished brain-capacity, at a period long
anterior to that of the palæolithic weapons, the earliest
works of art yet discovered. At a more remote epoch still,
we may perhaps find some indication of those successive
steps by which the erect form of Man was gradually acquired
by modification from that of an ape-like ancestor.

Hallam, in his 'Literature of Europe,' after indulging in
some profound reflections on 'the thoughts of Pascal,' and
the theological dogmas of his school respecting the fallen
nature of Man, thus speaks of Man's place in the creation :—
'It might be wandering from the proper subject of these
volumes if we were to pause, even shortly, to enquire whether,
while the creation of a world so full of evil must ever
remain the most inscrutable of mysteries, we might not be
led some way in tracing the connexion of moral and physical
evil in mankind, with his place in that creation, and especially, whether the law of continuity, which it has not
pleased his Maker to break with respect to his bodily structure, and which binds that, in the unity of one great type, to
the lower forms of animal life by the common conditions of
nourishment, reproduction, and self-defence, has not rendered necessary both the physical appetites and the propensities
which terminate in self; whether again, the superior endowments of his intellectual nature, his susceptibility of moral

emotion, and of those disinterested affections which, if not
exclusively, he far more intensely possesses than an inferior
being—above all, the gifts of conscience and a capacity to
know God, might not be expected, even beforehand, by their
conflict with the animal passions, to produce some partial
inconsistencies, some anomalies at least, which he could not
himself explain in so compound a being. Every link in the
long chain of creation does not pass by easy transition into
the next. There are necessary chasms, and, as it were, leaps
from one creature to another, which, though not exceptions
to the law of continuity, are accommodations of it to a new
series of being. If Man was made in the image of God, he
was also made in the image of an ape. The framework of
the body of Him who has weighed the stars and made the
lightning his slave, approaches to that of a speechless brute,
who wanders in the forests of Sumatra. Thus standing on
the frontier land between animal and angelic natures, what
wonder that he should partake of both!' *

The law of continuity here spoken of, as not being violated
by occasional exceptions, or by leaps from one creature to another, is not the law of variation and natural selection above
explained (Chap. XXI.), but that unity of plan supposed to
exist in the Divine Mind, whether realised or not materially
and in the visible creation, of which the 'links do not
pass by an easy transition' the one into the other, at least
as beheld by us.

Dr. Asa Gray, an eminent American botanist, to whom we
are indebted for a philosophical essay of great merit on the
Origin of Species by Variation and Natural Selection, has
well observed, when speaking of the axiom of Leibnitz,
'Natura non agit saltatim,' that nature secures her ends, and
makes her distinctions, on the whole, manifest and real, but

* Hallam, 'Introduction to the Literature of Europe,' &c., vol. iv. p. 162.

without any important breaks or long leaps. 'We need not wonder that gradations between species and varieties should occur, or that genera and other groups should not be absolutely limited, though they are represented to be so in our systems. The classifications of the naturalist define abruptly where nature more or less blends. Our systems are nothing if not definite.'

The same writer reminds us that 'plants and animals are so different, that the difficulty of the ordinary observer would be to find points of comparison, whereas, with the naturalist, it is all the other way. All the broad differences vanish one by one as we approach the lower confines of the animal and vegetable kingdoms, and no absolute distinction whatever is now known between them.' *

The late Mr. Hopkins of Cambridge, himself an accomplished geologist, when reviewing Darwin's 'Origin of Species' declared, that if we embrace the doctrine of the 'continuous variation of all organic forms from the lowest to the highest, including Man as the last link in the chain of being, there must have been a transition from the instinct of the brute to the noble mind of Man; and in that case, where,' he asks, 'are the missing links, and at what point of his progressive improvement did Man acquire the spiritual part of his being, and become endowed with the awful attribute of immortality?' †

Before we raise objections of this kind to a scientific hypothesis, it would be well to pause and enquire whether there are no analogous enigmas in the constitution of the world around us, some of which present even greater difficulties than that here stated. When we contemplate, for example, the many hundred millions of human beings who

---

* Natural Selection not inconsistent with Natural Theology, p. 55, by Dr. Asa Gray. Trübner & Co., London, 1861.

† Physical Theories of the Phenomena of Life, Fraser's Magazine, July 1860, p. 88.

now people the earth, we behold thousands who are doomed
to helpless imbecility, and we may trace an insensible
gradation between them and the half-witted, and from these
again to individuals of perfect understanding, so that tens
of thousands must have existed in the course of ages, who, in
their moral and intellectual condition, have exhibited a
passage from the irrational to the rational, or from the ir-
responsible to the responsible. Moreover, we may infer
from the returns of the Registrar-General of births and
deaths in Great Britain, and from Quetelet's statistics of
Belgium, that one fourth of the human race die in early
infancy, nearly one tenth before they are a month old ; so
that we may safely affirm that millions perish on the earth
in every century, in the first few hours of their existence.
To assign to such individuals their appropriate psychological
place in the creation, is one of the unprofitable themes on
which theologians and metaphysicians have expended much
ingenious speculation.

The philosopher, without ignoring these difficulties, does
not allow them to disturb his conviction that ' whatever is,
is right,' nor do they check his hopes and aspirations in
regard to the high destiny of his species ; but he also feels
that it is not for one who is so often confounded by the
painful realities of the present to test the probability of
theories respecting the past, by their agreement or want of
agreement with some ideal of a perfect universe which those
who are opposed to his opinions may have pictured to themselves.

Origin of Superior Races and Men of Genius.

We may demur to the assumption that the hypothesis
of variation and natural selection obliges us to assume
that there was an absolutely insensible passage from the
highest intelligence of the inferior animals to the improvable

N N

reason of Man. The birth of an individual of transcendent
genius, whose immediate progenitors are not known to have
displayed any intellectual capacity above the average standard of their age or race, is a phenomenon not to be lost
sight of, when we are conjecturing whether the successive
steps in advance, by which a progressive scheme has been
developed, may not admit of occasional strides constituting
apparent breaks in an otherwise continuous series of psychical changes.

The inventors of useful arts, the poets and prophets of
the early stages of a nation's growth, the promulgators of
new systems of religion, ethics, and philosophy, or of new
codes of laws, have often been looked upon as messengers
from Heaven, and after their death have had divine honours
paid to them, while fabulous tales have been told of the pro-
digies which accompanied their birth. Nor can we wonder
that such notions have prevailed when we consider what
important revolutions in the moral and intellectual world
such leading spirits have brought about; and when we
reflect that mental as well as physical attributes are trans-
missible by inheritance, so that we may possibly discern in
such cases of exceptional pre-eminence the origin of the
superiority of certain races of mankind.

Mr. Darwin has shown, when treating of inheritance, that
in Man, as in animals, the most trifling as well as the most
important characters are habitually transmitted from one
generation to another, and Mr. Francis Galton, in his work
on 'Hereditary Genius,' has brought together a large
number of facts to prove that in Man, even the higher kinds
of intellectual excellence are exhibited in members of the
same family more conspicuously than in others which are
unconnected with them, or which have sprung from stocks
notoriously inferior in capacity. Keeping in view this
principle of hereditary transmission, he has speculated on

the origin of distinct races and the prevalence of national
characters. 'The ablest race,' he says, 'of whom history
bears any record is unquestionably the ancient Greek, partly
because their master-pieces in the principal departments of
intellectual activity are still unsurpassed, and partly because
the population that gave birth to the creators of those
master-pieces was very small. Of the various Greek sub-
races, that of Attica was the ablest, and she was no doubt
largely indebted to the following cause for her superiority.
Athens opened her arms to immigrants, but not indiscriminately, for her social life was such that none but very able
men could take any pleasure in it; on the other hand, she
offered attractions such as men of the highest ability and
culture could find in no other city. Thus, by a system of
partly unconcious selection, she built up a magnificent
breed of human animals, which, in the space of one century
—viz. between 530 and 430 B.C., produced the following
illustrious persons, fourteen in number:—Themistocles, Miltiades, Aristides, Cimon, Pericles, Thucydides, Socrates,
Xenophon, Plato, Æschylus, Sophocles, Euripides, Aristo-
phanes, and Phidias.'* The population which produced
these men amounted, neglecting the resident aliens and
slaves, to less than 90,000 free-born individuals; allowing
three generations in a century, this would give 270,000 free-
born persons, or only 135,000 males, born in the century
named.†

Mr. Galton reasonably suggests the doubt whether, if
we could select the leading men from all the millions
which have flourished in Europe during the last eighteen
centuries, we could match the fourteen illustrious characters
which adorned that single century in Athens. We have
none, he says, to put by the side of Socrates and Phidias.

* Galton, 'Hereditary Genius,' 1869, p. 310.   † Ibid. p. 341.

N N 2

But it will be seen that in the list of fourteen the influence of kinship tells but little, being confined to two of them, viz. Miltiades and Cimon, who stand to each other in the relationship of father and son. Probably neither this principle of inheritance, nor the 'unconscious selection' before alluded to, are sufficient to account for the wonderful phenomenon into which we are now enquiring. Sparta, it should be remembered, during the period in question, failed to produce men of genius who could be pointed to as the rivals of their Athenian contemporaries. Yet Sparta shared with Athens a military ascendency, during the same stirring period of Greek history. If it be said that the Athenian aristocracy had the advantage of slaves, and so were saved the drudgery of servile work, it must be borne in mind that the Spartans were on an equality in this respect, having also their Helots. Nor should it be forgotten, that, in our own time, the Northern or Free States of the American Union far excelled the Southern or Slave States in literature and science. Grote has no doubt justly attributed the intellectual inferiority of Sparta to her 'perpetual drill,' which, while it favoured the cultivation of a severe code of morality, prompt obedience, and high power of endurance, and gave them strength in war, occupied so much time as to leave no leisure for literature, the fine arts, and philosophical contemplation.

Where, then, are we to look for the cause of Athenian pre-eminence in the century of her greatest renown? Herodotus has declared his opinion that equality of political rights was the main cause of Athenian greatness, and that so long as they were under the rule of a single individual, or tyrant, the citizens did not display that courage and patriotic devotion to their country by which they were afterwards distinguished ; but Herodotus is alluding chiefly to what led to their success in war, which followed close on the adoption of a more liberal form of government.   This, however, will not help us to explain the manner in which she eclipsed Sparta in so many other lines of excellence.

Perhaps the much more prominent part played by Athens should be ascribed, in addition to the causes at which Mr. Galton has hinted, viz. hereditary transmission and unconscious selection, to the freedom of thought and great liberty of expression of opinion which was enjoyed for three generations in the Senate and lecture-room soon after the introduction of a more democratic form of government.   The trial and condemnation of so great and good a man as Socrates may perhaps seem fatal to this speculation.   But it is well known that Socrates had already reached his seventieth year when for the first time he was arraigned before the Dikastery, and he was charged, amongst other capital offences, with corrupting the youth by teaching unsound theological doctrines, and even substituting divinities of his own for those which the city worshipped.   He had gone on for no less than thirty years unmolested, practising what he seems to have thought his divine mission, viz. to inspire his fellow-countrymen with more elevated and purer views of the sacredness of truth and of social and political morality.   The influence of what had been termed his 'colloquial magic' is admitted to have been immense, and Grote has pointed out how much his conversation had contributed to form the doctrines of Aristotle and Plato, and afterwards the theories of leading schools of philosophy which succeeded them.   In spite, therefore, of the capital sentence passed on this great teacher, it may be doubted whether the toleration extended to Socrates for thirty years would have been granted in any city or university of modern Europe, whether before or after the Reformation, to anyone whose opinions were as much opposed to those generally received as were the views of Socrates to the public opinion of Athens.

But we should do wrong to Greece generally if we dwelt too much on the exceptional excellence displayed in a very short period by the Athenians as if it represented the genius and taste of the whole Hellenic race. The steps by which this people came to excel the surrounding Asiatic and European tribes in intellectual activity and independence of character are a mystery. Even in the time of Homer, eight centuries before our era, they possessed a copious and rich language, and were able to appreciate the Homeric poetry and to hand it down to posterity. No doubt the growth of the high civilisation finally reached was gradual, and the unity of national character and its improvement in various kinds of excellence was promoted by the confluence of vast multitudes at such games as the Olympic and Pythian. During these sacred festivals the wars of neighbouring petty states were suspended, and prizes were given for excellence in poetry, philosophy, oratory, painting, sculpture, music, and architecture, as well as in chariot-racing and athletic exercises. A common pride was felt in works of Greek genius by all who spoke the same language, from whatever part of Hellas they came. As to the rapid decline of Athens after the three generations above alluded to, it may be easily traced to the corruption of a licentious democracy, and, as Mr. Galton points out, to an increasing laxity of social morality : 'marriage having become unfashionable and being avoided, many of the more ambitious and accomplished women were avowed courtesans and consequently infertile, and the mothers of the incoming population were of a heterogeneous class. In a small sea-bordered country,' says Mr. Galton, ' where emigration and immigration are constantly going on, and where the manners are as dissolute as were those of Greece in the period of which I speak, the purity of a race would necessarily fail.' *

* Galton, ' Hereditary Genius,' p. 343.

Transmutation and Natural Theology.

I cannot better conclude this volume than by reminding the reader that Dr. Asa Gray, in the excellent essay already cited (p. 544), has pointed out that there is no tendency in the doctrine of Variation and Natural Selection to weaken the foundations of Natural Theology; for, consistently with the derivative hypothesis of species, we may hold any of the popular views respecting the manner in which the changes of the natural world are brought about. We may imagine 'that events and operations in general go on in virtue simply of forces communicated at the first, and without any subsequent interference; or we may hold that now and then, and only now and then, there is a direct interposition of the Deity; or, lastly, we may suppose that all the changes are carried on by the immediate orderly and constant, however infinitely diversified, action of the intelligent, efficient Cause.' They who maintain that the origin of an individual, as well as the origin of a species or a genus, can be explained only by the direct action of the creative cause, may retain their favourite theory compatibly with the doctrine of transmutation.

Professor Agassiz, having observed that, 'while human thought is consecutive, divine thought is simultaneous,' Dr. Asa Gray has replied that, 'if divine thought is simultaneous, we have no right to affirm the same of divine action.'

The whole course of nature may be the material embodiment of a preconcerted arrangement; and if the succession of events be explained by transmutation, the perpetual adaptation of the organic world to new conditions leaves the argument in favour of design, and therefore of a designer, as valid as ever; 'for to do any work by an instrument must

require, and therefore presuppose, the exertion rather of
more than of less power, than to do it directly.' *

As to the charge of materialism brought against all forms
of the development theory, Dr. Gray has done well to remind us that 'of the two great minds of the seventeenth
century, Newton and Leibnitz, both profoundly religious as
well as philosophical, one produced the theory of gravita-
tion, the other objected to that theory, that it was subversive
of natural religion.' †

It may be said that, so far from having a materialistic
tendency, the supposed introduction into the earth at successive geological periods of life,—sensation,—instinct,—the
intelligence of the higher mammalia bordering on reason,—
and lastly the improvable reason of Man himself, presents
us with a picture of the over-increasing dominion of mind
over matter.

* 'Natural Selection not Inconsistent with Natural Theology,' p. 55, by
Dr. Asa Gray. Trübner, & Co., London, 1861.   † Ibid. p. 31.

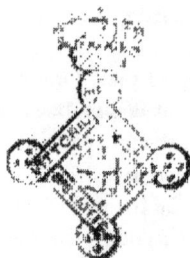

INDEX.

www.ingramcontent.com/pod-product-compliance
Lightning Source LLC
Chambersburg PA
CBHW020853210326
41598CB00018B/1655